APPLIED PARTIAL

DIFFERENTIAL

EQUATIONS

AN INTRODUCTION

APPLIED PARTIAL
DIFFERENTIAL
EQUATIONS

AN INTRODUCTION

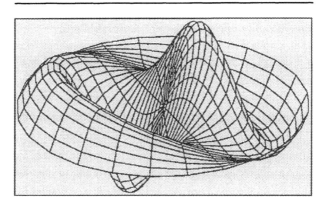

ALAN JEFFREY

University of Newcastle-upon-Tyne

ACADEMIC PRESS

An imprint of Elsevier Science

Amsterdam   Boston   London   New York   Oxford   Paris
San Diego   San Francisco   Singapore   Sydney   Tokyo

Senior Editor, Mathematics	Barbara Holland
Senior Project Manager	Angela Dooley
Editorial Coordinator	Tom Singer
Product Manager	Anne O'Mara
Cover Design	Gary Ragaglia
Copyeditor	Charles Lauder
Composition	International Typesetting and Composition
Printer	Maple-Vail

This book is printed on acid-free paper. ∞

Cover graphic: The illustration represents a snapshot of an asymmetric mode of vibration of the circular drum head considered in Example 6.9 of Chapter 6.
The time dependent vibration is the product of this static mode and a sinusoidal function of time. In general, the response of the drum head to an arbitrary disturbance is expressible as the linear superposition of all possible time dependent modes of vibration.

Academic Press
An imprint of Elsevier Science
525 B Street, Suite 1900, San Diego, California 92101-4495, USA
http://www.academicpress.com

Academic Press
An imprint of Elsevier Science
84 Theobald's Road, London WC1X 8RR, UK
http://www.academicpress.com

Academic Press
An imprint of Elsevier Science
200 Wheeler Road, Burlington, Massachusetts 01803, USA

Library of Congress Catalog Card Number: 2002109015

International Standard Book Number: 0-12-382252-1

Transferred to digital printing 2006

To Lisl

Contents

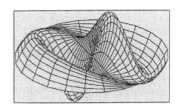

Preface

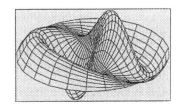

Many books deal with partial differential equations, some at an elementary level and others at more advanced levels, so it is necessary that some justification should be given for the publication of another introductory text. With few exceptions, existing texts written at a similar level restrict their subject matter to the study of the boundary and initial value problems associated with the three fundamental linear second-order partial differential equations of hyperbolic, parabolic and elliptic type, and to solutions obtained by the method of separation of variables. Although these fundamental linear second-order equations are extremely important, and have many classical applications, other more recent applications require familiarity with more general forms of partial differential equations, and also with some of the simpler properties of first-order systems of partial differential equations. Thus the purpose of this book is to attempt to cover all of the standard requirements expected of such a text, though sometimes using a slightly more general approach than usual to unify ideas, and also to introduce the fresh material necessary for understanding many new practical applications that involve systems of hyperbolic partial differential equations.

The text places emphasis on the importance of conservation laws in integral form, and on the fundamental role integral conservation laws play in the derivation of the partial differential equations used to model physical situations, particularly when discontinuities are involved. The three fundamental types of linear second-order partial differential equations are shown to arise as special cases of very simple systems of first order equations. In general a system of coupled first-order equations cannot be reduced to an equivalent single higher order equation as occurs, for example, with the heat equation, the wave equation and the Laplace equation, so because of this an introduction to the study of systems of partial differential equations has been provided.

The effect of nonlinearity on solutions of a hyperbolic equation or system is shown to have important consequences that make their solutions have quite different properties from those of linear hyperbolic equations. This is because when nonlinearity is present in a hyperbolic equation it is possible for a discontinuous solution to evolve from perfectly smooth initial data, and such discontinuous solutions have important properties of their own and propagate in a special way. An everyday physical example of a discontinuous solution occurs when an aircraft is in supersonic flight, because the abrupt change of pressure produced ahead of the aircraft produces a shock wave which is experienced by an observer on the ground as a sonic boom. Analogous phenomena are to be found in the study of water waves where hydraulic jumps can occur, in solids where shock waves can develop, in chromatography where an abrupt change in color or concentration occurs across an interface in a fluid, in continuum models of traffic flow when traffic jams cause vehicles to come to a standstill, and in many other situations.

The emphasis in this text is on the analytical and qualitative properties of partial differential equations, rather than on their numerical solution, which forms a quite separate field of study. However, before turning to the use of numerical methods of solution, it is desirable to have an understanding of the qualitative properties of the solutions of the different types of partial differential equations that occur, and essential to understand what forms of initial and boundary conditions are appropriate for different types of equations, all of which material is to be found in this book.

All but a few of the exercises found at the ends of chapters can be solved with pencil and paper, though access to a symbolic algebra package will simplify many of the calculations. The software package MAPLE V Release 5.1 was used when performing some of the routine calculations found in the examples, and also to generate many of the diagrams.

In addition to discussing standard topics like the classification of second-order equations, the reduction of partial differential equations to their standard form, and solutions obtained by the method of separation of variables, an introductory account is given of material that is becoming increasingly important in many current applications.

Special features of the book are as follows:

1. An emphasis on the role of integral conservation laws and their use when deriving the three fundamental types of linear second-order partial differential equations, followed by a systematic development of linear and nonlinear first-order equations, and a detailed study of the linear wave equation.

2. An introduction to the solution of all three fundamental types of linear partial differential equations by the method of separation of variables, using a unified approach that allows all types of equations to be studied simultaneously, followed by a study of the main properties of the associated Sturm-Liouville problems.

3. Because of the extensive use made of Fourier series in applications, the key ideas are described in detail, and as generalized Fourier series arise when applications depend on cylindrical and spherical polar coordinates, the reader is helped by the provision of detailed summaries of the most important properties of Legendre polynomials and Bessel functions.

4. Key qualitative properties of parabolic and elliptic equations are derived together with the fundamental solution of the heat equation and an explanation of Duhamel's principle. Quasilinear hyperbolic first-order systems of equations are introduced leading to the study of Riemann invariants and simple waves and, when conservation laws are involved, to shock solutions and to the solution of Riemann problems.

The physical applications used to illustrate the various methods of solution described in the text have been chosen because of the relative ease with which they can be understood, without the necessity for lengthy background introductions. However, the methods described are not limited to the applications found in this book, as they also apply to a far wider group of problems.

Many detailed worked examples are provided throughout the text. Their purpose is to illustrate some of the ways partial differential equations arise in applications, to provide examples of how methods developed in the text are to be applied, and also to show how some types of problems can be solved in very different ways. The exercises following most sections are designed to use ideas developed in the text, and to give the reader experience solving problems using different methods. Some of the exercises are of theoretical interest, while others are of practical interest and often require the use of cylindrical or spherical polar coordinate systems, and hence the use of Legendre polynomials and Bessel functions.

The necessity for the production of a *Student Solutions Manual* has been avoided by the inclusion of very detailed solutions to odd-numbered exercises, though an *Instructor's Manual* is available containing detailed solutions to all but the simplest exercises found in the book.

In conclusion, I wish to express my gratitude to the reviewers, Yuxi Zheng, Penn State University, William Moss, Clemson University, Chris Judge, Indiana University, Anthony Peressini, University of Illinois Urbana, Robert Fisher, University of Massachusets Amhurst, and Donald Hartig, California

Polytechnic State University. Their combined comments and helpful suggestions were invaluable, as they identified the need to add some new sections and to expand others. I also wish to express my gratitude to the accuracy checker Edgar Pechlaner, Simon Frazer University, for his comments and careful attention to detail. His valuable contribution has enabled me to free the book from some inconsistencies and various errors, and all errors that remain are my sole responsibility.

Alan Jeffrey

Introduction to Partial Differential Equations

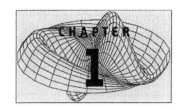

CHAPTER

1

1.1 What Is a Partial Differential Equation?

The behavior of scalar physical and mathematical quantities that can be represented by an unknown function u of two or more variables can often be characterized by an equation that relates some of the partial derivatives of u. A typical physical example is provided by the temperature $T(x, y, z, t)$ at time t inside a uniform block of metal at a point with the rectangular Cartesian coordinates (x, y, z). It will be seen later that in this case the temperature distribution as a function of time is described by the equation

$$\frac{\partial T}{\partial t} = \kappa \left(\frac{\partial^2 T}{\partial x^2} + \frac{\partial^2 T}{\partial y^2} + \frac{\partial^2 T}{\partial z^2} \right), \tag{1.1}$$

where κ is a constant that depends on the metal. This equation is called the *heat equation*, although it is also known as the *diffusion equation* because diffusive processes and heat conduction behave in a similar manner. As partial derivatives of the unknown function T with respect to the independent variables x, y, z, t are involved, the equation is called a partial differential equation for the function T involving three space variables and the time. This is often abbreviated by saying that (1.1) is the heat equation in $3 + 1$ independent variables, with the understanding that the number 3 refers to the number of space variables, without specifying the coordinate system to be used, while number 1 refers to the time variable. Thus the heat equation in $2 + 1$ independent variables refers to the heat equation in two space variables and time.

The presence of the time t in the temperature T means that (1.1) describes how the temperature evolves with time, so in this sense the temperature

1

distribution determined by (1.1) is said to be *time dependent* or *unsteady*. When the temperature T is independent of the time, so $T = T(x, y, z)$, the term on the left of (1.1) vanishes and T is then determined by the simpler partial differential equation

$$\frac{\partial^2 T}{\partial x^2} + \frac{\partial^2 T}{\partial y^2} + \frac{\partial^2 T}{\partial z^2} = 0. \tag{1.2}$$

A temperature determined by Eq. (1.2) is said to be a *steady-state* temperature distribution, and the partial differential equation (1.2) satisfied by T is called the three-dimensional *Laplace equation* expressed in terms of Cartesian coordinates. The Laplace equation only involves partial derivatives with respect to *space* coordinates, and never with respect to the time. We will see later that this important equation arises in many different ways and that it will need to be expressed in terms of other coordinate systems, like plane and cylindrical polar coordinates and spherical coordinates, which will be defined later.

Definition 1.1.   (Partial Differential Equation (PDE))
A partial differential equation involving an unknown function u of two or more independent variables, usually abbreviated to a PDE in u, is an equation that determines the behavior of u in terms of some of its partial derivatives, possibly also u itself, and the independent variables involved. The *order* of a PDE is the order of the highest partial derivative of u to appear in the equation.

The time-dependent heat equation (1.1) involves a first-order partial derivative of T with respect to the time t, and second-order partial derivatives of T with respect to each of the space variables x, y, and z, so it is a second-order PDE because the highest order partial derivatives to occur in the equation are of order 2.

For conciseness, and to simplify what is to follow, it is necessary to introduce some convenient notations. Partial derivatives will usually be denoted by means of suffixes so, for example, if $u(x, y, z, t)$ is a suitably differentiable function we will write

$$u_t = \frac{\partial u}{\partial t}, \quad u_x = \frac{\partial u}{\partial x}, \quad u_{yy} = \frac{\partial^2 u}{\partial y^2}, \quad u_{yz} = \frac{\partial^2 u}{\partial y \partial z} = \frac{\partial}{\partial z}\left(\frac{\partial u}{\partial y}\right), \dots.$$

Furthermore, when u is twice continuously differentiable, it will often be necessary to make use of the *equality of mixed derivatives*, which permits the interchange of operations of partial differentiation so that, for example, $u_{xy} = u_{yx}$ and $u_{yz} = u_{zy}$.

Sometimes when working with PDEs it is convenient to replace the variables x, y, z and t by x_1, x_2, x_3 and x_4, respectively, causing partial derivatives like u_x, u_{yy}, u_{yz} and u_t to be written $u_{x_1}, u_{x_2 x_2}, u_{x_2 x_3}$ and u_{x_4}. Another notation with certain advantages involves denoting these same partial derivatives by $\partial_{x_1} u, \partial_{x_2 x_2} u, \partial_{x_2 x_3} u$ and $\partial_{x_4} u$, respectively.

Using the first of these notations, a general first-order PDE for an unknown function $u(x, y)$ defined in some region D of the (x, y) plane can be written

$$F(x, y, u, u_x, u_y) = 0, \tag{1.3}$$

where F is an arbitrary continuous function of its arguments. A convenient shorthand notation used to show that the independent variables x and y in a function $u(x, y)$ are to be confined to a region D in the (x, y) plane involves writing $(x, y) \in D$, where the symbol $\in$ taken from set theory is to be read "belongs to."

An immediate extension of the notation for a general first-order PDE used in (1.3) involves denoting a general second-order PDE by writing

$$G(x, y, u, u_x, u_y, u_{xx}, u_{xy}, u_{yx}, u_{yy}) = 0, \tag{1.4}$$

where $u(x, y)$ is a suitably differentiable function defined in some region D of the (x, y) plane and G is an arbitrary continuous function of its arguments.

The bracketed expression on the right of (1.1), which also appears on the left of (1.2), is called the *Laplacian* of T when expressed in rectangular Cartesian coordinates. In what follows, the Laplacian of a suitably differentiable function u of several independent variables will be denoted by Δu, to be read "Laplacian u." Thus, when (1.1) is written in this concise manner, which omits any reference to the coordinate system involved, it becomes

$$T_t = \kappa \Delta T. \tag{1.1$'$}$$

In engineering applications, and much of applied mathematics, the Laplacian of u is often written $\nabla^2 u$, to be read "del squared u." The two notations are equivalent, and the symbol ∇^2 comes from the way the Laplacian of u is derived from vector analysis. In terms of the rectangular Cartesian coordinates x, y, and z the *Laplacian differential operator* Δ, equivalently ∇^2, becomes

$$\Delta \equiv \nabla^2 \equiv \frac{\partial^2}{\partial x^2} + \frac{\partial^2}{\partial y^2} + \frac{\partial^2}{\partial z^2}, \tag{1.5}$$

and this operator only gives rise to a function when it operates on a twice differentiable function of the variables x, y and z. Hence in terms of the suffix notation, the Laplacian of u becomes

$$\Delta u \equiv \nabla^2 u = u_{xx} + u_{yy} + u_{zz}. \tag{1.6}$$

If the variables x, y, and z in (1.6) are replaced by x_1, x_2, and x_3, respectively, we can take advantage of the numerical suffixes to write $\Delta u = \sum_{i=1}^{3} u_{x_i x_i}$. If the summation convention is used, where a repeated suffix implies summation over that suffix, this result contracts still more to $\Delta u = u_{x_i x_i}$, with $i = 1, 2, 3$.

Sometimes, when the Laplacian symbol Δ is used, it is necessary to show the number of independent variables entering into the Laplacian, and this is made clear by writing either $\Delta_2 u$ or $\Delta_3 u$, where the suffix 2 indicates a two-dimensional Laplacian and the suffix 3 a three-dimensional one.

When the region D in space to which the solution of a PDE is confined is either finite or semi-infinite, the *boundary* of D and its precise shape become important. To clarify the notion of a boundary, let us consider an arbitrary two-dimensional region D, and let P be a point belonging to D. Then point P will be called a *boundary point* of D if every circle with its center at P, however small its radius, always contains points that belong to D and points that do not. A point Q will be called an *interior point* of D if a circle with its center at Q can be constructed such that it only contains points of D. Analogously, a point R will be called an *exterior point* of D if a circle with its center at R that contains no points of D can be constructed. These definitions of internal, boundary, and external points of a region D extend immediately to three-dimensional regions if the word "circle" is replaced by "sphere." The *boundary* of a region D is the set of all of the boundary points of D and this will be denoted by ∂D. Our concern will mainly be with two-dimensional regions D for which the set of boundary points forms a continuous piecewise smooth curve with finitely many segments.

Thus we will always assume that with the exception of at most a finite number of points P on ∂D it will be possible to construct a unique *outward drawn unit normal* $\mathbf{n}(P)$ to ∂D. Here, the term *outward drawn* is used to refer to a normal $\mathbf{n}(P)$ to a point P on the boundary ∂D that is directed *away* from the interior of D. Thus if, for example, the boundary ∂D of region D is a sphere, the outward drawn normal to any point P on ∂D will be directed *away* from the interior of the sphere. The relationship between ∂D, the interior and exterior points of a plane region D, and the outward drawn unit normal to ∂D at P is illustrated in Fig. 1.1. The boundary point Q is exceptional, because no tangent to ∂D can be defined at Q, so no outward drawn normal to ∂D can be constructed at this point.

A PDE will be considered to be *solved* when all functions u that satisfy it throughout D have been found, so there may be more than one solution. In most physical applications only one solution is expected, and this is chosen from all possible solutions by requiring the solution u to satisfy both the PDE and certain conditions which are prescribed on all or part of ∂D. These auxiliary conditions to be satisfied by u on ∂D are called *boundary conditions*. When working with physical problems, the choice of coordinate system to be used and the boundary conditions that must be satisfied usually arise naturally from the problem itself, although this is not always the case.

In most physical problems, in addition to requiring a solution u to *exist*, and there to be only *one* solution, the solution is not expected to be excessively sensitive to small changes in the boundary conditions. This means

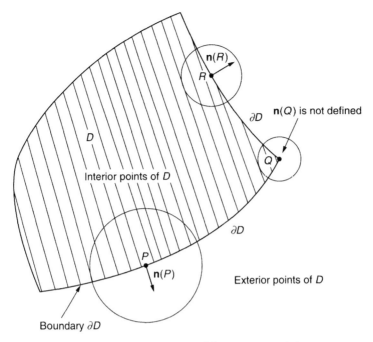

Figure 1.1   A region D, its boundary ∂D, and outward drawn unit normals $\mathbf{n}(P)$ and $\mathbf{n}(R)$.

that in a sense still to be defined, a small change in the boundary conditions must only produce a correspondingly small change in the solution itself. These two essential features of a solution of a PDE describing some physical phenomenon, which are suggested by both intuition and experiment, are expressed by saying that the solution of a physical problem must *exist* and also be *stable*.

When a solution $u(x, y)$ satisfies suitable boundary conditions on the boundary ∂D of D, and a unique smooth solution exists, it can be represented in the form of a surface $u = u(x, y)$ called an *integral surface*, or a *solution surface*, which is defined for all $(x, y) \in D$. Then, if P' is any point in D, at each corresponding point $u(P)$ on the integral surface it is possible to construct a unit vector $\mathbf{n}(P)$ normal to the surface, where $\mathbf{n}(P)$ is unique apart from its sign, since it can be directed away from either side of the surface. This is illustrated in Fig. 1.2, where the choice of orientation of $\mathbf{n}(P)$ has been chosen arbitrarily.

In applications like gas dynamics, electromagnetic theory, and elasticity several unknown scalar functions $u_1, u_2, \ldots, u_n$ occur, each a function of m independent variables, and they are related by N coupled PDEs. A simultaneous set of PDEs of this type is called a *system* of PDEs, and it can happen that $n > N, n = N$, or $n < N$. A system for which $n > N$ is said to be *underdetermined*, since it contains more unknowns than equations. A system in which $n = N$ is said to be *properly determined*, because the number of unknowns

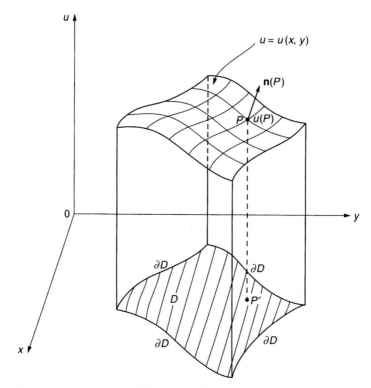

Figure 1.2   A normal $\mathbf{n}(P)$ at point P on the surface $u = u(x, y)$.

equals the number of equations involved, while a system for which $n < N$ is said to be *overdetermined*, because the n unknowns are required to satisfy more than n equations. Only properly determined systems will be considered in what is to follow. A typical example of a properly determined system is provided by the coupled first-order equations $u_x = v_y$, $u_y = -v_x$. It will be recognized that these are the *Cauchy–Riemann equations* from complex analysis, where they are shown to be the equations that must be satisfied simultaneously by the real part $u(x, y)$ and imaginary part $v(x, y)$ of a complex analytic function $w = u + iv$.

A general qualitative description of individual PDEs, which is useful because it reflects the order of the complexity of solutions to be expected, is provided by the following definition.

Definition 1.2.   (Linear and Nonlinear PDEs)

(i) A PDE satisfied by a suitably differentiable function u of the independent variables $x_1, \ldots, x_n$ is said to be *linear* if u and its partial derivatives only occur linearly and, possibly, with coefficients that are functions of the independent variables. This PDE may, or may not, contain a function f that depends only on the independent variables. The PDE is said to be *homogeneous* when the function $f(x_1, \ldots, x_n) \equiv 0$; otherwise it is said to be *nonhomogeneous*.

(ii) A PDE satisfied by a suitably differentiable function u of the independent variables $x_1, \ldots, x_n$ is said to be *semilinear* when all derivatives of u occur linearly, with coefficients that may be functions of the independent variables, but u itself occurs nonlinearly.

(iii) A PDE of order k satisfied by a suitably differentiable function u of the independent variables $x_1, \ldots, x_n$ is said to be *quasilinear* if its partial derivatives of order k appear linearly, possibly with coefficients that are functions of u and derivatives of u of order less than k and also the independent variables.

(iv) A PDE that belongs to none of the above categories is said to be *nonlinear*.

These definitions extend in an obvious way to systems of PDEs, and the following examples illustrate some of the most important linear, semilinear, and quasilinear equations and systems that occur in applications to physical problems.

1. *Linear transport (advection) equation in 3 + 1 dimensions*

$$u_t = \sum_{i=1}^{3} a_i u_{x_i},$$

   with the a_i constants (first-order linear homogeneous PDE).

2. *The Laplace equation in two space dimensions*

$$\Delta_2 u = 0 \quad \text{(second-order linear homogeneous PDE)}.$$

3. *The Poisson equation in three space dimensions*

$$\Delta_3 u = -f(x_1, x_2, x_3) \quad \text{(second-order linear nonhomogeneous PDE)}.$$

4. *The Poisson equation in three space dimensions with $f(u)$ being a nonlinear function*

$$\Delta_3 u = -f(u) \quad \text{(second-order semilinear PDE)}.$$

5. *The heat equation in 3 + 1 dimensions with a source term Q*

$$u_t = \kappa \Delta_3 u + Q \quad \text{(second-order linear nonhomogeneous PDE)}.$$

6. *The wave equation in 2 + 1 dimensions*

$$u_{tt} = c^2 \Delta_2 u, \quad c = \text{constant} \quad \text{(second-order linear homogeneous PDE)}.$$

7. *The one-dimensional vibrating beam equation with a distributed load*

$$u_{tt} + a^2 u_{xxxx} = F \quad \text{(fourth-order linear nonhomogeneous PDE)}.$$

8. *Generalized Burgers' equation without dissipation*

$$u_t + f(u)u_x = 0 \quad \text{(first-order quasilinear PDE)}.$$

9. *Generalized Burgers' equation with dissipation*

$$u_t + f(u)u_x = u_{xx} \quad \text{(second-order quasilinear PDE)}.$$

10. *Korteweg–de Vries (KdV) equation*

$$u_t + uu_x + u_{xxx} = 0 \quad \text{(third-order quasilinear PDE)}.$$

11. *Maxwell's equations in a vacuum*

$$(1/c)\mathbf{E}_t = \text{curl } \mathbf{H}, \qquad (1/c)\mathbf{H}_t = -\text{curl } \mathbf{E}, \qquad \text{div } \mathbf{H} = \text{div } \mathbf{E} = 0,$$

with $\mathbf{E}$ being the electric vector, $\mathbf{H}$ the magnetic vector, and c the speed of light (a first-order properly determined linear homogeneous system for the three-dimensional vectors $\mathbf{E}$ and $\mathbf{H}$).

12. *Euler's equations for a compressible fluid*

$$\rho_t + \text{div }(\rho\mathbf{u}) = 0, \qquad \mathbf{u}_t + (\mathbf{u} \cdot \text{grad})\mathbf{u} + (1/\rho)\text{grad } p = 0,$$

with $\mathbf{u}$ being the fluid velocity vector, ρ the fluid density, and $p = p(\rho)$ a known function relating the pressure p and density ρ (the *constitutive equation*) (a first-order properly determined quasilinear system for $\mathbf{u}$ and ρ).

13. *Telegraph equation*

$$u_{tt} = c^2 u_{xx} - au_t - bu \quad \text{(second-order linear homogeneous PDE)}.$$

14. *Minimal surface equation*

$$\left(1 + u_y^2\right)u_{xx} - 2u_x u_y u_{xy} + \left(1 + u_x^2\right)u_{yy} = 0$$
$$\text{(second-order nonlinear PDE)}.$$

This equation describes the shape of a soap film spanning a wire loop, and in mathematical terms a *minimal surface* is one for which the average of the maximum and minimum curvature of the surface (its *mean curvature*) is zero at each point of the surface.

Inspection of the above equations shows the Poisson equation to be a non-homogeneous Laplace equation, while setting $f(u) = c$ (constant) in Burgers' equation without dissipation reduces it to the one-dimensional form of the transport equation. The wave equation in 6 is linear and second-order in both space and time, while the vibrating beam equation, which is also linear, is second order in time but fourth order in the space variable x. This equation describes the transverse vibrations of a beam, where the nature of the vibrations depends on the way the ends of the beam are fixed, how the vibrations are started, and the forcing function F. The systems in 11 and 12 are both properly determined, because the vector form of Maxwell's equations is equivalent to six scalar equations for the three components of the electric vector $\mathbf{E}$ and the three components of the magnetic vector $\mathbf{H}$, while the Euler equations reduce to four scalar equations for the three components of the velocity $\mathbf{u}$ and the density ρ (equivalently the pressure p).

The telegraph equation in 13 is a one-dimensional wave equation that takes into account the effect of two important effects called dispersion and dissipation. Dispersion causes the shape of a disturbance to change as it propagates, while dissipation causes it to decay with time.

The minimal surface equation in 14 is nonlinear and second order in the two space variables x and y. It describes, for example, the shape of a soap film spanning a twisted loop of wire.

It is to be expected that the solution of a partial differential equation is simplified if a coordinate system that matches as closely as possible to the geometry of the problem is used. For example, if the temperature distribution is required in a rectangular block of metal, it is natural to use rectangular Cartesian coordinates with axes parallel to the edges of the block, while when seeking the electric potential in a cylindrical cavity it is natural to use cylindrical polar coordinates with the z axis coinciding with the center line of the cavity.

This being so, in addition to knowing the relationship between these different coordinate systems, it is necessary to know how the Laplacian in rectangular Cartesian coordinates given in (1.6) transforms into the *cylindrical polar coordinate system* (r, θ, z) and the *spherical polar coordinate system* (r, ϕ, θ). These, and other matters, are taken up later in Section 1.6, but because of their considerable importance we repeat here the notations used and the forms taken by the Laplacian in these two different coordinate systems.

To avoid confusion caused by differing notations, where in spherical coordinates the angles θ and ϕ are often interchanged, the convention adopted here for cylindrical polar coordinates is shown in Fig. 1.3, and for spherical polar coordinates in Fig. 1.4. When referring to a point P in cylindrical coordinates the notation (r, θ, z) is used, in this order, with r being the radial distance of the projection of P onto the plane $z = 0$, θ the azimuthal angle measured counterclockwise from the x axis to the radial line r, and z the ordinate from the plane $z = 0$ to the point P, as shown in Fig. 1.3. Thus r is a nonnegative quantity, $0 \leq \theta < 2\pi$.

The convention is different when working with spherical polar coordinates, because there a point P is identified by using the notation (r, θ, ϕ), in this order, where now r is the radial distance from the origin to the point P, ϕ is the azimuthal angle measured counterclockwise from the x axis to the projection of

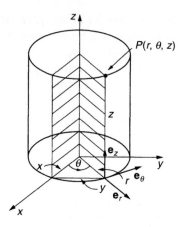

Figure 1.3 Cylindrical polar coordinates (r, θ, z).

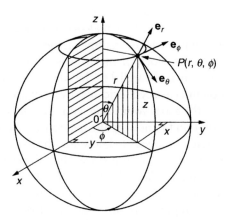

Figure 1.4 Spherical polar coordinates (r, θ, ϕ).

r onto the plane $z = 0$, and θ is the polar angle, measured as shown in Fig. 1.4. Thus in spherical polar coordinates, as before r is a nonnegative quantity, $0 \leq \phi < 2\pi$ and $0 \leq \theta \leq \pi$.

The connections between these two coordinate systems and rectangular Cartesian coordinates are given below.

Cylindrical polar coordinates (r, θ, z):

$$x = r \cos\theta, \qquad y = r \sin\theta, \qquad z = z \quad (0 \leq \theta < 2\pi). \quad (1.7)$$

Spherical polar coordinates (r, θ, ϕ):

$$x = r \sin\theta \cos\phi, \quad y = r \sin\theta \sin\phi, \quad z = r \cos\theta$$
$$(0 \leq \phi < 2\pi, \; 0 \leq \theta \leq \pi). \quad (1.8)$$

In vector calculus the gradient operator written grad, and also denoted by ∇ and read "del", takes the form

$$\text{grad} \equiv \nabla \equiv \mathbf{i}\frac{\partial}{\partial x} + \mathbf{j}\frac{\partial}{\partial y} + \mathbf{k}\frac{\partial}{\partial z}, \tag{1.9}$$

where $\mathbf{i}$, $\mathbf{j}$, and $\mathbf{k}$ are unit vectors along the respective x, y, and z coordinate axes, so if $u(x, y, z)$ is a suitably differentiable scalar function,

$$\text{grad}\, u \equiv \nabla u = \mathbf{i}\frac{\partial u}{\partial x} + \mathbf{j}\frac{\partial u}{\partial y} + \mathbf{k}\frac{\partial u}{\partial z}. \tag{1.10}$$

In vector calculus, the vector $(\text{grad}\, u)_0$ at a point (x_0, y_0, z_0) is a vector pointing in the direction in which u changes most rapidly, and the magnitude of this vector measures the rate of change of u at the point. A useful result is that if $\hat{\mathbf{n}}$ is an arbitrary unit vector, the quantity $\hat{\mathbf{n}} \cdot \text{grad}\, u$ is the *directional derivative* of u in the direction $\hat{\mathbf{n}}$, or the rate of change of the derivative of u in the direction $\hat{\mathbf{n}}$.

For reference purposes we now quote the forms taken by grad u and Δu in cylindrical and spherical polar coordinates, but more information concerning other vector operators is to be found in Section 1.6.

Cylindrical polar coordinates (r, θ, z):

$$\text{grad}\, u \equiv \nabla u = \frac{\partial u}{\partial r}\mathbf{e}_r + \frac{1}{r}\frac{\partial u}{\partial \theta}\mathbf{e}_\theta + \frac{\partial u}{\partial z}\mathbf{e}_z, \tag{1.11}$$

where $\mathbf{e}_r$, $\mathbf{e}_\theta$, and $\mathbf{e}_z$ form a right-handed set of unit vectors in the respective directions in which r, θ, and z increase (see Fig. 1.3), and

$$\Delta u = \frac{\partial^2 u}{\partial r^2} + \frac{1}{r}\frac{\partial u}{\partial r} + \frac{1}{r^2}\frac{\partial^2 u}{\partial \theta^2} + \frac{\partial^2 u}{\partial z^2}. \tag{1.12}$$

A special case of cylindrical polar coordinates arises when $u = u(r, \theta)$, in which case u experiences no change in the z direction. As a result, (1.7) reduces to $x = r\cos\theta$, $y = r\sin\theta$, while in (1.11) the term $\partial u/\partial z \equiv 0$, and in (1.12) the term $\partial^2 u/\partial z^2 \equiv 0$. Coordinates of this type, which involve only r and θ, are called **plane polar coordinates**, or sometimes simply **polar coordinates**. This coordinate system involving only two independent variables can be considered to describe a position in the plane $z = 0$ in terms of r and θ. A different interpretation involves regarding the coordinates as describing the situation in three-dimensional space when there is no change in the z direction, in which case the point (r, θ) can be considered to refer to a point in any plane $z = $ constant.

Spherical polar coordinates (r, θ, ϕ):

$$\text{grad}\, u \equiv \nabla u = \frac{\partial u}{\partial r}\mathbf{e}_r + \frac{1}{r}\frac{\partial u}{\partial \theta}\mathbf{e}_\theta + \frac{1}{r\sin\theta}\frac{\partial u}{\partial \phi}\mathbf{e}_\phi, \tag{1.13}$$

where e_r, e_θ, and e_ϕ form a right-handed set of unit vectors in the respective directions in which r, θ, and ϕ increase (see Fig. 1.3), and

$$\Delta u = \frac{1}{r^2} \frac{\partial}{\partial r}\left(r^2 \frac{\partial u}{\partial r}\right) + \frac{1}{r^2 \sin\theta} \frac{\partial}{\partial \theta}\left(\sin\theta \frac{\partial u}{\partial \theta}\right) + \frac{1}{r^2 \sin^2\theta} \frac{\partial^2 u}{\partial \phi^2}. \qquad (1.14)$$

EXERCISES 1.1

1. Use the coordinate transformation (1.7) in the Cartesian form of the Laplacian to show that in cylindrical polar coordinates

$$\Delta u = \frac{\partial^2 u}{\partial r^2} + \frac{1}{r} \frac{\partial u}{\partial r} + \frac{1}{r^2} \frac{\partial^2 u}{\partial \theta^2} + \frac{\partial^2 u}{\partial z^2}.$$

2. Use the coordinate transformation (1.8) in the Cartesian form of the Laplacian to show that in spherical polar coordinates

$$\Delta u = \frac{1}{r^2} \frac{\partial}{\partial r}\left(r^2 \frac{\partial u}{\partial r}\right) + \frac{1}{r^2 \sin\theta} \frac{\partial}{\partial \theta}\left(\sin\theta \frac{\partial u}{\partial \theta}\right) + \frac{1}{r^2 \sin^2\theta} \frac{\partial^2 u}{\partial \phi^2}.$$

3. Write down the form taken by the Laplace equation in spherical polar coordinates when the solution u depends only on r and θ, and explain the geometrical implication for the solution.

4. Show that by a suitable scaling of the space coordinates, the heat equation,

$$u_t = \kappa(u_{xx} + u_{yy} + u_{zz}),$$

can be reduced to the standard form $v_t = \Delta v$, where u becomes v after scaling.

5. By using suitable scaling of the solution u and the time t, show that the Korteweg–de Vries (KdV) equation,

$$u_t + uu_x + \mu u_{xxx} = 0, \quad \text{with } \mu \neq 1 \text{ a constant,}$$

can be reduced to the standard form $v_\tau + vv_x + v_{xxx} = 0$, where u becomes v and t becomes τ after scaling.

1.2 Representative Problems Leading to PDEs, Initial and Boundary Conditions

Only in special cases is it possible to find a general solution of a PDE. Consequently, instead of seeking a general solution, it becomes necessary to devise methods for solving specific problems whose solution satisfies a PDE. Here a specific problem means finding the solution of a given PDE subject to some auxiliary conditions which, together with the PDE, lead to a unique solution. The auxiliary conditions are often suggested by the problem itself, in which case they can usually be given a simple physical interpretation.

To discover the sort of auxiliary conditions that must be imposed when solving a PDE we will derive some PDEs that model several important though very different physical situations. In the process, the nature of the problem will be seen to suggest appropriate auxiliary conditions. The various types of auxiliary conditions that arise will be found to be associated with PDEs, which exhibit very different characteristics, and this is one of the reasons why it is necessary to classify PDEs according to type, so when a PDE is encountered, the correct form of auxiliary conditions can be used. An explanation of the basis for the classification of second-order PDEs, and a discussion of the three essentially different types that arise, is given later in Chapter 3.

The traffic flow problem

In the first physical problem to be considered we derive a simple mathematical model for the one-way flow of traffic along a long straight road between a point A at one end and a point B at the other, on the assumption that vehicles move to the right and can neither enter nor leave the road between A and B. This situation is illustrated in Fig. 1.5 where the x axis lies along the road, with the sense of increasing x taken in the direction of the traffic flow.

To arrive at a simple mathematical model of traffic flow, instead of examining the behavior of individual vehicles, we will assume them to be sufficiently numerous that they can be considered to be distributed continuously from A to B. Accordingly, we define the continuous and differentiable function $\rho(x, t)$ to be the number of vehicles in a unit length of the road at time t and position x, and call this the *vehicle density*. A quantity that is easily recorded by observation of traffic flow along a road is the *flux* of vehicles $q(x, t)$, defined as the number of vehicles that at time t pass a given point with coordinate x in a unit of time.

Let us now consider the flow of vehicles past any two fixed points x_1 and x_2 located between A and B, with $x_1 < x_2$, as shown in Fig. 1.5. As vehicles can neither enter nor leave the road between A and B, it follows that the increase in the number of vehicles in this interval in a unit of time must be the difference between the number of vehicles entering at x_1 and leaving at x_2. From the vehicle density it then follows that at time t this number must be $\int_{x_1}^{x_2} \rho(x, t)\, dx$, so the rate of change of this quantity is $\frac{d}{dt}\int_{x_1}^{x_2} \rho(x, t)\, dx$. In terms of the flux $q(x, t)$ of vehicles at time t, which we assume to be continuous and differentiable, the

Figure 1.5   Traffic flow between A and B.

difference between the number of vehicles entering at x_1 and leaving at x_2 in a unit of time is $q(x_1, t) - q(x_2, t)$, so as this difference must equal the previous result, we can write

$$\frac{d}{dt} \int_{x_1}^{x_2} \rho(x, t)dx = q(x_1, t) - q(x_2, t). \tag{1.15}$$

Provided $q(x, t)$ is considered to be a differentiable function of x, it follows from the fundamental theorem of calculus that

$$\int_{x_1}^{x_2} \frac{\partial}{\partial x} q(x, t)dx = q(x_2, t) - q(x_1, t). \tag{1.16}$$

Taking account of the difference in sign between (1.15) and (1.16), and assuming $\rho(x, t)$ to be differentiable with respect to t, after taking the time derivative under the integral sign in (1.15), result (1.16) becomes

$$\int_{x_1}^{x_2} \{\rho_t(x, t) + q_x(x, t)\}dx = 0. \tag{1.17}$$

A result of this type, which describes the balance between certain quantities in terms of an integral, is called an *integral conservation law*. Conservation laws such as this play a fundamental role throughout applied mathematics, physics, and engineering. As x_1 and x_2 are arbitrary, and the integral in (1.17) is true irrespective of their values, this can only follow if the integrand is identically zero. As a result, (1.17) implies the *PDE*

$$\rho_t(x, t) + q_x(x, t) = 0, \tag{1.18}$$

which is called a *PDE conservation law*. Note that the second term in (1.18) occurs in the form of a *divergence*, although here it only involves one space dimension. The presence of such a divergence term is characteristic of all PDE conservation laws. In this case, it is appropriate to refer to (1.18) as the *PDE traffic conservation law*.

To make further progress it is necessary to postulate a functional relationship between the vehicle flux q and the traffic density ρ, so when q is substituted into (1.18) it becomes a first-order PDE in ρ. In practice the choice of function connecting q and ρ is determined by observation of traffic patterns, and no one relationship is suitable for all situations.

In the simplest case, when the vehicles all move to the right at a constant speed $c > 0$, the flux $q = c\rho$, and then $q_x = c\rho_x$. Using this in (1.18) we arrive at the simple first-order linear PDE for the traffic density function $\rho(x, t)$

$$\rho_t + c\rho_x = 0. \tag{1.19}$$

This PDE is seen to be a one-dimensional example of the transport equation given in the list of PDEs in Section 1.1, where it occurs as entry number 1.

The form of (1.19) is sufficiently simple for it to be seen to have the *general solution*

$$\rho(x, t) = f(x - ct), \tag{1.20}$$

where f is an arbitrary differentiable function of its argument $x - ct$. This is easily confirmed by substituting (1.20) into (1.19), although a direct derivation of this general solution can be obtained by using the method of characteristics which will be discussed later. To discover the connection between the arbitrary function f and the traffic flow in the special solution described by (1.20), we must consider how the traffic flow started. Suppose at time $t = 0$ the traffic density distribution was $\rho(x, 0) = F(x)$, where $F(x)$ is a known function of x found by observation at time $t = 0$. Then the condition $\rho(x, 0) = F(x)$ imposed on the traffic density at $t = 0$ is what is called an *initial condition* for $\rho(x, t)$. Setting $t = 0$ in (1.20) shows $f(x) = F(x)$, so replacing f by F and x by $x - ct$, leads to the following solution of this initial value problem,

$$\rho(x, t) = F(x - ct). \tag{1.21}$$

To understand the nature of this solution we make a change of variable and set $\zeta = x - ct$, when we can write $\rho(x, t) = F(\zeta)$. This shows that the traffic flow density $\rho(x, t) = F(\zeta)$ is constant along each straight line, $x - ct =$ constant. All of the straight lines $\zeta =$ constant are parallel with slope $dx/dt = c$, but as x is a distance and t is the time, this confirms that c must be a speed, since it has the dimensions of length/time. The physical significance of this solution is that, whatever the initial traffic density distribution, the traffic density distribution at a time $t > 0$ is obtained by shifting the initial distribution to the right through a distance ct, without any change of shape. Expressed differently, this result says that if one of the parallel straight lines intersects the x axis at $x = \xi$, then everywhere along the line $x - ct = \xi$ it must follow that $\rho = \rho(\xi, 0) = F(\xi)$. Thus the value of $\rho(x, t)$ remains constant along each of these lines with its value being equal to $F(\xi)$ where the line passes through the point $(\xi, 0)$ on the x axis. As the constant value of ρ depends on ξ, it follows that, in general, ρ has different constant values along different lines. This situation is illustrated graphically in Fig. 1.6, where the initial traffic density distribution in any interval $x_1 \leq x \leq x_2$ at time $t = 0$ is seen to move to the right with speed c. Because of this property, the solution of PDE (1.19) can be considered to represent a *wave* moving to the right at a constant speed c without change of shape. This result is to be expected, because as the vehicles all move at the same constant speed c, the distances between vehicles must remain constant, causing the shape of the initial distribution to be preserved as it moves to the right.

A more interesting situation arises when vehicle speeds are not constant. If $u(x, t)$ is the traffic speed at point x and time t, then as the flux $q = u\rho$, we see that when the density ρ vanishes the flux q must also vanish. Also, when the density is large, causing the vehicles to be very close together, the

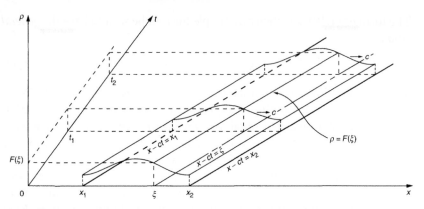

Figure 1.6   The initial density distribution moves to the right with constant speed c without change of shape.

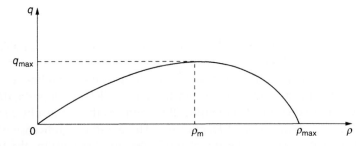

Figure 1.7   A realistic vehicle density distribution.

vehicles can again only move very slowly, as is familiar to drivers in dense city traffic. In the limiting case when the density ρ attains its maximum value ρ_{max}, corresponding to vehicles almost touching one another, the flux q must again be zero. Between the densities $\rho = 0$ and $\rho = \rho_{max}$, there will be some value $\rho = \rho_m$ where the flux q attains a maximum q_{max}. Intuition suggests that a typical relationship of this type should be like that in Fig. 1.7.

A possible flux function with these properties is

$$q = \begin{cases} 0, & \rho < 0 \\ \rho \frac{q_{max}(\rho_{max}-\rho)}{\rho_m(\rho_{max}-\rho_m)}, & 0 \le \rho \le \rho_{max}. \\ 0, & \rho > \rho_{max} \end{cases} \qquad (1.22)$$

As $q = q(\rho)$, it follows that $q_x = (\partial q/\partial \rho)\rho_x$, where $\partial q/\partial \rho$ is now a function of ρ, so that the traffic conservation law (1.18) becomes the first-order quasilinear PDE

$$\rho_t + (\partial q/\partial \rho)\rho_x = 0. \qquad (1.23)$$

This is seen to be in the form of the dissipationless generalized Burgers' equation given in the list of PDEs in Section 1.1, where it occurs as entry number 8.

By analogy with the linear case in (1.16), it is reasonable to suppose that when the initial condition $\rho(x, 0) = F(x)$ is imposed on (1.23) this will again suffice to determine a unique solution. However, the solution cannot now be as simple as that in (1.21), and when we come to examine the solution of first-order quasilinear PDEs it will be seen that although such equations can still be considered to propagate an initial condition in the form of a *wave*, this time the shape of the wave will be distorted as it advances. Furthermore, it will be seen that this distortion can produce a dramatic change in behavior of the wave from that found in the linear case. This is because at a finite time $t = t_c$ after the start of the traffic flow, the solution can distort to the point where it becomes nonunique. When the distortion reaches this stage partial derivatives of the solution can no longer be defined, so the evolution of the solution beyond this critical time can no longer be described by the original PDE.

The heat equation with a source term

Let us now derive the PDE that determines the time-dependent temperature distribution in a rather general heat-conducting solid occupying a volume V of space under the following assumptions, where t is the time and $\mathbf{r}$ is the position vector of an arbitrary point in the solid:

- the solid is *isotropic*, so at any given point $\mathbf{r}$ the heat-conducting properties are independent of direction;

- the density ρ of the solid may be a function of position, so $\rho = \rho(\mathbf{r})$;

- the thermal conductivity k of the solid is a function of position, so $k = k(\mathbf{r})$;

- there is a distributed time-dependent heat source (or sink) H throughout the solid that is a function of position and time, so $H = H(\mathbf{r}, t)$;

- the specific heat c of the solid may be a function of position, so $c = c(\mathbf{r})$.

These conditions are satisfied by all ordinary metals, for which the situation is somewhat simpler, because for such materials the density, thermal conductivity, and specific heat can be considered to be absolute constants.

Let S be an arbitrary smooth surface drawn inside V, and $T = T(\mathbf{r}, t)$ be the temperature at a point with position vector $\mathbf{r}$ on surface S. We now need to appeal to an experimental law called the *Fourier heat conduction law*, which asserts that the rate of flow of heat energy through a unit area in a unit of time (the *heat flux*) is proportional to the gradient of the temperature T normal to the area, with the heat flow in the direction of decreasing temperature. Assuming the *Fourier heat conduction law* holds in the material to be considered (it does in all practical situations involving metals), the amount of heat dq

flowing through an element dS of the surface located at $\mathbf{r}$ in an interval of time dt must be proportional to dt, dS, and the gradient of T normal to S at $\mathbf{r}$. Introducing the proportionality factor $k(\mathbf{r}) > 0$, called the *thermal conductivity* of the solid at the point $\mathbf{r}$, allows us to write

$$dq = -k(\mathbf{r})\hat{\mathbf{n}} \cdot \text{grad } T \, dS \, dt, \tag{1.24}$$

where $\hat{\mathbf{n}}$ is the unit normal to S at $\mathbf{r}$ in the direction of decreasing temperature, and the negative sign is introduced because heat flows from a higher temperature to a lower one. It follows from (1.24) that if q is the *heat flux* passing in unit time through a unit area of surface with unit normal $\hat{\mathbf{n}}$ located at a point with position vector $\mathbf{r}$, we can write

$$q = -k(\mathbf{r})\hat{\mathbf{n}} \cdot \text{grad } T. \tag{1.25}$$

Integrating (1.25) over the entire volume V occupied by the solid, the boundary surface of which is ∂V, shows that the amount of heat q_c present in volume V due to conductivity from time t_1 to time t_2 is

$$q_c = -\int_{t_1}^{t_2}\int_{\partial V} k(\mathbf{r})\hat{\mathbf{n}} \cdot \text{grad } T \, dS \, dt$$

$$= -\int_{t_1}^{t_2}\int_{\partial V} k(\mathbf{r})(\text{grad } T) \cdot d\mathbf{S} \, dt, \tag{1.26}$$

where $d\mathbf{S} = \hat{\mathbf{n}} \, dS$ is now the vector element of area of ∂V with its unit normal $\hat{\mathbf{n}}$ directed toward the *interior* of ∂V.

If a distributed time-dependent heat source (or sink) exists in the solid, possibly due to chemical or electrical effects, and at point $\mathbf{r}$ it produces (or removes) an amount of heat $H(\mathbf{r}, t)$ in a unit volume in a unit time, then integrating H over the volume of the solid shows the heat q_s generated (removed) from the solid due to the source (sink) in the interval of time from t_1 to t_2 is

$$q_s = \int_{t_1}^{t_2}\int_V H \, dV \, dt. \tag{1.27}$$

The total amount of heat entering V in this interval of time is $q_c + q_s$, and this must lead to a change in temperature throughout the solid.

To find the temperature change let dV be an element of volume V located at a point $\mathbf{r}$, and suppose that during the interval of time from t_1 to t_2 the temperature of the volume element dV changes from $T(\mathbf{r}, t_1)$ to $T(\mathbf{r}, t_2)$. Then the amount of heat dq required to produce this change is

$$dq = [T(\mathbf{r}, t_2) - T(\mathbf{r}, t_1)]c(\mathbf{r})\rho(\mathbf{r}) \, dV, \tag{1.28}$$

where $c(\mathbf{r})$ is the specific heat of the material at position vector $\mathbf{r}$. Assuming that $T(\mathbf{r}, t)$ is continuous and differentiable, it follows from the fundamental theorem of calculus that we can set

$$T(\mathbf{r}, t_2) - T(\mathbf{r}, t_1) = \int_{t_1}^{t_2} T_t \, dt,$$

so using this result in (1.28) and integrating over V shows the amount of heat q required to produce this temperature change throughout V must be

$$q = \int_{t_1}^{t_2} \int_V c(\mathbf{r}) \rho(\mathbf{r}) T_t \, dV \, dt. \tag{1.29}$$

When the temperature change is due only to the combined effects of conduction determined by (1.26) and a distributed heat source (sink) determined by (1.27), and the solid undergoes no phase change during which it can absorb or release heat without a temperature change, we must have $q = q_c + q_s$, so

$$\int_{t_1}^{t_2} \int_V c(\mathbf{r}) \rho(\mathbf{r}) T_t \, dV \, dt = - \int_{t_1}^{t_2} \int_{\partial V} k(\mathbf{r})(\mathrm{grad}\ T) \cdot d\mathbf{S} \, dt + \int_{t_1}^{t_2} \int_V H \, dV \, dt.$$
$$\tag{1.30}$$

An application of the Gauss divergence theorem (see Section 1.6) to the first term on the right, followed by a reversal of sign because the unit normal $\hat{\mathbf{n}}$ to a surface element d is directed *into* V, whereas in the divergence theorem it is required to be directed *out of* V, enables all the terms of (1.30) to be combined into the single integral

$$\int_{t_1}^{t_2} \int_V [c(\mathbf{r}) \rho(\mathbf{r}) T_t - \mathrm{div}(k(\mathbf{r})\mathrm{grad}\ T) - H] \, dV \, dt = 0. \tag{1.31}$$

This result is an integral form of the *conservation law for heat* in the solid. As the integrand is continuous, and V, t_1, and t_2 are arbitrary, (1.31) can only be true if the integrand is identically zero, from which it follows that

$$c(\mathbf{r}) \rho(\mathbf{r}) T_t = \mathrm{div}(k(\mathbf{r})\mathrm{grad}\ T) + H. \tag{1.32}$$

This is the PDE form of the *conservation law for heat flow* in the solid.

In the simplest case when c, ρ, and k are all constants, introducing the constant $\kappa = k/c\rho$, called the *diffusivity* of the solid, and using the result that $\mathrm{div}(k\ \mathrm{grad}\ T) = k\Delta T$ causes (1.32) to simplify to

$$T_t = \kappa \Delta T + H/(c\rho), \tag{1.33}$$

which is the form of the *heat equation* given in Section 1.1 with $Q = H/(c\rho)$.

When applying either (1.32) or (1.33) to an actual problem, the shape of the boundary ∂V of the solid must be taken into account together with the conditions to be satisfied by the temperature on ∂V. Some typical conditions that can be applied to the temperature on the boundary are as follows:

- The temperature T is specified on a boundary. A **boundary condition** of this type, where the value of the solution of a PDE is specified on a boundary, is called a **Dirichlet condition.**

- The rate at which heat is introduced (or removed) from the solid through a boundary is specified. The heat flux is $q = -k(\mathbf{r})\hat{\mathbf{n}} \cdot \text{grad } T$, so the specification of q over a boundary is seen to be equivalent to specifying $\hat{\mathbf{n}} \cdot \text{grad } T$ over a boundary. A boundary condition of this type, where the **normal derivative** of the solution of a PDE is specified over a boundary (that is, the derivative of the solution *normal* to ∂V), is called a **Neumann condition.**

- When the surface temperature T of the solid is close to that of the surrounding medium (usually air) at a constant temperature T_0, the transfer of heat from (or to) the solid by convection and radiation can be modeled by **Newton's law of cooling.** In this approximation, the heat flux at a point $\mathbf{r}_s$ on the surface is proportional to the difference in temperature $T - T_0$ at $\mathbf{r}_s$ between the surface temperature T and temperature T_0 of the surrounding medium. If h is the **heat transfer coefficient** between the surface and the surrounding medium, and we consider h to be constant, a condition of this type takes the form

$$h(T - T_0) = -k(\mathbf{r}_s)\hat{\mathbf{n}} \cdot \text{grad } T \tag{1.34}$$

at each point $\mathbf{r}_s$ on the surface. A more general boundary condition of this type, which includes (1.34) as a special case, is the requirement that on the boundary

$$\alpha u + \beta\, \hat{\mathbf{n}} \cdot \text{grad } u = f. \tag{1.35}$$

This is called a **Robin condition,** or sometimes a **mixed boundary condition,** and the condition is said to be **nonhomogeneous** when $f \neq 0$; otherwise it is **homogeneous.**

If there are no internal heat sources (or sinks) in the solid, the only external means by which the temperature distribution (equivalently the heat flow) within a solid can be influenced is through its boundary surface. Consequently, when time is involved, it is also necessary to specify the temperature distribution on the surface at some initial time, which for convenience is usually taken to be $t = 0$. A condition of this type applied to a solution of a PDE is another example of an **initial condition.**

It becomes necessary to impose an **internal boundary condition** on the solution of the heat equation if the solid is a composite material for which perfect contact is made across a surface Σ between two different heat-conducting materials with thermal conductivities k_1 and k_2. The form taken by the internal boundary conditions is easily found from the physical requirements that both the temperature and heat flux must be continuous across Σ. The first condition follows from the fact that the solids are in perfect contact across the surface Σ, so there can be no jump in temperature across it. The second condition comes from the fact that if the heat flux was not continuous across Σ, an unbounded accumulation of heat would occur at the surface. If the temperatures in the respective materials are T_1 and T_2, and a unit normal to Σ is $\hat{\mathbf{n}}$, the continuity of temperature requires that

$$T_1(\Sigma) = T_2(\Sigma), \tag{1.36}$$

while the continuity of heat flux requires that

$$k_1(\mathbf{r}_\Sigma)(\mathbf{n} \cdot \operatorname{grad} T_1)_\Sigma = k_2(\mathbf{r}_\Sigma)(\mathbf{n} \cdot \operatorname{grad} T_2)_\Sigma. \tag{1.37}$$

In the simple case that Σ is the plane $x = a$, and the thermal conductivities to the left and right of $x = a$ are constant, these internal boundary conditions become

$$T_1(a) = T_2(a), \tag{1.38}$$

and

$$k_1\left(\frac{\partial T_1}{\partial x}\right)_{x=a} = k_2\left(\frac{\partial T_2}{\partial x}\right)_{x=a}. \tag{1.39}$$

In summary, possible auxiliary conditions for the heat equation, which can be expected to lead to a unique solution, involve one or more of the above boundary conditions together with an initial condition.

If after a suitably long time the temperature distribution reaches a steady state, the function T becomes independent of the time t, causing the heat equation (1.32) to reduce to

$$\operatorname{div}(k(\mathbf{r})\operatorname{grad} T) + H = 0. \tag{1.40}$$

When k is a constant, (1.40) becomes the *Poisson equation* (see the list of PDEs in Section 1.1, where it occurs as entry number 3),

$$\Delta T + H/k = 0, \tag{1.41}$$

while if $H = 0$ the equation simplifies still further to the *Laplace equation* $\Delta T = 0$.

To understand why not all combinations of initial and boundary conditions are appropriate it will be sufficient for us to examine two quite different situations. Consider the steady-state heat equation with a source term H given in (1.41). In this case no initial condition can be imposed, because time is absent from the PDE. Now suppose the volume V occupied by the solid is finite and the Neumann condition $\hat{\mathbf{n}} \cdot \operatorname{grad} T = 0$ is imposed all over the surface of V.

Such a boundary condition is inappropriate for this steady-state temperature distribution since there can be no solution, because while the Neumann condition asserts no heat can flow across the boundary, the term H will add heat (remove it if there is a sink), causing the temperature to increase (decrease) without bound. In the study of heat conduction, because the Neumann condition $\hat{\mathbf{n}} \cdot \text{grad } T = 0$ applied to a surface means no heat can pass through it, a condition of this type corresponds to a surface that is (thermally) *insulated*.

A different example of an inappropriate set of auxiliary conditions for the heat equation that leads to a problem for which no solution is possible can be formulated as follows. Suppose, in addition to specifying any one of the above boundary conditions and the initial temperature distribution, the temperature distribution at a subsequent time $t = t_1$ is required to satisfy the condition $T(\mathbf{r}, t_1) = \hat{T}(\mathbf{r})$, where $\hat{T}(\mathbf{r})$ is an arbitrarily prescribed function (temperature distribution). Then, in general, this problem has no solution, because we will see later that when a solution exists, satisfying an initial condition and one of the above boundary conditions, it will be unique. Consequently, the boundary condition and initial condition will *determine* the solution $T(\mathbf{r}, t_1)$ at the time t_1, although this temperature distribution will not normally be equal to the prescribed temperature distribution $\hat{T}(\mathbf{r})$ when $t = t_1$.

The reason why PDEs (1.32) and (1.33) are also known as the *diffusion equations* is because whereas the conduction of heat is due to a temperature gradient, in a diffusion process a corresponding situation arises where a flow of material is caused by a concentration gradient. A simple example of a diffusion process is provided by a storage tank for liquid chemical waste, when the base is made of a thick layer of slightly porous concrete. If at time $t = 0$ the tank is filled with a certain volume of liquid chemical waste, the waste will immediately start to diffuse into the concrete. Provided the tank is large, the concentration C of waste in the concrete, represented by the amount of the liquid waste present in a unit volume of concrete, can be assumed to depend only on the depth x in the concrete and the time t, so we can write $C = C(x, t)$.

To illustrate matters further, for the sake of simplicity let us consider one-dimensional diffusion, where $C(x, t)$ is the concentration at position x and time t, and $j(x, t)$ is the rate of flow of waste matter diffusing through a unit area at position x and time t. An experimental result called *Fick's law*, analogous to the *Fourier law* in heat flow, relates $j(x, t)$ and $C(x, t)$ through the condition $j = -DC_x$, where the constant $D > 0$ is called the *diffusivity* of the medium through which diffusion takes place, and in general $D = D(C)$ is a function of the concentration.

It follows from the principle of conservation of mass that $C_t + j_x = 0$, so after combining this result with Fick's law, the concentration C is seen to be the solution of the diffusion equation $C_t = \partial_x\{D(C)C_x\}$, where $\partial_x \equiv \partial/\partial x$. This result extends at once to three dimensions and time, when the *diffusion equation* becomes

$$C_t = \text{div}\{D(C)\text{grad } C\}. \tag{1.42}$$

This PDE is a conservation law of the same form as the heat equation in (1.32), and once again the divergence operator is seen to appear in the law. If the diffusivity D is constant, (1.42) reduces to the *diffusion (heat)* equation

$$C_t = D\Delta C. \tag{1.43}$$

If in suitable circumstances the diffusion can reach a steady-state condition, the function C becomes independent of the time t and (1.43) reduces to the Laplace equation $\Delta C = 0$.

The auxiliary conditions that can be specified for the diffusion equation, either in its more general form (1.42) or in the simpler one given in (1.43), are similar to those for the heat equation.

Transverse vibrations of a string

Let the equilibrium position of a stretched elastic string of length L and tension T_0 coincide with the x axis, with the ends fixed at the origin and at $x = L$. If the distribution of mass along the string is variable, let the line density of the string (its mass per unit length) be $\rho(x)$, and suppose the string performs small *transverse vibrations* in a plane containing the x axis, in which an external distributed force of density $\hat{F}(x, t)$ acts on the string in a direction normal to the x axis. If the displacement of the string normal to the x axis is $u(x, t)$, its equation of motion can be derived by considering the motion of an arbitrary segment of the string at a time t when the string is displaced from its equilibrium position. In terms of *Newton's second law*, this means we must equate the force acting on the segment to its rate of change of momentum. The displaced string is shown in Fig. 1.8, where α is the angle between the tangent to the string at P and the x axis.

Before resolving forces it is necessary to determine the tension T in the string at a point P which must act along the tangent to the string at P. An elementary result from the calculus asserts that if δx is a small change in x, then the corresponding length δl along the string in its displaced position can be approximated by $\delta l \approx (1 + u_x^2)^{1/2}\delta x$. We now make the assumption that the displacement $u(x, t)$ of the string and its derivative u_x are sufficiently small

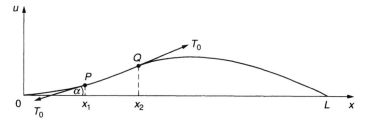

Figure 1.8   Internal forces acting on a segment of a string with ends clamped at $x = 0$ and $x = L$.

that squares and products of these quantities can be neglected. An immediate consequence of this assumption is that $\delta l = \delta x$ so, to this order of approximation, the small transverse displacement of the string has not caused the original element of length δx to change. Consequently, assuming that the elastic string obeys *Hooke's law* (string tension is proportional to the extension beyond its equilibrium length), it follows that the tension T at the displaced point P must be independent of the time t, and so $T = T_0$.

At point P the tension T_0 acting along the tangent to the string in the direction shown in Fig. 1.8 will have a component T_u in the u direction given by $T_u = T_0 \sin \alpha$. As $u(x, t)$ is small, $\sin \alpha \approx \tan \alpha$, but $\tan \alpha = u_x$ so to this order of approximation $T_u = T_0 u_x$. In an arbitrary interval of time from t_1 to t_2, the integral with respect to time of the force produced by T_u, which acts in opposite directions at each end of an arbitrary segment PQ of the string, is $T_0 \int_{t_1}^{t_2} [u_x(x_2, \tau) - u_x(x_1, \tau)] d\tau$, where P and Q have the respective x coordinates x_1 and x_2. Integrating the external force with density $\hat{F}(x, t)$ over the segment shows the external force acting on the segment is $\int_{x_1}^{x_2} \hat{F}(\xi, t) d\xi$. As the transverse speed of point P is $u_t(x, t)$, after taking into account the variable mass distribution $\rho(x)$ along the string, the change of momentum of the element from time t_1 to t_2 is $\int_{x_1}^{x_2} \rho(\xi)[u_t(\xi, t_2) - u_t(\xi, t_1)] d\xi$. Equating the change of momentum in the u direction to the integral with respect to time of the combined forces acting on the segment in the u direction gives

$$\int_{x_1}^{x_2} \rho(\xi)[u_t(\xi, t_2) - u_t(\xi, t_1)] d\xi = T_0 \int_{t_1}^{t_2} [u_x(x_2, \tau) - u_x(x_1, \tau)] d\tau$$
$$+ \int_{t_1}^{t_2} \int_{x_1}^{x_2} \hat{F}(\xi, \tau) d\xi \, d\tau.$$

Provided $u_x(x, t)$ and $u_t(x, t)$ are suitably differentiable, the terms in this equation can all be combined under a single double integral to give

$$\int_{t_1}^{t_2} \int_{x_1}^{x_2} [\rho(\xi)u_{\tau\tau} - T_0 u_{\xi\xi} - \hat{F}(\xi, \tau)] d\xi \, d\tau = 0. \qquad (1.44)$$

This is the integral form of the *conservation law*, which describes the motion of the oscillating string. As the integrand of (1.44) is continuous and t_1, t_2, x_1, and x_2 are arbitrary, it follows that (1.44) can only be possible if its integrand vanishes identically, so returning to the variables x and t the PDE form of the *conservation law* governing the transverse vibrations of the string is seen to be

$$\rho(x)u_{tt} = T_0 u_{xx} + \hat{F}(x, t). \qquad (1.45)$$

Dividing by $\rho(x)$ setting $T_0/\rho(x) = c^2$ and $F(x, t) = \hat{F}(x, t)/\rho(x)$, brings (1.45) into the form

$$u_{tt} = c^2 u_{xx} + F(x, t), \quad \text{with } c = c(x). \qquad (1.46)$$

This PDE is called the *one-dimensional nonhomogeneous wave equation,* with the nonhomogeneous term $F(x, t)$. When $c = $ constant, and no external force acts on the string, (1.46) simplifies to the one-dimensional homogeneous wave equation introduced in Section 1.1. The string line density $\rho(x)$ has the dimensions (mass/length), and the tension T_0 has the dimensions (mass × length)/ (time)^2, so c in (1.46) has the dimensions (length/time), thereby confirming that it is a *speed.*

The problem considered here involves a string of finite length L clamped at its ends, so

$$u(0, t) = u(L, t) = 0, \quad \text{for } t \geq 0. \tag{1.47}$$

The conditions in (1.47) form a suitable set of *boundary conditions* for the wave equation. In this case the problem is said to be set in a *bounded domain,* because the solution is confined to the semi-infinite strip $0 \leq x \leq L, t \geq 0$ in the (x, t) plane.

To describe the start of the transverse vibrations it is necessary to specify the initial shape of the string when it is displaced at $t = 0$, and the speed with which each point of the string starts to move in the direction of the u axis at this initial time. This can be accomplished by assigning the *initial conditions*

$$u(x, 0) = \hat{u}(x) \quad \text{and} \quad u_t(x, 0) = \hat{v}(x), \tag{1.48}$$

where $\hat{u}(x)$ and $\hat{v}(x)$ are arbitrary smooth functions subject only to the conditions that $\hat{u}(0) = \hat{u}(L) = 0$ and $\hat{v}(0) = \hat{v}(L) = 0$, which are necessary because the ends of the string are clamped.

One of these boundary conditions must be replaced by a more general one if an end is not clamped. For example, if the end at the origin is clamped but the end at $x = L$ is made to move in the u direction so its displacement at time t is $a \sin \omega t$, the boundary conditions (1.47) must be replaced by

$$u(0, t) = 0 \quad \text{and} \quad u(L, t) = a \sin \omega t. \tag{1.49}$$

Other more general boundary conditions are also possible, although they will not be considered here.

As with the heat equation, an internal boundary condition must be applied if the string is a composite formed by joining two different strings with the respective line densities ρ_1 and ρ_2, with the join occurring at a point with coordinate $x = a$. At $x = a$ the transverse displacement $u_1(x, t)$ of the part of the string with line density ρ_1, and the transverse displacement $u_2(x, t)$ of the part of the string with line density ρ_2 must be continuous. The internal forces can only act *along* the string, because otherwise a component of force at $x = a$ would act on an element of the string with an arbitrarily small mass causing an infinite acceleration. Consequently, the tangents to the string to the immediate left and right of $x = a$ must be continuous. So the internal boundary conditions to be applied at $x = a$ become

$$u_1(a, t) = u_2(a, t) \quad \text{and} \quad \left(\frac{\partial u_1}{\partial x}\right)_{x=a} = \left(\frac{\partial u_2}{\partial x}\right)_{x=a}. \tag{1.50}$$

Different physical situations can give rise to the one-dimensional wave equation, in which case the functions $u(x, t)$, $\rho(x)$, and $\hat{F}(x, t)$ will have different meanings, while the quantity corresponding to T_0 may be a function $T(x)$ of x. When this occurs, the arguments leading to the integral form of the conservation equation in (1.44) must be slightly modified, with the result that it becomes

$$\int_{t_1}^{t_2} \int_{x_1}^{x_2} \{\rho(\xi)u_{\tau\tau} - [T(\xi)u_\xi]_\xi - \hat{F}(\xi, \tau)\}d\xi \, d\tau = 0. \tag{1.51}$$

The PDE form of this conservation equation corresponding to (1.42) then takes on the more general form in terms of x and t

$$\rho(x)u_{tt} = [T(x)u_x]_x + \hat{F}(x, t), \tag{1.52}$$

although the boundary and initial conditions introduced earlier are still appropriate.

Vibrations of a membrane

A *membrane* is a thin flexible sheet of elastic material, like a metal panel in an aircraft wing, or the diaphragm forming the head of a drum. Suppose that when in equilibrium a membrane is stretched with tension T_0 over a plane area S and fastened to the boundary ∂S of S. If the membrane is displaced perpendicular to its equilibrium position, and then released, it will perform vibrations like those produced when the head of a drum is struck. The term *tension* used here has the following meaning. Let a straight line L of unit length be drawn in S. Then the material on one side of L will produce a force $\mathbf{T}$ (the tension in the membrane) on the material on the other side, where the vector representing $\mathbf{T}$ is tangential to the displaced surface of the membrane and perpendicular to L.

The displacement of point $P(x, y)$ of the membrane normal to S at time t will be denoted by $u(x, y, t)$, the area density of the membrane will be denoted by $\rho(x, y)$ (its mass per unit area), a distributed external force of density $\hat{F}(x, y, t)$ will be assumed to act on the membrane at time t, and the membrane will be assumed to be *isotropic*, so the tension T at any point P of the membrane will be the same in all directions through P.

Let the area A in Fig. 1.9 with boundary ∂A be an arbitrary part of membrane S with boundary ∂S when in its equilibrium position, and let area A' with boundary $\partial A'$ located vertically above A be the corresponding part of the membrane when in its displaced position at time t.

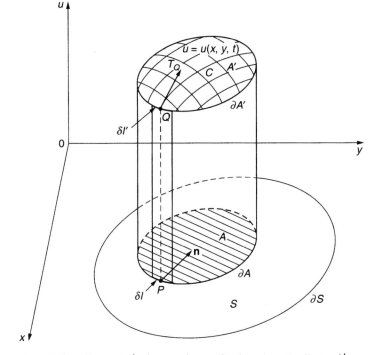

Figure 1.9   Element A of a membrane displaced vertically to A'.

From elementary calculus, the area A' is given by

$$A' = \int_A \left(1 + u_x^2 + u_y^2\right)^{1/2} dS.$$

Provided the vibrations are small, so squares and products of u, u_x, and u_y can be ignored, it follows that $(1 + u_x^2 + u_y^2)^{1/2} \approx 1$, so to this order of approximation we see that $A' = A$. If, in addition, the membrane obeys Hooke's law, the fact that $A' = A$ implies that the tension in the displaced membrane remains unchanged at T_0.

The force T_Q tangent to the membrane at Q and perpendicular to line element $\delta l'$ due to the tension T_0 acting on element $\delta l'$ in Fig. 1.9 is $T_Q = T_0 \delta l'$. The argument used to show that $A' = A$ also implies that $\delta l' = \delta l$, so $T_Q = T_0 \delta l$. In Fig. 1.10 the line of action of the force T_Q, which is perpendicular to $\delta l'$, must lie in the same plane as the unit vector $\mathbf{n}$ in A, which is normal to δl. The plane of T_Q and $\mathbf{n}$ is shown in Fig. 1.10, from which it can be seen that the component T_u of T_Q in the u direction is $T_u = T_Q \sin \alpha$.

As the displacement $u(x, y, t)$ is small, the angle α will also be small, so we may use the approximation $\sin \alpha \approx \tan \alpha = \mathbf{n} \cdot \operatorname{grad} u$, showing that to this order of approximation $T_u = \mathbf{n} \cdot \operatorname{grad} u \, \delta l$.

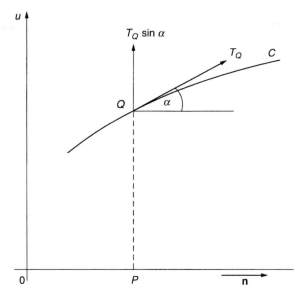

Figure 1.10   The force T_Q acting tangentially to the membrane at Q.

Thus, integrating T_u around $\partial A'$ to find the force Φ_u on A' in the u direction due to the tension is equivalent to integrating T_u around ∂A, so

$$\Phi_u = T_0 \int_{\partial A} \mathbf{n} \cdot \operatorname{grad} u \, dl. \qquad (1.53)$$

Applying the two-dimensional form of the Gauss divergence theorem (Green's theorem) to this result gives

$$\Phi_u = T_0 \int_A \operatorname{div}(\operatorname{grad} u) \, dS = T_0 \int_A \Delta u \, dS.$$

So the integral of Φ_u over an arbitrary interval of time from t_1 to t_2 is given by

$$\int_{t_1}^{t_2} \Phi_u \, dt = T_0 \int_{t_1}^{t_2} \int_A \operatorname{div}(\operatorname{grad} u) \, dS \, dt. \qquad (1.54)$$

The distributed external force of density $\hat{F}(x, y, t)$ acting over the displaced membrane A' in the u direction produces a total force $\Psi_u = \int_{A'} \hat{F}(x, y, t) dA'$, but as $A' = A$ we have $dA' = dS$, so integrating Ψ_u over the interval of time from t_1 to t_2 gives

$$\int_{t_1}^{t_2} \Psi_u \, dt = \int_{t_1}^{t_2} \int_A \hat{F}(x, y, t) \, dS \, dt. \qquad (1.55)$$

Then taking into account the variable area density of the membrane, and the fact that $A' = A$, shows the momentum of membrane in its displaced position to be

$$M_u = \int_A \rho(x, y) u_t \, dS.$$

In terms of this result the change in the momentum in the interval of time from t_1 to t_2 is given by

$$\int_A M_u \, dS = \int_A \rho(x, y)[u_t(x, y, t_2) - u_t(x, y, t_1)] \, dS. \tag{1.56}$$

Provided u is twice differentiable with respect to t this last result can be written

$$\int_A M_u \, dS = \int_{t_1}^{t_2} \int_A \rho(x, y) u_{tt} \, dS \, dt. \tag{1.57}$$

The form of argument used when deriving the wave equation for the string now shows that (1.57) must equal the sum of results (1.54) and (1.55), so the integral form of the *conservation equation* governing the oscillations of a membrane is

$$\int_{t_1}^{t_2} \int_A [\rho(x, y) u_{tt} - T_0 \Delta u - \hat{F}(x, y, t)] \, dS \, dt = 0. \tag{1.58}$$

As the integrand in (1.54) is continuous, and A, t_1, and t_2 are arbitrary, this can only be true if the integrand vanishes identically, so the PDE *conservation law* governing the vibrations of the membrane is

$$\rho(x, y) u_{tt} = T_0 \Delta u + \hat{F}(x, y, t). \tag{1.59}$$

Dividing this result by $\rho(x, y)$, setting $T_0/\rho(x, y) = c^2$ and $\hat{F}(x, y, t)/\rho(x, y) = F(x, y, t)$ the PDE governing the vibrations becomes

$$u_{tt} = c^2 (u_{xx} + u_{yy}) + F(x, y, t), \tag{1.60}$$

where $c = c(x, y)$ again has the dimensions of a speed. This PDE is called the *two-dimensional nonhomogeneous wave equation* with the nonhomogeneous term $F(x, y, t)$.

Like the PDE governing the vibrations of a stretched string, the PDE in (1.60) also requires both boundary and initial conditions if a unique solution is to be obtained. When the membrane is clamped around the boundary ∂S of S there can be no displacement on ∂S, so the *boundary condition* to be applied there becomes

$$u(x, y, t)|_{(x,y) \text{ on } \partial S} = 0, \tag{1.61}$$

while appropriate *initial conditions* are

$$u(x, y, 0) = \hat{u}(x, y) \quad \text{and} \quad u_t(x, y, 0) = \hat{v}(x, y), \tag{1.62}$$

where the first condition describes the initial shape of the membrane, and the second one the speed with which points on the membrane start moving in the u direction, so that $\hat{v}(x, y)$ will be zero if the membrane starts from rest. As the boundary of the membrane is assumed to be clamped, $\hat{u}(x, y)$ and $\hat{v}(x, y)$ must vanish at all points on the boundary ∂S of S. As with the vibrating string, a boundary condition more general than (1.61) must be applied if the boundary is not rigidly clamped.

The telegraph equation

Consider a transmission line formed by a pair of parallel wires with x being the distance measured along them, and let $V(t)$ be the voltage difference between the wires at a point x and time t, when the currents in the respective wires are $\pm i(x, t)$. Let the characteristic parameters of the transmission line be uniformly distributed along the line, with C being the capacitance of a unit length of the line, L the self-inductance, $\frac{1}{2}R$ the resistance on each of the wires, and G the leakage conductance for a unit length of the line (the reciprocal of the leakage resistance between the lines).

Figure 1.11 shows an element of the transmission line of length δx, and to determine the equations governing the current i and voltage V we need to make use of the following laws:

> *Kirchhoff's current law*: The current at a junction is conserved, so that the sum of the currents entering a junction equals the sum of the currents leaving it.

> *Kirchhoff's voltage law*: The algebraic sum of the voltage drops around any closed circuit is zero, provided a voltage increase is regarded as a negative voltage drop.

> *Ohm's law*: The current i flowing through a resistor R across which a voltage V is applied is determined by the law $i = V/R$.

Applying these laws at a time t to the element of the transmission line shown in Fig. 1.11 situated at a position x, followed by the cancelation of δx, shows

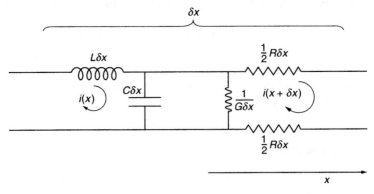

Figure 1.11 An element of the transmission line of length δx.

that in the limit as $\delta x \to 0$

$$\frac{\partial}{\partial t} Li + Ri + \frac{\partial V}{\partial x} = 0 \quad \text{and} \quad C\frac{\partial V}{\partial t} + GV + \frac{\partial i}{\partial x} = 0. \quad (1.63)$$

Setting $c = \frac{1}{\sqrt{LC}}$, $p = \frac{R}{L} + \frac{G}{C}$, and $q = \frac{RG}{LC}$, and eliminating either i or V, we find that in each case we obtain the homogeneous second-order constant coefficient PDE

$$\frac{\partial^2 u}{\partial t^2} = c^2 \frac{\partial^2 u}{\partial x^2} - p\frac{\partial u}{\partial t} - qu, \quad (1.64)$$

where u is either i or V. This PDE is called the *telegraph equation*, because it first arose when determining the current and voltage distribution along telegraph land lines.

Longitudinal vibrations of a free elastic rod with a variable cross section

The top diagram in Fig. 1.12 shows a linearly elastic rod with constant density ρ and a slowly varying cross-sectional area $S(x)$, with x measured along the axis of the rod from O. Let us consider *longitudinal vibrations* along the length of the rod, instead of the transverse vibrations that occur in stretched strings.

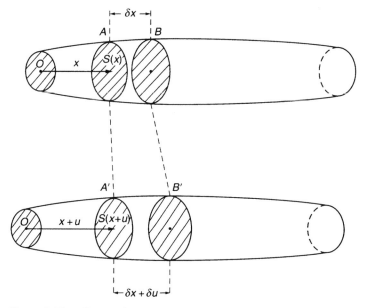

Figure 1.12   The top diagram shows the rod in equilibrium, and the bottom one shows it in a displaced condition at time t due to a longitudinal vibration.

A longitudinal vibration will displace each cross-sectional element of the rod along the axis of the rod. As a result the element AB of length δx and mass $\rho S(x)\delta x$ in the top diagram, with A at a distance x from O, will become the element $A'B'$ in the bottom diagram. The mass of element $A'B'$ will be the same as that of element AB, but its length will now be $\delta x + \delta u$, with A' at a distance $x + u$ from O.

If the material of the rod is linearly elastic it will obey *Hooke's law*, which asserts that the force necessary to change by x length units the length of an elastic rod or spring from its natural length is proportional to x. When Hooke's law is applied to this rod of varying cross section the tension $T_{A'}$ at A' is given by

$$T_{A'} = \lambda S(x) \cdot (\text{extension/natural length})$$

$$= \lambda S(x) \lim_{\delta x \to 0} \frac{\delta x + \delta u - \delta x}{\delta x}$$

$$= \lambda S(x)\frac{\partial u}{\partial x},$$

where λ is a constant that describes the elastic nature of the material.

The acceleration of element $A'B'$ is $\partial^2 u/\partial t^2$, so the difference in the tension $T_{B'}$ at B' and the tension $T_{A'}$ at A' must equal the product of the mass $\rho\delta x S(x)$ of the element and its acceleration, so

$$\rho S(x)\delta x\frac{\partial^2 u}{\partial t^2} = T_{B'} - T_{A'}$$

$$= \frac{\partial T_{P'}}{\partial x}\delta x = \frac{\partial}{\partial x}\left(\lambda S(x)\frac{\partial u}{\partial x}\right)\delta x.$$

Canceling δx, setting $c^2 = \lambda/\rho$, and expanding the expression on the right gives the PDE that describes the displacement u of the rod at point x and time t

$$\frac{1}{c^2}\frac{\partial^2 u}{\partial t^2} = \frac{\partial^2 u}{\partial x^2} + \frac{1}{S(x)}\frac{dS}{dx}\frac{\partial u}{\partial x}. \tag{1.65}$$

When $S(x)$ is constant, this generalization of the wave equation reduces to the ordinary one-dimensional wave equation that was found to govern the transverse vibrations of an elastic string, although here it describes the *longitudinal* vibrations of a rod.

Electromagnetic wave propagation in free space

The electric vector $\mathbf{E}$ and the magnetic vector $\mathbf{H}$ in an electromagnetic field in free space (a vacuum) obey the *Maxwell equations*

$$(1/c)\mathbf{E}_t = \text{curl } \mathbf{H} \quad \text{and} \quad (1/c)\mathbf{H}_t = -\text{curl } \mathbf{E} \quad \text{with div } \mathbf{E} = \text{div } \mathbf{H} = 0, \tag{1.66}$$

where the constant c is the speed of light in a vacuum in appropriate units.

Differentiating the first equation partially with respect to the time t gives $(1/c)\mathbf{E}_{tt} = \mathrm{curl}\,\mathbf{H}_t$, and after substituting for $\mathbf{H}_t$ from the second equation we find that $\mathbf{E}_{tt} = -c^2\mathrm{curl}(\mathrm{curl}\,\mathbf{E})$. However, from vector identity (4) in Section 1.6, $\mathrm{curl}(\mathrm{curl}\,\mathbf{E}) = \mathrm{grad}(\mathrm{div}\,\mathbf{E}) - \Delta\mathbf{E}$, but $\mathrm{div}\,\mathbf{E} = 0$, so the electric vector $\mathbf{E}$ is seen to satisfy the equation $\mathbf{E}_{tt} = c^2\Delta\mathbf{E}$. A similar argument shows the magnetic vector $\mathbf{H}$ satisfies this same equation, so in a vacuum the components of both $\mathbf{E}$ and $\mathbf{H}$ all satisfy the scalar *three-dimensional wave equation*

$$u_{tt} = c^2\Delta u. \tag{1.67}$$

In Cartesian coordinates (1.67) becomes

$$u_{tt} = c^2(u_{xx} + u_{yy} + u_{zz}). \tag{1.68}$$

If this PDE describes the solution in a finite region of space the components of $\mathbf{E}$ and $\mathbf{H}$ must satisfy *boundary conditions*, while *initial conditions* must also be given to specify the way in which electromagnetic waves start. If, however, the region of space in which electromagnetic waves propagate is unbounded then although initial conditions for $\mathbf{E}$ and $\mathbf{H}$ must still be given, in this case the *boundary conditions* usually take the form of the requirement that the components of $\mathbf{E}$ and $\mathbf{H}$, or some functions of them, are bounded at infinity. A condition of this form is called a *boundary condition at infinity*.

Acoustic waves in a gas

A different physical phenomenon also governed by the wave equation in one or more space variables is the propagation of *acoustic waves* (*sound waves*) in a gas. The *nonlinear* fluid equations involved are

$$\mathbf{v}_t + \mathbf{v}\cdot\mathrm{grad}\,\mathbf{v} + (1/\rho)\mathrm{grad}\,p = 0 \quad \text{(the momentum equation)} \tag{1.69}$$

and

$$\rho_t + \mathrm{div}(\rho\mathbf{v}) = 0 \quad \text{(the mass conservation equation),} \tag{1.70}$$

where $\mathbf{v}$ is the gas velocity vector, ρ is the gas density, and $p = p(\rho)$ is the gas pressure with the function $p(\rho)$ being a specified function of the density ρ, called the *constitutive equation* for the gas. In air, for example, $p(\rho) = k\rho^{1.4}$, where k is a constant. In sound waves in a gas, the gas velocity $\mathbf{v}$ is such that the term $\mathbf{v}\cdot\mathrm{grad}\,\mathbf{v}$ in the momentum equation can be neglected. In these circumstances the pressure and density changes are small relative to the ambient conditions p_0 and ρ_0, so we may set $p = p_0 + p'$ and $\rho = \rho_0 + \rho'$, where p' and ρ' are small quantities representing changes from the constant ambient conditions. When these substitutions are made in the momentum and mass conservation equations, and vector identity (6) from Section 1.6 is used, they reduce to the respective equations for ρ' and p'

$$\rho_t' + \rho_0\mathrm{div}\,\mathbf{v} = 0 \quad \text{and} \quad \mathbf{v}_t + (1/\rho_0)\mathrm{grad}\,p' = 0. \tag{1.71}$$

In a sound wave we can use the approximation $p' = (\partial p/\partial \rho_0)\rho'$, in which case the mass conservation equation becomes

$$p'_t + \rho_0(\partial p/\partial \rho)_0 \, \text{div } \mathbf{v} = 0. \tag{1.72}$$

Acoustic wave propagation is then governed by the two simplified equations

$$p'_t + \rho_0(\partial p/\partial \rho)_0 \, \text{div } \mathbf{v} = 0 \text{ (linearized mass conservation equation)} \tag{1.73}$$

and

$$\mathbf{v}_t + (1/\rho_0)\text{grad } p' = 0 \quad \text{(linearized momentum equation).} \tag{1.74}$$

Introducing a scalar quantity ϕ called the **velocity potential**, which is related to the velocity by $\mathbf{v} = \text{grad } \phi$, the linearized momentum equation becomes

$$\text{grad } (\phi_t + (1/\rho_0)\, p') = \mathbf{0}.$$

Thus $p' = -\rho_0 \partial \phi/\partial t$, and after substituting this into the linearized mass conservation equation and using vector identity (5) of Section 1.6 the velocity potential ϕ is seen to be a solution of the three-dimensional wave equation

$$\phi_{tt} = c^2 \Delta \phi, \tag{1.75}$$

where $c = \sqrt{(\partial p/\partial \rho)_0}$ is the speed of sound under ambient conditions.

The heat and wave equations, which are linear second-order equations, were each derived by elimination of a dependent variable from a simple system of two first-order equations. It must be recognized that this situation is very special, and it comes about because of a peculiarity in the structure of the first-order equations involved. In general, if a system of n first-order equations arises, it is *not* possible to eliminate all but one of the n-dependent variables, and then to replace the system by a single nth-order equation in the remaining dependent variable. A case in point is the two first-order equations (1.69) and (1.70) from which Eq. (1.75) governing acoustic wave propagation was derived. These are quasilinear equations, and the derivation of the wave equation satisfied by the velocity potential ϕ was only possible because the equations were linearized. It will be seen in Chapter 8 how in order to study problems without linearization, when important new physical phenomena can be described, it will be necessary to work with the full system of first-order quasilinear equations.

EXERCISES 1.2

1. If f is an arbitrary once differentiable function of its argument $x - ct$, where $c = \text{constant}$, show by differentiation that $u(x, t) = f(x - ct)$ is a solution of the first-order homogeneous equation $u_t + cu_x = 0$.

2. Use the result of Exercise 1 to find the solution of $u_t - 4u_x = 0$, subject to the initial condition $u(x, 0) = \tanh x$.

3. Explain why requiring the equation $u_t + 2u_x = 0$ to satisfy the initial condition $u_t(x, 0) = x$ will *not* lead to a unique solution.

4. Explain why no solution will exist if the initial conditions $u(x, 0) = \sin x$ and $u_t(x, 0) = x$ are imposed on the equation $u_t + 3u_x = 0$.

5. If f and g are arbitrary twice differentiable functions of their respective arguments $x - ct$ and $x + ct$, where $c = $ constant, show by differentiation that $u(x, t) = f(x - ct) + g(x + ct)$ is a solution of the second-order homogeneous equation $u_{tt} = c^2 u_{xx}$.

6. Use the form of solution in Exercise 5, together with the corresponding expression for $u_t(x, t)$, to find the solution of $9u_{tt} = u_{xx}$ subject to the initial conditions $u(x, 0) = x$ and $u_t(x, 0) = 1$.

7. Show how the arguments leading to the integral form of the conservation law (1.44) and its PDE form in (1.45) must be modified to obtain the corresponding more general forms (1.51) and (1.52) when the tension in the string is *not* constant and $T = T(x)$.

8. Prove by contradiction that if $f(\mathbf{x}, t)$ is a continuous function of $\mathbf{x}$ and t in some region of space D for $0 \leq t \leq T$, and such that $\int_{t_1}^{t_2} \int_{\hat{D}} f(\mathbf{x}, t) \, d\hat{D} \, dt = 0$ for all subintervals $t_1 \leq t \leq t_2$ of $0 \leq t \leq T$, including the complete time interval itself, and all subregions $\hat{D}$ of D including the region D itself and on its boundary, then $f(\mathbf{x}, t)$ is identically zero in D and on its boundary for all for $0 \leq t \leq T$.

9. Explain why, when deriving the PDE in (1.65), it is necessary to impose the condition that $S(x)$ must be a slowly varying function of x.

10. The wave equation in three space dimensions for a function $u(x, y, z, t)$ has the form $\frac{\partial^2 u}{\partial t^2} = c^2 (\frac{\partial^2 u}{\partial x^2} + \frac{\partial^2 u}{\partial y^2} + \frac{\partial^2 u}{\partial z^2})$. Confirm by differentiation that $u = f(lx + my + nz - ct) + g(lx + my + nz + ct)$ with $l^2 + m^2 + n^2 = 1$ is a solution, where f and g are arbitrary twice differentiable functions of their arguments. Explain why solutions of this type represent **plane waves**.

11. Show by differentiation that $\Phi(x, t) = \frac{1}{\sqrt{4\pi\kappa t}} e^{-x^2/4\kappa t}$, with $\kappa > 0$ is a solution of the heat equation $\Phi_t = \kappa \Phi_{xx}$ for $t > 0$ and $-\infty < x < \infty$ (Φ is meaningless if $t = 0$).

12. Show that if $u(x, t)$ is a solution of the diffusion equation $u_t = \kappa u_{xx}$, then so also is $u(x - k, t)$ where k is a constant. This shows that the solution on the entire real line can be **translated** (shifted) in space while still remaining a solution.

13. Show that if $u(x, t)$ is a solution of the diffusion equation $u_t = \kappa u_{xx}$, then so also is $u(x\sqrt{k}, kt)$ with $k > 0$ being a constant. This shows the solution on the entire real line can be **dilated** (magnified or diminished) while still remaining a solution.

14. Show by differentiation that $u(x, y) = \sin x \sinh y$ is a solution of the Laplace equation $u_{xx} + u_{yy} = 0$ in the rectangle $0 \leq x \leq \pi/2, 0 \leq y \leq 1$. Confirm that the maxima and minima of u occur on the boundary of this rectangle (for more information about this property of the Laplace equation see Theorem 7.6).

15. The Laplace equation satisfied by a function $u(r, \theta, z)$ in the cylindrical polar coordinates (r, θ, z) (see Fig. 1.16 and the Laplacian in (1.105)) is $\frac{1}{r}\frac{\partial}{\partial r}\left(r\frac{\partial u}{\partial r}\right) + \frac{1}{r^2}\frac{\partial^2 u}{\partial \theta^2} + \frac{\partial^2 u}{\partial z^2} = 0$. Find the form of this equation when u is radially symmetric, and so is independent of θ and z. Hence find the general solution $u(r)$ for $r > 0$.

16. Let $u(x, y)$ be a solution of the Laplace equation $u_{xx} + u_{yy} = 0$ in a region D of the (x, y) plane. By making the change of variable $x = X - x_0$, $y = Y - y_0$, where (x_0, y_0) is an arbitrary point in D, show that if $u(x, y)$ becomes the function $U(X, Y)$, then U is also a solution of the Laplace equation so $U_{XX} + U_{YY} = 0$. This shows that a solution of the Laplace equation remains a solution when the origin is **translated** (shifted) to some other point in D. Verify this by first showing $u(x, y) = x \sin x \cosh y - y \cos x \sinh y$ satisfies the Laplace equation and then that this remains true after the variable change $x = X + 2, y = Y - 4$.

1.3 What Is a Solution of a PDE?

Let an nth-order PDE involving a real unknown scalar function $u(x_1, \ldots, x_m)$ be defined in a region D of $(x_1, \ldots, x_m)$ space. Then u will be said to be a *classical solution* of the PDE in D if it has continuous partial derivatives of all orders up to and including those of order n with respect to the m real independent variables $x_1, \ldots, x_m$, and the substitution of u into the PDE reduces it to an identity.

To illustrate this definition it will be sufficient to consider the two-dimensional Laplace equation in Cartesian coordinates and the function $u(x, y) = \sin x \cosh y$. This function is continuous, twice differentiable with continuous second-order derivatives for all x and y, and it satisfies the Laplace equation $\Delta_2 u = 0$. Consequently, since $u(x, y)$ satisfies the forgoing criteria, it is a classical solution of the two-dimensional Laplace equation in any region D of the (x, y) plane.

In two-dimensional (x, y) space a solution $u = u(x, y)$ of a PDE defines an *integral surface*, while in m-dimensional $(x_1, x_2, \ldots, x_m)$ space a solution

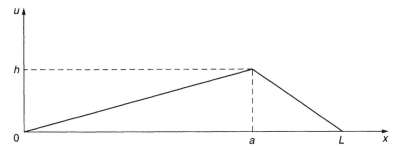

Figure 1.13 The initial shape of a stretched string when the point at $x = a$ is displaced perpendicular to its length.

$u = u(x_1, \ldots, x_m)$ of a PDE defines an *integral manifold* (an *m*-dimensional solution surface).

In applications of mathematics giving rise to PDEs, it sometimes happens that while a solution u is continuous, some of the partial derivatives of u entering into the PDE may be discontinuous. This situation means that if such problems possess solutions, they cannot be classical solutions. A typical example of this type occurs when considering the transverse vibrations of a stretched string with fixed ends, if it released from rest after a point on the string has been pulled aside through a small distance.

This situation is illustrated in Fig. 1.13, where a string of length L has its ends clamped at $x = 0$ and $x = L$, with the point at $x = a$ displaced perpendicular to the x axis through a small distance h, and then maintained in this position until the string is released. The boundary conditions for the one-dimensional wave equation that will determine the subsequent vibrations of the string are the usual ones

$$u(0, t) = u(L, t) = 0,$$

but the initial conditions are

$$u(x, 0) = \begin{cases} hx/a, & 0 \leq x \leq a \\ h(L - x)/(L - a), & a \leq x \leq L \end{cases} \quad \text{and} \quad u_t(x, 0) = 0.$$

Thus the solution of the second-order wave equation is required to satisfy an initial condition that is continuous for $0 \leq x \leq L$, but such that $u_x(x, 0)$ is discontinuous at $x = a$. It will be seen later that a discontinuity of this type introduces no fundamental difficulty when seeking a solution, because methods that apply when the initial conditions are smooth can also be used when the initial conditions are continuous, but have a discontinuity in a derivative with respect to x.

It is for reasons such as this that the strict differentiability requirements of a classical solution need to be relaxed to admit more general solutions, called *generalized solutions*. To be specific, we will consider a function to be a generalized solution of a PDE if it is has the following properties:

Definition 1.3. (Generalized Solutions)
A function u will be called a *generalized solution* of a PDE if it possesses the following properties:

(i) u is continuous with continuous derivatives up to and including those of the order of the PDE, and it satisfies the PDE at all but a finite number of points;

(ii) at each of the finite number of points in (i) where u fails to satisfy the PDE, one or more of its derivatives of order equal to or less than the order of the PDE may be discontinuous.

More will be said about this matter later, but for the moment it will be sufficient to remark that generalized solutions are not unusual when PDEs arise in connection with physical problems.

A further extension of the concept of a solution involves defining what is called a *weak solution* of a PDE, which allows the solution itself to be *discontinuous*. Weak solutions arise in various circumstances as, for example, in the study of the fluid mechanics of compressible fluids where discontinuous solutions called *shocks* can arise. A physical manifestation of a shock encountered in everyday life is the sonic bang heard when an aircraft passing overhead exceeds the speed of sound. Something of the nature of shocks, and the fact that in physical applications not all discontinuous solutions are necessarily physically realizable, will be examined when first-order quasilinear PDEs are considered.

In what is to follow, most attention will be devoted the study of linear PDEs and, in particular, to the class of *second-order linear* PDEs which arise in many applications. A general nonhomogeneous second-order linear PDE for the real scalar unknown function $u(x_1, \ldots, x_n)$ has the form

$$\sum_{i,j=1}^{n} a_{ij} u_{x_i x_j} + \sum_{i=1}^{n} b_i u_{x_i} + cu = f, \tag{1.76}$$

where the coefficients a_{ij}, b_i, c and the nonhomogeneous term f may be functions of the real independent variables $x_1, \ldots, x_n$, but *not* of the solution u itself. When the second-order partial derivatives are continuous it follows that $u_{x_i x_j} = u_{x_j x_i}$ so, unless stated to the contrary, PDEs will always be simplified by making use of this equality of mixed derivatives by setting $a_{ij} = a_{ji}$, because this allows the terms $a_{ij} u_{x_i x_j}$ and $a_{ji} u_{x_j x_i}$ to be combined.

If the *linear differential operator* L is defined as

$$L \equiv \sum_{i,j=1}^{n} a_{ij} \frac{\partial^2}{\partial x_i \partial x_j} + \sum_{i=1}^{n} b_i \frac{\partial}{\partial x_i} + c, \tag{1.77}$$

then $L[u]$ will mean

$$L[u] = \sum_{i,j=1}^{n} a_{ij} u_{x_i x_j} + \sum_{i=1}^{n} b_i u_{x_i} + cu. \tag{1.78}$$

This notation allows us to write (1.76) in the concise form

$$L[u] = f. \tag{1.79}$$

It will be recalled from Section 1.1 that when the nonhomogeneous term $f \equiv 0$, the PDE in (1.76) becomes a *homogeneous* equation.

To see why L in (1.78) is called a *linear* differential operator, it is necessary to recall from elementary calculus that when $k =$ constant and u, v are suitably differentiable functions of x_i and x_j,

$$\partial(ku)/\partial x_i = ku_{x_i}, \qquad \partial^2(ku)/\partial x_i \partial x_j = ku_{x_j x_i}, \quad \text{and}$$
$$\partial^2(u+v)/\partial x_i \partial x_j = u_{x_j x_i} + v_{x_j x_i} \quad \text{for } i, j = 1, \dots, n.$$

When applied to L, these basic properties of the operation of partial differentiation, taken together with the fact that the coefficients of L are functions only of the independent variables, lead to the following fundamental properties of L:

- When the operator L acts on the product ku, where k is a constant and u is a suitably differentiable function, the constant k may be taken through the operator L to become a multiplicative constant of $L[u]$, so that

$$L[ku] = kL[u]. \tag{1.80}$$

- When the operator L acts on the sum $u + v$ of two suitably differentiable functions u and v, the result is the sum of operator L acting first on u and then on v, so

$$L[u+v] = L[u] + L[v]. \tag{1.81}$$

These properties of a linear operator have two important consequences for *homogeneous* equations. The first is that if u is a solution of the homogeneous equation $L[u] = 0$, property (1.80) tells us that u may be scaled by a constant k and the result ku will also be a solution of the same homogeneous PDE.

The second consequence is that if u and v are solutions of a homogeneous PDE, so that $L[u] = 0$ and $L[v] = 0$, then from (1.81) the sum $u + v$ will also be a solution of the same homogeneous PDE, because $L[u+v] = L[u] + L[v] = 0$.

This last result extends to the sum of any finite number of solutions $u_1, \dots, u_n$ of the homogeneous PDE $L[u_i] = 0$, for $i = 1, \dots, n$, because

$$L[u_1 + \dots + u_n] = L[u_1] + L[u_2] + \dots + L[u_n] = 0.$$

When an infinite sequence of classical solutions $\{u_n\}$ of a homogeneous linear PDE can be found, and $a_1, a_2, \dots$ is any infinite set of constants, the sum

$\sum_{n=1}^{\infty} a_n u_n$ is said to be a *formal solution* of the PDE. This is because, to be mathematically rigorous, it is necessary to prove that the infinite series converges uniformly to a sum function with continuous second-order partial derivatives, before the sum function can be called a classical solution of $L[u] = 0$. Conversely if the series converges uniformly to a sum function, but the sum function does not have continuous second-order partial derivatives, the sum function will be a generalized solution of $L[u] = 0$.

The additive properties of solutions of homogeneous linear second-order PDEs demonstrated above extend in an obvious manner to homogeneous nth-order linear PDEs, and they will be used later when examining methods of solution for both homogeneous and nonhomogeneous equations.

EXERCISES 1.3

1. Show that $u(x, y) = \frac{x + x^2 + y^2}{1 + 2x + x^2 + y^2}$ is a classical solution of the two-dimensional Laplace equation $\Delta_2 u = 0$ where $y \neq 0$ or $x + 1 \neq 0$.

2. Show that $u(x, y) = \frac{4xy + x^2 y + y^3}{4 + 4x + x^2 + y^2}$ is a classical solution of the two-dimensional Laplace equation $\Delta_2 u = 0$ where $y \neq 0$ or $x + 2 \neq 0$.

3. Show that $u(x, y) = x \cosh x \cos y - y \sinh x \sin y + 3x^2$ is a classical solution of the two-dimensional Poisson equation $\Delta_2 u = 6$.

4. Show that $u(x, y) = x^2 + y^3 + e^x (\sin y \cos x \cosh y - \cos y \sin x \sinh y)$ is a classical solution of the two-dimensional Poisson equation $\Delta_2 u = 2 + 6y$.

5. Show that if u is a classical solution of the nonhomogeneous equation $L[u] = f$ and v is a classical solution of the homogeneous equation $L[v] = 0$, then $w = u + v$ is a classical solution of the nonhomogeneous equation $L[w] = f$.

6. Show that if u and v are the respective classical solutions of the nonhomogeneous equations $L[u] = g$ and $L[v] = h$, then $w = u + v$ is a classical solution of the nonhomogeneous equation $L[w] = g + h$.

1.4 The Cauchy Problem

Section 1.2 showed how, when deriving PDEs modeling physical situations, in order to find the solution of a specific problem it was necessary to impose certain auxiliary conditions that usually arise naturally from the problem itself. Collectively, these auxiliary conditions are called *boundary conditions*. If, however, one of the independent variables involved is the time t, or an independent variable that can be regarded as being time-like, then the condition to be satisfied at the start (usually when $t = 0$) is called an *initial condition*.

The determination of the solution of a PDE subject only to an initial condition is called a *pure initial value problem.*

When a time or a time-like independent variable is involved, as with the vibrating string problem, it is often the case that the solution of a PDE in a region D is required to satisfy both some initial conditions and conditions on the physical boundary of D. Finding the solution of a given PDE subject to an initial condition and conditions on the physical boundary of D is called an *initial boundary value problem,* often abbreviated to a problem of IBVP type.

To develop ideas further, we now consider the following semi-linear second-order PDE for the twice differentiable unknown function $u(x, y)$

$$A(x, y)u_{xx} + 2B(x, y)u_{xy} + C(x, y)u_{yy} = \Phi(x, y, u, u_x, u_y), \quad (1.82)$$

where $\Phi(x, y, u, u_x, u_y)$ is some nonlinear function of u. The assumption that $u(x, y)$ is twice differentiable allows the mixed derivative terms u_{xy} and u_{yx} to be combined to form the second term on the left of (1.82), where the factor 2 has been introduced for convenience as it simplifies the analysis that follows.

Note that PDE (1.82) reduces to the general second-order nonhomogeneous linear PDE in two independent variables when Φ depends linearly on u, u_x, and u_y, because then

$$\Phi = p(x, y)u + q(x, y)u_x + r(x, y)u_y + f(x, y).$$

Let the curve ∂D in the (x, y) plane shown in Fig. 1.14, forming the boundary of a region D, have the parametric equations

$$x = x_D(s), \qquad y = y_D(s), \qquad (1.83)$$

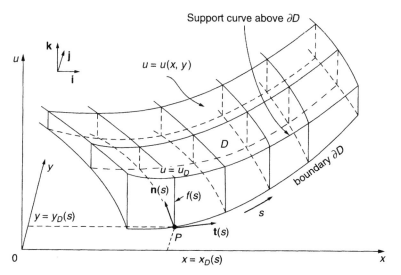

Figure 1.14   Cauchy conditions on the boundary ∂D and the support of $u(x, y)$.

where $x_D(s)$ and $y_D(s)$ are single-valued differentiable functions of the variable s, which can always be taken to be the arc length along ∂D measured from some convenient reference point. Then a fundamental problem for PDE (1.82), called the *Cauchy problem*, involves finding the solution $u(x, y)$ in D subject to the conditions

$$u(x, y)\big|_{(x,y)\in\partial D} = f(s) \quad \text{and} \quad \mathbf{n}(s) \cdot \text{grad } u\big|_{(x,y)\in\partial D} = g(s), \quad (1.84)$$

where $\mathbf{n}(s)$ is the unit normal to ∂D in the (x, y) plane at the point $(x_D(s)$, $y_D(s))$ shown as point P in Fig. 1.14, with $\mathbf{n}(s)$ directed *into* region D, and the unit vectors $\mathbf{i}$, $\mathbf{j}$, and $\mathbf{k}$ directed along the x, y, and u axes, respectively. When applied to the PDE in (1.82), conditions (1.84) are called *Cauchy conditions*.

When expressed in words, the imposition of Cauchy conditions on the PDE in (1.82) means that the solution u must be equal to the given function $f(s)$ on ∂D, and the normal derivative $\mathbf{n}(s) \cdot \text{grad } u$ of the solution u on ∂D (that is, the derivative of u along the unit normal $\mathbf{n}(s)$ in the (x, y) plane) must be equal to the given function $g(s)$. The curve ∂D in the (x, y) plane on which the Cauchy conditions are prescribed is called the *support curve* of the solution, while the function $f(s) = u(x_D(s), y_D(s))$ determining the value of u on ∂D is called the *support* of the solution.

One way of seeking a solution of this Cauchy problem is to try to determine it in the form of a Taylor series expansion about points on ∂D. To discover how this can be achieved, we take as our starting point the Cauchy conditions in (1.84). When doing so, it is first necessary to express the directional derivative $\mathbf{n}(s) \cdot \text{grad } u$ in terms of the parametric description of ∂D given in (1.83). If $\mathbf{t}(s)$ is the unit tangent to ∂D at P in Fig. 1.14, then since s is the arc length along ∂D it follows that dx_D/ds and dy_D/ds are the direction cosines of $\mathbf{t}(s)$, so we may set $\mathbf{t}(s) = (dx_D/ds)\mathbf{i} + (dy_D/ds)\mathbf{j}$. The vector $\mathbf{n}(s)$ directed into the region D is $\mathbf{n}(s) = \mathbf{k} \times \mathbf{t}(s)$, so $\mathbf{n}(s) = -\mathbf{i}(dy_D/ds) + \mathbf{j}(dx_D/ds)$. When this result is used with the second Cauchy condition in (1.84) it yields

$$-u_x(dy_D/ds) + u_y(dx_D/ds) = g(s). \quad (1.85)$$

This equation relates u_x and u_y to the known Cauchy condition $g(s)$ at P on ∂D, although it determines neither u_x nor u_y. A second equation relating u_x and u_y can be found by determining the directional derivative of u in the direction $\mathbf{t}(s)$ at P. To do this we take the dot product of the grad u, found from the first Cauchy condition, and $\mathbf{t}(s) = (dx_D/ds)\mathbf{i} + (dy_D/ds)\mathbf{j}$, and as a result arrive at $\mathbf{t}(s) \cdot \text{grad } u = u_x(dx_D/ds) + u_y(dy_D/ds)$. However, the derivative of $f(s)$ in the direction $\mathbf{t}(s)$ is df/ds, so a second equation relating u_x and u_y, this time to the first Cauchy condition at P, is

$$u_x(dx_D/ds) + u_y(dy_D/ds) = df/ds. \quad (1.86)$$

Equations (1.85) and (1.86) will always have a solution at any point P on ∂D, because s is the arc length along ∂D so the determinant of the coefficients of Eqs. (1.85) and (1.86) will always be nonsingular because $(dx_D/ds)^2 + (dy_D/ds)^2 = 1$.

Solving (1.85) and (1.86) for u_x and u_y on ∂D in terms of the Cauchy conditions gives

$$u_x(x, y)|_{(x,y)\in\partial D} = (df/ds)(dx_D/ds) - g(s)(dy_D/ds) \qquad (1.87)$$

and

$$u_y(x, y)|_{(x,y)\in\partial D} = g(s)(dx_D/ds) + (df/ds)(dy_D/ds). \qquad (1.88)$$

As (x, y) in (1.87) and (1.88) lies on ∂D, the points of which are parameterized in terms of s, we can set $u_x(x, y)|_{(x,y)\in\partial D} = U(s)$ and $u_y(x, y)|_{(x,y)\in\partial D} = V(s)$.

In order to develop a Taylor series for $u(x, y)$ it is necessary to determine all higher-order partial derivatives of $u(x, y)$. To find the second-order partial derivatives, assuming the necessary differentiability we differentiate (1.87) and (1.88) along ∂D at P using $\mathbf{t}(s) \cdot \operatorname{grad} u_x = dU/ds$ and $\mathbf{t}(s) \cdot \operatorname{grad} u_y = dV/ds$ to obtain

$$\begin{aligned} u_{xx}(dx_D/ds) + u_{xy}(dy_D/ds) &= dU/ds \quad \text{and} \\ u_{yx}(dx_D/ds) + u_{yy}(dy_D/ds) &= dV/ds, \end{aligned} \qquad (1.89)$$

where u_{xx}, u_{xy}, u_{yx}, and u_{yy} are all evaluated at P. As the existence of continuous second-order partial derivatives has been assumed it follows that $u_{xy} = u_{yx}$. So supplementing the two equations in (1.89) by the original PDE (1.82) at P gives the following three linear equations from which to determine u_{xx}, u_{xy}, and u_{yy} at P:

$$\begin{aligned} u_{xx}(dx_D/ds) + u_{yx}(dy_D/ds) &= dU/ds, \\ u_{xy}(dx_D/ds) + u_{yy}(dy_D/ds) &= dV/ds, \\ Au_{xx} + 2Bu_{xy} + Cu_{yy} &= \Phi. \end{aligned} \qquad (1.90)$$

It follows from elementary linear algebra that it will only be possible to solve for the second-order partial derivatives if the determinant Δ of the coefficients of this system is nonsingular (Δ must be nonzero), where

$$\Delta = \begin{vmatrix} dx_D/ds & dy_D/ds & 0 \\ 0 & dx_D/ds & dy_D/ds \\ A & 2B & C \end{vmatrix}$$

$$= C\left(\frac{dx_D}{ds}\right)^2 - 2B\left(\frac{dx_D}{ds}\right)\left(\frac{dy_D}{ds}\right) + A\left(\frac{dy_D}{ds}\right)^2. \qquad (1.91)$$

Provided $\Delta \neq 0$, and the coefficients $A(x, y)$, $B(x, y)$, $C(x, y)$ and the functions $f(s)$ and $g(s)$ are analytic, and so can be differentiated with respect to x and y arbitrarily many times, successive differentiation of system (1.89) will lead to the determination of all partial derivatives of $u(x, y)$ with respect to x and y. So, if (x_0, y_0) is any point on ∂D, the solution $u(x, y)$ of the PDE in some neighborhood of this point can be represented in the form of the Taylor series expanded about the (x_0, y_0)

$$u(x, y) = u(x_0, y_0) + (x - x_0)u_x(x_0, y_0) + (y - y_0)u_y(x_0, y_0)$$

$$+ \frac{1}{2!}[(x - x_0)^2 u_{xx}(x_0, y_0) + 2(x - x_0)(y - y_0)u_{xy}(x_0, y_0)$$

$$+ (y - y_0)^2 u_{yy}(x_0, y_0)] + \cdots, \tag{1.92}$$

where (x, y) is a point in D, in some neighborhood of (x_0, y_0). The series in (1.92) can only be regarded as a *formal solution* of the PDE until the Taylor series has been shown to be uniformly continuous in some subregion $\hat{D}$ of D, in which case $u(x, y)$ then becomes a classical solution in $\hat{D}$.

The convergence of (1.92) in some suitably small disc centered on point P on ∂D can be established by overestimating (*majorizing*) the magnitude of terms in series (1.92), and using the result to find the radius of the disc containing points in D for which (1.92) converges. This argument can be applied to a sequence of points along ∂D, chosen so the circles of convergence about successive points overlap. As a result, the Taylor series solution for u can be developed in a strip inside D, bounded by ∂D. If another support curve is drawn in this strip, this same form of argument can be repeated, enabling the solution to be extended further into D. The formal proof of this result, called the *Cauchy–Kovalevskaya theorem* will not be given here, but it is mentioned because it establishes the *existence* of a solution of PDE (1.82) subject to the Cauchy data in (1.84), and so it forms an *existence theorem* for the solution. This very general result is mainly of theoretical importance, rather than for practical use. Its practical limitations as a method of solution are obvious, and it does not apply to generalized solutions, so other more effective methods of solution must be derived.

Although this theorem is not useful when seeking a practical method of solution of the PDE in (1.82) subject to Cauchy conditions, the expression Δ in (1.91) turns out to be extremely important. Indeed, in the next section, the condition corresponding to $\Delta = 0$ will be used to find a simple and extremely important criterion for the classification of second-order semilinear PDEs, of which a second-order linear PDE is a special case.

1.5 Well-Posed and Improperly Posed Problems

In the following chapters the only problems to be considered will be those known as well-posed problems. The term *well-posed* problem is used here to refer to a problem involving a PDE defined in a given domain with associated auxiliary conditions that ensure the solution possesses the following properties:

- *Existence*: At least one solution that satisfies the PDE and its auxiliary conditions can be found.

- *Uniqueness*: There is only one solution that satisfies the PDE and its auxiliary conditions.

- *Stability*: The solution must depend on the auxiliary conditions in a continuous manner in such a way that a small change in the auxiliary conditions only produces a small change in the solution.

The *existence* of a solution depends on the nature of the PDE and the auxiliary conditions. The imposition of too many auxiliary conditions, or an auxiliary condition that cannot be satisfied, can lead to there being no solution, whereas the imposition of too few conditions can lead to the existence of more than one solution, and so to the *nonuniqueness* of solutions.

Suppose, for example, when seeking a solution of $u_x^2 + u_y^2 = 1$ an auxiliary condition that requires the solution to satisfy the condition $u_y > 1$ on a boundary is imposed. Then clearly no such solution exists, because if $u_y > 1$ there is no real function u_x such that on the boundary $u_x^2 + u_y^2 = 1$, showing that in this case the auxiliary condition is too restrictive.

A simple example of the *nonuniqueness* of a solution caused by the imposition of too few auxiliary conditions can be seen by considering the oscillating string problem introduced in Section 1.2. If, instead of imposing Cauchy conditions, only the initial shape of the string is given the solution cannot be unique, because the imposition of different initial transverse speed distributions along the string will lead to different solutions.

The *stability* of a solution of a PDE subject to its associated auxiliary conditions is a normal expectation when modeling most physical problems. Previously, when considering the stability of a solution, the requirement that "a small change in the auxiliary conditions should only produce a small change in the solution" was too vague to be useful. To overcome this difficulty, the precise way in which the change is to be measured must be defined. To illustrate ideas, we consider a vibrating string problem, where the transverse displacement of a string of length L clamped at its ends is governed by the one-dimensional wave equation $u_{tt} = c^2 u_{xx}$, subject to the boundary conditions

$$u(0, t) = u(L, t) = 0$$

and the initial conditions

$$u(x, 0) = \hat{u}(x) \quad \text{and} \quad u_t(x, 0) = \hat{v}(x).$$

We will define small changes in the auxiliary conditions from $u(x, 0) = \hat{u}(x)$ to $u(x, 0) = \hat{u}_1(x)$, and from $u_t(x, 0) = \hat{v}(x)$ to $u_t(x, 0) = \hat{v}_1(x)$, to mean that $|\hat{u}(x) - \hat{u}_1(x)| < \varepsilon$ and $|\hat{v}(x) - \hat{v}_1(x)| < \varepsilon$ for $0 \le x \le L$ and some suitably small number $\varepsilon > 0$. If the solution corresponding to the changed initial conditions is $u_1(x, t)$, we will consider it to have changed by a small amount if the quantity $|u(x, t) - u_1(x, t)| < \delta$ for $0 \le x \le L$ and $t > 0$, where $\delta = \delta(\varepsilon)$ is some suitably small positive quantity related to ε. Other measures of change are possible, but this one is the simplest.

A problem involving a PDE defined in a given domain with associated auxiliary conditions that is not well posed is said to be *improperly posed*. The following is a classical example of an improperly posed problem for the two-dimensional Laplace equation, and it is based on an example due to the French mathematician J. Hadamard. Specifically, the example demonstrates that the Laplace equation subject to certain Cauchy and Neumann conditions is improperly posed.

Consider the Laplace equation $u_{xx} + u_{yy} = 0$ in the half-strip $0 \le x \le \pi$ and $y > 0$, subject to the Dirichlet conditions $u|_{x=0} = u|_{x=\pi} = u|_{y=0} = 0$, and the *additional* Neumann condition $u_y|_{y=0} = e^{-\sqrt{n}} \sin nx$, where n is an integer. Hence, whereas Dirichlet conditions are specified on the semi-infinite sides of the strip, Neumann conditions are imposed on the finite line $y = 0$, $0 \le x \le \pi$.

The function $u_n(x, y) = (1/n)e^{-\sqrt{n}} \sin nx \sinh ny$ is easily shown to be a classical solution of the Laplace equation subject to the stated auxiliary conditions, and it is unique. However, as $n \to \infty$, the solution $u_n(x, y)$ and all its derivatives tend uniformly to 0 throughout the half-strip in the (x, y) plane. However, when $y \ne 0$, the factor $\sinh ny$ in the solution causes the amplitude of the solution to become arbitrarily large, even though $u_y|_{y=0} = e^{-\sqrt{n}} \sin nx$ becomes arbitrarily small, showing that the stability requirement of a well-posed problem is not satisfied. This has shown by example the unsuitability of Cauchy conditions for the Laplace equation.

1.6 Coordinate Systems, Vector Operators, and Integral Theorems

The task of solving a partial differential equation is simplified if an appropriate coordinate system is used with the property that a boundary in a physical problem coincides with a constant value of a coordinate variable or that we seek a solution taking constant values on one family of coordinate surfaces. Thus in regions with rectangular boundaries it is natural to use Cartesian coordinates, while in cylindrical and spherical regions cylindrical and spherical polar coordinates are appropriate.

It is the purpose of this section to collect together for reference the vector operators used in this book when working with Cartesian, cylindrical, and

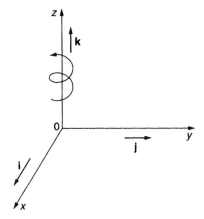

Figure 1.15   A right-handed
system of rectangular Cartesian
coordinates with the associated
unit vectors **i**, **j**, and **k**.

spherical polar coordinates, some of which were introduced in Section 1.1.
The section closes by recording the most important vector integral theorems
that prove useful when working with partial differential equations.

Cartesian coordinates

The right-handed system of rectangular Cartesian coordinates $O\{x, y, z\}$ is
illustrated in Fig. 1.15, together with the right-handed set of unit vectors **i**, **j**, and
k directed, respectively, in the positive sense along the mutually perpendicular
x, y, and z axes. Denoting a scalar product (inner product) of two vectors by a
dot, and a vector product (cross product) of two vectors by a cross, these unit
vectors have the property that $\mathbf{i} \cdot \mathbf{i} = \mathbf{j} \cdot \mathbf{j} = \mathbf{k} \cdot \mathbf{k} = 1, \mathbf{i} \cdot \mathbf{j} = \mathbf{i} \cdot \mathbf{k} = \mathbf{j} \cdot \mathbf{k} = 0$,
$\mathbf{i} \times \mathbf{j} = \mathbf{k}, \mathbf{j} \times \mathbf{k} = \mathbf{i}, \mathbf{k} \times \mathbf{i} = \mathbf{j}, \mathbf{i} \times \mathbf{i} = \mathbf{j} \times \mathbf{j} = \mathbf{k} \times \mathbf{k} = \mathbf{0}$. Reversing the
order of vectors in a vector product changes its sign so, for example, $\mathbf{k} \times \mathbf{j} =$
$-\mathbf{i}$, but reversing the order of vectors in a scalar product leaves the result
invariant, so if **a** and **b** are any two vectors, $\mathbf{a} \cdot \mathbf{b} = \mathbf{b} \cdot \mathbf{a}$.

The gradient operator

The **gradient operator** is denoted either by grad or by ∇, and when it acts on
a continuously differentiable function $\Phi(x, y, z)$ it is defined as the vector

$$\text{grad } \Phi \equiv \nabla \Phi \equiv \frac{\partial \Phi}{\partial x}\mathbf{i} + \frac{\partial \Phi}{\partial y}\mathbf{j} + \frac{\partial \Phi}{\partial z}\mathbf{k}. \tag{1.93}$$

The directional derivative

Let $\mathbf{n} = n_1\mathbf{i} + n_2\mathbf{j} + n_3\mathbf{k}$ be a unit vector; then the derivative of a continuously differentiable scalar function $\Phi(x, y, z)$ in the direction $\mathbf{n}$, called the **directional derivative** of Φ in the direction $\mathbf{n}$, is the scalar function

$$\mathbf{n} \cdot \text{grad } \Phi \equiv n_1 \frac{\partial \Phi}{\partial x} + n_2 \frac{\partial \Phi}{\partial y} + n_3 \frac{\partial \Phi}{\partial z}. \tag{1.94}$$

The divergence operation

Let $\mathbf{f} = f_1(x, y, z)\,\mathbf{i} + f_2(x, y, z)\,\mathbf{j} + f_3(x, y, z)\,\mathbf{k}$ be a continuously differentiable vector function of x, y, and z. Then the **divergence** of $\mathbf{f}$, written either div $\mathbf{f}$ or $\nabla \cdot \mathbf{f}$, is defined as

$$\text{div } \mathbf{f} \equiv \nabla \cdot \mathbf{f} \equiv \left(\frac{\partial}{\partial x}\mathbf{i} + \frac{\partial}{\partial y}\mathbf{j} + \frac{\partial}{\partial z}\mathbf{k} \right) \cdot (f_1\mathbf{i} + f_2\mathbf{j} + f_3\mathbf{k})$$

$$= \frac{\partial f_1}{\partial x} + \frac{\partial f_2}{\partial y} + \frac{\partial f_3}{\partial z}. \tag{1.95}$$

The curl operation

Let $\mathbf{f} = f_1(x, y, z)\mathbf{i} + f_2(x, y, z)\mathbf{j} + f_3(x, y, z)\mathbf{k}$ be a continuously differentiable vector function of x, y, and z. Then the **curl** of $\mathbf{f}$, written either curl $\mathbf{f}$ or $\nabla \times \mathbf{f}$, is defined as

$$\text{curl } \mathbf{f} \equiv \nabla \times \mathbf{f} \equiv \left(\frac{\partial f_3}{\partial y} - \frac{\partial f_2}{\partial z} \right)\mathbf{i} + \left(\frac{\partial f_1}{\partial z} - \frac{\partial f_3}{\partial x} \right)\mathbf{j} + \left(\frac{\partial f_2}{\partial x} - \frac{\partial f_1}{\partial y} \right)\mathbf{k} \tag{1.96}$$

or, more conveniently, in terms of the symbolic determinant

$$\text{curl } \mathbf{f} \equiv \begin{vmatrix} \mathbf{i} & \mathbf{j} & \mathbf{k} \\ \partial/\partial x & \partial/\partial y & \partial/\partial z \\ f_1 & f_2 & f_3 \end{vmatrix}, \tag{1.97}$$

which is expanded in the usual manner.

The Laplacian

The **Laplacian operator**, denoted here by Δ, but often by ∇^2, when acting on a twice continuously differentiable scalar function $\Phi(x, y, z)$ is defined as the scalar function

$$\Delta \Phi \equiv \nabla^2 \Phi \equiv \frac{\partial^2 \Phi}{\partial x^2} + \frac{\partial^2 \Phi}{\partial y^2} + \frac{\partial^2 \Phi}{\partial x^2}. \tag{1.98}$$

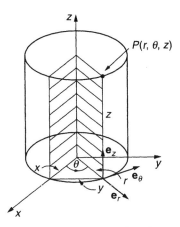

Figure 1.16   The cylindrical polar coordinate system.

If the Laplacian acts on a vector function $\mathbf{f} = f_1\mathbf{i} + f_2\mathbf{j} + f_3\mathbf{k}$, it acts on each component of $\mathbf{f}$, so that

$$\Delta f \equiv \Delta f_1\mathbf{i} + \Delta f_2\mathbf{j} + \Delta f_3\mathbf{k}. \tag{1.99}$$

Cylindrical polar coordinates and plane polar coordinates

The system of cylindrical polar coordinates (r, θ, z) of a point P, always given in this order when defining a point, is illustrated in Fig. 1.16. The variable r is the radial distance from the origin in the $O\{x, y\}$ plane of the projection of point P onto the plane, θ is the azimuth angle measured counterclockwise from the x axis to r, so $0 \le \theta < 2\pi$, and z is the ordinate of P from the $O\{x, y\}$ plane. The mutually perpendicular unit vectors in the direction of increasing r, θ, and z are, respectively, $\mathbf{e}_r$, $\mathbf{e}_\theta$, and $\mathbf{e}_z$, where $\mathbf{e}_z = \mathbf{k}$.

The connection between the (x, y, z) and (r, θ, z) coordinate systems

Cartesian and cylindrical polar coordinates are related by

$$x = r\cos\theta, \quad y = r\sin\theta, \quad \text{and} \quad z = z, \quad \text{with } 0 \le \theta < 2\pi. \tag{1.100}$$

The gradient operator

The **gradient operator** acting on a continuously differentiable scalar function $\Phi(r, \theta, z)$ is defined as the vector

$$\operatorname{grad}\Phi \equiv \nabla\Phi \equiv \frac{\partial\Phi}{\partial r}\mathbf{e}_r + \frac{1}{r}\frac{\partial\Phi}{\partial\theta}\mathbf{e}_\theta + \frac{\partial\Phi}{\partial z}\mathbf{e}_z. \tag{1.101}$$

The directional derivative

Let $\mathbf{n} = n_1\mathbf{e}_r + n_2\mathbf{e}_\theta + n_3\mathbf{e}_z$ be a unit vector. Then the **directional derivative** in the direction $\mathbf{n}$ of a continuously differentiable scalar function $\Phi(r, \theta, z)$ is the scalar function

$$\mathbf{n} \cdot \mathrm{grad}\ \Phi \equiv \mathbf{n} \cdot \nabla\Phi \equiv n_1\frac{\partial\Phi}{\partial r} + n_2\frac{1}{r}\frac{\partial\Phi}{\partial\theta} + n_3\frac{\partial\Phi}{\partial z}. \tag{1.102}$$

The divergence operation

Let $\mathbf{f} = f_1(r, \theta, z)\mathbf{e}_r + f_2(r, \theta, z)\mathbf{e}_\theta + f_3(r, \theta, z)\mathbf{e}_z$, then the **divergence** of a continuously differentiable function $\mathbf{f}$ of r, θ, and z is defined as the scalar function

$$\mathrm{div}\ \mathbf{f} \equiv \nabla \cdot \mathbf{f} \equiv \frac{1}{r}\frac{\partial(rf_1)}{\partial r} + \frac{1}{r}\frac{\partial f_2}{\partial\theta} + \frac{\partial f_3}{\partial z}. \tag{1.103}$$

The curl operation

Let $\mathbf{f} = f_1(r, \theta, z)\mathbf{i} + f_2(r, \theta, z)\mathbf{j} + f_3(r, \theta, z)\mathbf{k}$ be a continuously differentiable vector function of r, θ, and z. Then the **curl** of $\mathbf{f}$ is defined as the vector

$$\mathrm{curl}\ \mathbf{f} \equiv \nabla \times \mathbf{f} \equiv \left[\frac{1}{r}\frac{\partial f_3}{\partial\theta} - \frac{\partial f_2}{\partial z}\right]\mathbf{e}_r + \left[\frac{\partial f_1}{\partial z} - \frac{\partial f_3}{\partial r}\right]\mathbf{e}_\theta + \frac{1}{r}\left[\frac{\partial(rf_2)}{\partial r} - \frac{\partial f_1}{\partial\theta}\right]\mathbf{e}_z. \tag{1.104}$$

The Laplacian

The **Laplacian operator** when acting on a twice continuously differentiable scalar function $\Phi(r, \theta, z)$, written either as $\Delta\Phi$ or as $\nabla^2\Phi$, is the scalar function

$$\Delta\Phi \equiv \nabla^2\Phi \equiv \frac{1}{r}\frac{\partial}{\partial r}\left(r\frac{\partial\Phi}{\partial r}\right) + \frac{1}{r^2}\frac{\partial^2\Phi}{\partial\theta^2} + \frac{\partial^2\Phi}{\partial z^2}. \tag{1.105}$$

A related coordinate system involving only the two variables r and θ in the (x, y) plane is called a **plane polar coordinate system**, and the corresponding vector operators are derived from those above by omitting the independent variable z and its derivatives.

Spherical polar coordinates

The system of spherical polar coordinates (r, θ, ϕ) of a point P used throughout this book, and always given in this order when defining a point, is illustrated in Fig. 1.17. Unlike the cylindrical polar coordinate system, in the spherical polar coordinate system the variable r is the radial distance from the origin in the Cartesian coordinate system $O\{x, y, z\}$ to the point P, ϕ is the azimuth angle

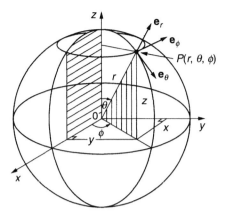

Figure 1.17   The spherical polar coordinate system.

measured counterclockwise for the x axis to the projection of r on the $O\{x, y\}$ plane, so $0 \leq \phi < 2\pi$ and θ is the polar angle measured from the positive z direction to r, so $0 \leq \theta \leq \pi$. Care must always be taken when using a spherical polar coordinate system because different conventions are in use, and in the one used here the polar angle θ in spherical polar coordinates should *not* be confused with the azimuth angle θ in cylindrical polar coordinates.

    The mutually perpendicular unit vectors in the direction of increasing r, ϕ, and θ are, respectively, $\mathbf{e}_r, \mathbf{e}_\theta$, and $\mathbf{e}_\phi$, as shown in Fig. 1.17.

The connection between the (x, y, z) and (r, θ, ϕ) coordinate systems

$$x = r \sin \theta \cos \phi, \quad y = r \sin \theta \cos \phi, \quad z = r \cos \theta, \quad \text{with}$$
$$0 \leq \theta \leq \pi, \ 0 \leq \phi < 2\pi. \tag{1.106}$$

The gradient operator

The **gradient operator** acting on a continuously differentiable function $\Phi(r, \phi, \theta)$ is defined as the vector

$$\text{grad } \Phi \equiv \frac{\partial \Phi}{\partial r} \mathbf{e}_r + \frac{1}{r} \frac{\partial \Phi}{\partial \theta} \mathbf{e}_\theta + \frac{1}{r \sin \theta} \frac{\partial \Phi}{\partial \phi} \mathbf{e}_\phi. \tag{1.107}$$

The directional derivative

Let $\mathbf{n} = n_1 \mathbf{e}_r + n_2 \mathbf{e}_\theta + n_3 \mathbf{e}_\phi$ be a unit vector; then the **directional derivative** of Φ in the direction $\mathbf{n}$ of a continuously differentiable scalar function $\Phi(r, \theta, \phi)$

in the direction of **n**, is the scalar function

$$\mathbf{n} \cdot \text{grad } \Phi \equiv n_1 \frac{\partial \Phi}{\partial r} + n_2 \frac{1}{r} \frac{\partial \Phi}{\partial \theta} + n_3 \frac{1}{r \sin \theta} \frac{\partial \Phi}{\partial \phi}. \tag{1.108}$$

The divergence operation

Let $\mathbf{f} = f_1(r,\theta,\phi)\mathbf{e}_r + f_2(r,\theta,\phi)\mathbf{e}_\phi + f_3(r,\theta,\phi)\mathbf{e}_\theta$; then the divergence of a continuously differentiable function $\mathbf{f}$ of r, ϕ, and θ is defined as

$$\text{div } \mathbf{f} \equiv \frac{1}{r^2} \frac{\partial (r^2 f_1)}{\partial r} + \frac{1}{r \sin \theta} \frac{\partial (f_2 \sin \theta)}{\partial \theta} + \frac{1}{r \sin \theta} \frac{\partial f_3}{\partial \phi}. \tag{1.109}$$

The curl operation

Let $\mathbf{f} = f_1(r,\theta,\phi)\mathbf{e}_r + f_2(r,\theta,\phi)\mathbf{e}_\phi + f_3(r,\theta,\phi)\mathbf{e}_\theta$ be a continuously differentiable vector function of r, θ, and z. Then the **curl** of $\mathbf{f}$ is defined as the vector

$$\text{curl } \mathbf{f} = \frac{1}{r \sin \theta} \left[\frac{1}{r} \frac{\partial (f_3 \sin \theta)}{\partial \theta} - \frac{\partial f_2}{\partial \phi} \right] \mathbf{e}_r + \left[\frac{1}{r \sin \theta} \frac{\partial f_1}{\partial \phi} - \frac{1}{r} \frac{\partial (r f_3)}{\partial r} \right] \mathbf{e}_\theta$$

$$+ \left[\frac{1}{r} \frac{\partial (r f_2)}{\partial r} - \frac{1}{r} \frac{\partial f_1}{\partial \theta} \right] \mathbf{e}_\phi. \tag{1.110}$$

The Laplacian

The **Laplacian operator** when acting on a twice continuously differentiable scalar function $\Phi(r, \theta, \phi)$ takes the form

$$\Delta \Phi = \frac{1}{r^2} \frac{\partial}{\partial r} \left(r^2 \frac{\partial \Phi}{\partial r} \right) + \frac{1}{r^2 \sin \theta} \frac{\partial}{\partial \theta} \left(\sin \theta \frac{\partial \Phi}{\partial \theta} \right) + \frac{1}{r^2 \sin^2 \theta} \frac{\partial^2 \Phi}{\partial \phi^2}. \tag{1.111}$$

The Gauss divergence theorem

Let Ω be a closed bounded region in space with a boundary surface S formed by m piecewise smooth surfaces $S_1, S_2, \ldots, S_m$, each enclosed by a space curve. At each point of the surfaces S_i, with the exception of points on their boundaries, let there be a unit normal $\mathbf{n}$ directed *out* of Ω. Furthermore, let $\mathbf{F}(x, y, z)$ be a vector function that is continuous with continuous partial derivatives at all points inside Ω and on S, with the exception of points on the space curve boundaries of $S_1, S_2, \ldots, S_m$. Then

$$\iiint_\Omega \text{div } \mathbf{F} \, dV = \iint_S \mathbf{F} \cdot \mathbf{n} \, dA, \tag{1.112}$$

where dV is a volume element of Ω and dA is a surface element of S.

Green's integral theorems

Let the region Ω with the boundary surface S and outward drawn unit normal $\mathbf{n}$ satisfy the conditions of the Gauss divergence theorem. In addition, let $f(x, y, z)$ and $g(x, y, z)$ be two twice continuously differentiable scalar functions of position in Ω, with $d\mathbf{A} = \mathbf{n}\, dA$ the vector element of surface area S.

Green's first theorem

$$\iiint_\Omega (f \Delta g + \operatorname{grad} f \cdot \operatorname{grad} g)\, dV = \iint_S f \operatorname{grad} g \cdot d\mathbf{A}$$
$$= \iint_S f \frac{\partial g}{\partial n}\, dA,$$

where $\partial g / \partial n = \mathbf{n} \cdot \operatorname{grad} g$ is the directional derivative of g normal to S.

Green's second theorem

$$\iiint_\Omega (f \Delta g - g \Delta f)\, dV = \iint_S (f \operatorname{grad} g - g \operatorname{grad} f) \cdot d\mathbf{A}.$$

Stokes' theorem

Let S be an oriented piecewise smooth surface in space with a piecewise smooth boundary Γ in the form of a simple closed space curve, and let the position vector $\mathbf{r}$ on Γ be parametrized in terms of the arc length σ around Γ by $\mathbf{r} = \mathbf{r}(\sigma)$. Take $\mathbf{F}(x, y, z)$ to be a continuous vector function with continuous first-order partial derivatives on S, and let the vector $\mathbf{n}$ be the unit normal to S, oriented with respect to the direction of integration around Γ as shown in Fig. 1.18.

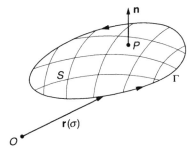

Figure 1.18 The orientation of $\mathbf{n}$ with respect to the direction of integration round Γ.

Then

$$\iint_S (\text{curl } \mathbf{F}) \cdot \mathbf{n} \, dA = \oint_\Gamma \mathbf{F} \cdot \left(\frac{d\mathbf{r}}{d\sigma} \right) d\sigma.$$

Useful identities involving vector operators

Entries 1 through 10 are standard identities involving the vector operators div, grad, and curl and the Laplacian Δ. They are included mainly for reference purposes, although proving them will provide revision for the most useful vector operator identities. For convenience, the identities are expressed both in terms of grad, div, and curl and the Laplacian Δ, and also using only the symbols ∇ and Δ. The reason the Laplacian Δ is written ∇^2 in many texts is because in operator notation $\Delta \equiv \text{div(grad)} \equiv \nabla \cdot \nabla \equiv \nabla^2$. The vector functions $\mathbf{F}$ and $\mathbf{G}$ and scalar functions V and W that occur in the identities are all assumed to be sufficiently differentiable for the stated results to be true.

Proofs of the identities can be found in any standard calculus text and they can be established in many different ways, often by using one identity when establishing another. The most elementary way of establishing an identity is first to decompose a vector expression like curl $\mathbf{F}$ or grad V into its scalar components, then to perform the necessary operation on each component, and finally to recombine the components to form a vector.

An expression like $\mathbf{F} \cdot$ grad, or equivalently $\mathbf{F} \cdot \nabla$, is a scalar operator, because if $\mathbf{F} = F_1\mathbf{i} + F_2\mathbf{j} + F_3\mathbf{k}$, then $\mathbf{F} \cdot \nabla \equiv F_1\frac{\partial}{\partial x} + F_2\frac{\partial}{\partial y} + F_3\frac{\partial}{\partial z}$. Thus if $\mathbf{F} \cdot \nabla$ operates on a scalar function the result will be a scalar function, whereas if it operates on a vector function the result will be a vector function. Consequently, as $\mathbf{F}/|\mathbf{F}|$ is a unit vector in the direction of $\mathbf{F}$, the expression $(\mathbf{F} \cdot \text{grad } V)/|\mathbf{F}|$ is the **directional derivative** of grad V in the direction of the *unit vector* $\mathbf{F}/|\mathbf{F}|$ (see Eq. (1.94)).

1. $\text{div(curl } \mathbf{F}) \equiv \nabla \cdot (\nabla \times \mathbf{F}) \equiv 0.$

2. $\text{curl(grad } V) \equiv \nabla \times (\nabla V) \equiv \mathbf{0}.$

3. $\text{grad}(VW) \equiv V\text{grad } W + W\text{grad } V \equiv V\nabla W + W\nabla V.$

4. $\text{curl(curl } \mathbf{F}) \equiv \text{grad(div } \mathbf{F}) - \Delta\mathbf{F} \equiv \nabla(\nabla \cdot \mathbf{F}) - \Delta\mathbf{F}.$

5. $\text{div(grad } V) \equiv \nabla \cdot (\nabla V) \equiv \Delta V.$

6. $\text{div}(V\mathbf{F}) \equiv V\text{div } \mathbf{F} + \mathbf{F} \cdot \text{grad } V \equiv \nabla \cdot (V\mathbf{F}) \equiv V\nabla \cdot \mathbf{F} + \mathbf{F} \cdot \nabla V.$

7. $\text{curl}(V\mathbf{F}) \equiv V\text{curl } \mathbf{F} - \mathbf{F} \times \text{grad } V \equiv V\nabla \times \mathbf{F} - \mathbf{F} \times \nabla V.$

8. $\text{grad}(\mathbf{F} \cdot \mathbf{G}) \equiv \mathbf{F} \times \text{curl } \mathbf{G} + \mathbf{G} \times \text{curl } \mathbf{F} + (\mathbf{F} \cdot \text{grad})\mathbf{G} + (\mathbf{G} \cdot \text{grad})\mathbf{F}$
   $\equiv \mathbf{F} \times (\nabla \times \mathbf{G}) + \mathbf{G} \times (\nabla \times \mathbf{F}) + (\mathbf{F} \cdot \nabla)\mathbf{G} + (\mathbf{G} \cdot \nabla)\mathbf{F}.$

9. $\text{div}(\mathbf{F} \times \mathbf{G}) \equiv \mathbf{G} \cdot \text{curl } \mathbf{F} - \mathbf{F} \cdot \text{curl } \mathbf{G} \equiv \mathbf{G} \cdot (\nabla \times \mathbf{F}) - \mathbf{F} \cdot (\nabla \times \mathbf{G}).$

10. $\text{curl}(\mathbf{F} \times \mathbf{G}) \equiv \mathbf{F} \text{ div } \mathbf{G} - \mathbf{G} \text{ div } \mathbf{F} + (\mathbf{G} \cdot \text{grad})\mathbf{F} - (\mathbf{F} \cdot \text{grad})\mathbf{G}$
$$\equiv \mathbf{F}(\nabla \cdot \mathbf{G}) - \mathbf{G}(\nabla \cdot \mathbf{F}) + (\mathbf{G} \cdot \nabla)\mathbf{F} - (\mathbf{F} \cdot \nabla)\mathbf{G}.$$

Examples of PDE applications of vector integral theorems

The Gauss divergence theorem

An application of the Gauss divergence theorem has already been encountered in Section 1.2 when proceeding from Eq. (1.30) to (1.31). There it enabled the integral of the normal component of a vector over a surface (the second term on the left of (1.30)) to be transformed into the integral of the divergence of the vector over a volume. As a result the sum of integrals in (1.30) over both a surface S and a volume V was expressed in (1.31) as a single integral over a volume V, from which there then followed the heat equation (1.32).

Green's theorems

Green's theorems have many uses, ranging from an application of the two-dimensional form when deriving the Cauchy integral formula in complex analysis, to many and diverse applications to PDEs. A useful special case of Green's first theorem follows by setting $f = g = u$, when it reduces to the following relationship between a volume integral and a surface integral:

$$\iiint_D (\nabla u)^2 \, dV = \iint_\Gamma (u\nabla u) \cdot dA.$$

If, for example, it is known that $u = 0$ on the boundary Γ of volume D, it follows at once that

$$\iiint_D (\nabla u)^2 \, dV = 0.$$

The integrand is nonnegative, so this can only be possible if $\nabla u \equiv 0$, and this in turn can only happen if $u \equiv 0$ throughout D. This simple result is used in Section 7.2 to prove the uniqueness of a solution of the Laplace equation in a region D by considering the difference of two supposedly different solutions, each of which assumes the same values on the boundary Γ of D.

Stokes' theorem

It is shown in calculus texts that a physical interpretation of the curl of a vector quantity $\mathbf{v}$ at a point P, namely curl $\mathbf{v}$, can be given in terms of the angular velocity of a uniformly rotating body where $\mathbf{v}$ is the linear velocity of the body at point P. To be precise, the angular velocity of the body is one-half of curl $\mathbf{v}$, so the curl of a vector provides a measure of its *rotational* effect at a point. It is for this reason that in some books curl $\mathbf{v}$ is written rot $\mathbf{v}$.

At a point P in a fluid where the velocity is $\mathbf{v}$, the swirling effect of the fluid is measured by curl $\mathbf{v}$ and called the **circulation** of the fluid at P, and also the **vorticity** of the fluid at P. In fluid mechanics the vorticity, or circulation, of the fluid moving with velocity $\mathbf{v}$ is usually denoted by ζ, so that $\zeta = $ curl $\mathbf{v}$.

With this idea in mind Stokes' theorem, which originated from the study of fluid mechanics, is capable of a simple and useful interpretation. Suppose S is an open surface like a hemispherical cap, bounded by a closed curve Γ. Then the total circulation of $\mathbf{v}$ normal to the surface S will be the integral over S of (curl $\mathbf{v}$) $\cdot$ $\mathbf{n}$, where $\mathbf{n}$ is the unit normal to S. To avoid ambiguity due to the fact that at any point P of S the unit normal $\mathbf{n}$ may be directed toward either side of S, the direction of $\mathbf{n}$ is always chosen to be such that it points in the direction in which a right-handed screw advances as the closed curve Γ is traversed (see Fig. 1.18).

Thus the total circulation over S is given by the surface integral $\iint_S$(curl $\mathbf{v}$) $\cdot$ $\mathbf{n}\, dA$, where dA is a surface element of S. If, as is often the case, S is required to be an *arbitrary* surface with only its bounding curve Γ known, this integral cannot be evaluated as it stands. It is here, however, that Stokes' theorem is invoked, because it asserts that

$$\iint_S (\text{curl } \mathbf{v}) \cdot \mathbf{n}\, dA = \oint_\Gamma \mathbf{F} \cdot \left(\frac{d\mathbf{r}}{d\sigma}\right) d\sigma,$$

where $d\sigma$ is an element of the closed curve Γ bounding S. Unlike the integral on the left, the line integral on the right can always be evaluated, either analytically or numerically, because $\mathbf{F}$ is known, as is the parametrization $\mathbf{r} = \mathbf{r}(\sigma)$ of Γ. Thus Stokes' theorem enables the circulation to be calculated independently of the nature of surface S, when only its bounding curve Γ is known.

An important consequence of this last result follows when vector $\mathbf{v}$ is the gradient of a scalar function ϕ called a **potential function**, so that $\mathbf{v} = $ grad ϕ. From entry (2) of the vector operator identities in Section 1.6 it follows that curl(grad ϕ) $\equiv 0$, so that Stokes' theorem then shows that the circulation (vorticity) is identically zero throughout the region where $\mathbf{v}$ is defined. Because of this a vector field of the form $\mathbf{v} = $ grad ϕ is said to be **irrotational**, a name derived from the early use of rot $\mathbf{v}$ for curl $\mathbf{v}$.

An irrotational vector field is a very important special case of a vector field, because it is shown in calculus texts that if $\mathbf{v} = $ grad ϕ in a region D, so no circulation is present in D, when the vector $\mathbf{v}$ represents a force, the force field grad ϕ is **conservative**, in the sense that the work done moving a mass from one point to another in a conservative gravitational field is fully recovered when it is moved back to its original point, irrespective of the paths taken. In Newtonian mechanics, when no dissipative effects are involved, it is this property which ensures that the sum of the potential and kinetic energies of a body is an absolute constant. Put another way, a conservative gravitational field has the property that, in the absence of dissipative effects like air resistance, when a stone is projected vertically upward against gravity with speed V, on its return to the same point its speed will again be V, though in the downward direction.

Stokes' theorem finds many other applications, as for example in electromagnetic theory, where the vector $\mathbf{v}$ may be either the electric vector $\mathbf{E}$ or the magnetic vector $\mathbf{H}$.

Linear and Nonlinear First-Order Equations and Shocks

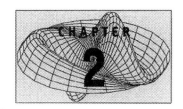

2.1 Linear and Semilinear Equations in Two Independent Variables

The most general *semilinear* first-order PDE for a scalar function $u(x, y)$ of the two independent variables x and y is

$$a(x, y)u_x + b(x, y)u_y = f(x, y, u), \qquad (2.1)$$

where $f(x, y, u)$ is a continuous function of its arguments that depends nonlinearly on u, while the coefficients $a(x, y)$, $b(x, y)$ are assumed to be continuous functions of the independent variables x and y.

In the special case that $f(x, y, u) = c(x, y)u + d(x, y)$, Eq. (2.1) reduces to the general *linear* first-order PDE

$$a(x, y)u_x + b(x, y)u_y = c(x, y)u + d(x, y), \qquad (2.2)$$

where now the functions $c(x, y)$ and $d(x, y)$ are also assumed to be continuous functions of their arguments. The linear equation (2.2) is *nonhomogeneous* when $d(x, y)$ is not identically zero; otherwise, it is *homogeneous*. A particularly simple form of linear equation arises when the coefficients $a(x, y)$, $b(x, y)$, and $c(x, y)$ in (2.2) are all absolute constants, because then the PDE becomes a *constant coefficient equation*.

The purpose of this section will be to arrive at a method of solution for the general semilinear equation (2.1), from which will follow the solution of linear equation (2.2) as a special case. The auxiliary condition to be satisfied by u will be taken to be that it must equal a given function $h(x, y)$ on a specified curve Γ in the (x, y) plane with the equation $g(x, y) = 0$. The curve Γ and the function h on Γ can always be defined parametrically in terms of a parameter σ, and although a parametric representation of the curve Γ is not unique, this is unimportant for what is to follow.

Define the points (x, y) on the curve Γ in terms of a parameter σ by

$$x = X(\sigma) \quad \text{and} \quad y = Y(\sigma), \tag{2.3}$$

where $X(\sigma)$ and $Y(\sigma)$ are given continuous and differentiable functions of σ. Then, in terms of σ, on the curve Γ the condition $u = h(x, y)$ becomes

$$u(\Gamma) = h(X(\sigma), Y(\sigma)), \tag{2.4}$$

where we will assume h to be a continuous and piecewise differentiable function of its arguments. Condition (2.4) is simply a *Cauchy condition* for solution u of (2.1), so the curve Γ in the (x, y) plane on which the Cauchy data for u are specified will be called the *Cauchy data curve.*

Let us now turn our attention to solving the semilinear equation (2.1). To accomplish this, let $\tilde{C}$ be a general curve in the (x, y) plane (not the Cauchy data curve Γ), which is described parametrically in terms of s by

$$x = x(s), \quad y = y(s), \tag{2.5}$$

where the functions $x(s)$ and $y(s)$ are assumed to be continuous and differentiable.

Then, on $\tilde{C}$, a differentiable function $u(x, y)$ becomes the function $u = u(x(s), y(s))$, so

$$\frac{du}{ds} = \frac{\partial u}{\partial x}\frac{dx}{ds} + \frac{\partial u}{\partial y}\frac{dy}{ds} \tag{2.6}$$

is the directional derivative of u along $\tilde{C}$.

Comparing the terms on the left of (2.1) with those on the right of (2.6) shows that along a special family of curves we will denote by C, which are found by integrating the *ordinary differential equations* (ODEs)

$$\frac{dx}{ds} = a(x, y), \quad \text{and} \quad \frac{dy}{ds} = b(x, y), \tag{2.7}$$

the original PDE in (2.1) can be replaced by the ODE

$$\frac{du}{ds} = f(x, y, u). \tag{2.8}$$

This family of curves C in the (x, y) plane are called the *characteristic curves,* and often simply the *characteristics,* of the PDE in (2.1), namely

$$a(x, y)u_x + b(x, y)u_y = f(x, y, u).$$

Equations (2.7) are called the parametric form of the *characteristic equations* of PDE (2.1), when (2.8) becomes an ODE for the solution u along any characteristic C as a function of the independent variable s. The ODE in (2.8) is called the *compatibility condition* for PDE (2.1) with respect to the family of characteristic curves C. Note that in linear and semilinear PDEs the characteristics can be determined independently of the solution u, since they only depend on the known functions $a(x, y)$ and $b(x, y)$.

When Eqs. (2.7) are integrated, the constants of integration introduced will correspond to an arbitrary point $(X(\sigma_0), Y(\sigma_0))$ on the Cauchy data curve Γ. Hence, from the initial condition to be satisfied by u on Γ, it follows that in terms of σ the initial condition for u along the characteristic through the point $(X(\sigma_0), Y(\sigma_0))$ will be $u = h(X(\sigma_0), Y(\sigma_0))$.

To see why the compatibility condition (2.9) is an ODE along a characteristic, and how the ODE can be solved, it is only necessary to recognize that once the equations in (2.7) have been integrated, the variables x and y become known functions of s. Consequently, the compatibility condition in (2.8) becomes an ODE for u along a characteristic curve with s as the independent variable.

Denoting the characteristic through the point $(X(\sigma_0), Y(\sigma_0))$ on Γ by C_{σ_0}, and using the fact that at this point $u = h(X(\sigma_0), Y(\sigma_0))$, integrating the compatibility condition along the characteristic C_{σ_0} shows that in terms of s, the solution on this characteristic can be written

$$u = U(X(\sigma_0), Y(\sigma_0), h(\sigma_0), s). \tag{2.9}$$

The point $(X(\sigma_0), Y(\sigma_0))$ on Γ was arbitrary, so this method will determine the solution of (2.1) in any domain of the (x, y) plane traversed by characteristics emanating from points on Γ.

This method of solution is called *the method of characteristics*, and it is illustrated in Fig. 2.1. The curve C_{σ_0} in Fig. 2.1 is a typical characteristic originating from the point $(X(\sigma_0), Y(\sigma_0))$ on the Cauchy data curve Γ, at which point $u_0 = h(X(\sigma_0), Y(\sigma_0))$.

This has established that when the method of characteristics is applied to a first-order PDE, the solution $u(x, y)$ is determined at every point along a characteristic curve C_{σ_0} emanating from a representative point $(X(\sigma_0), Y(\sigma_0))$ on the Cauchy data curve Γ, where $u_0 = h(X(\sigma_0), Y(\sigma_0))$. In general, this can be repeated at each point of Γ, so the solution $u(x, y)$ is determined at each point of the (x, y) plane traversed by nonintersecting characteristics emanating from points on Γ. Because of this property, it is appropriate to call the point $(X(\sigma_0), Y(\sigma_0))$ on Γ the *point of dependence* of the solution along the characteristic C_{σ_0}, and the characteristic itself the *line of determinacy* of the solution with respect to the point $(X(\sigma_0), Y(\sigma_0))$. This means, for example, that the solution in a region of the (x, y) plane traversed by characteristics emanating from a segment Γ_0 of Γ will depend *only* on the behavior of the boundary condition for u on Γ_0. This important

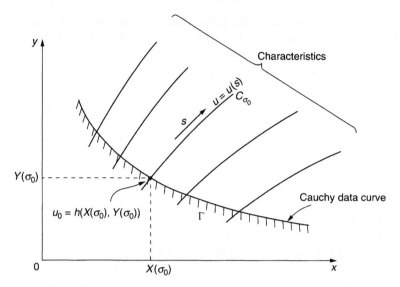

Figure 2.1 Characteristics emanating from a Cauchy data curve Γ.

property will be needed later when solving the boundary value problem in Example 2.2.

When the second-order wave equation for $u(x, t)$ is considered it will be seen that two families of characteristic curves arise. The notions of a point of dependence and a line of determinacy then generalize to a *domain of dependence* in the form of a line segment of Γ and a *domain of determinacy* in the (x, t) plane contained between characteristics through the end points of the domain of dependence.

Definition 2.1. (A Cauchy Problem for a Semilinear Equation)
A Cauchy problem for the semilinear PDE

$$a(x, y)u_x + b(x, y)u_y = f(x, y, u) \qquad (2.10)$$

is the determination of the solution $u(x, y)$ of the PDE in some domain of the (x, y) plane containing a Cauchy data curve Γ defined in terms of σ by

$$x = X(\sigma) \quad \text{and} \quad y = Y(\sigma),$$

such that on the data curve Γ the solution u satisfies the Cauchy condition

$$u(\Gamma) = h(X(\sigma), Y(\sigma)),$$

with h being a given continuous and differentiable function of σ.

When required, the parameter s can be eliminated from Eqs. (2.7) and (2.8) by dividing each by dx/ds to obtain an equivalent set of ODEs, in which the characteristics C are determined by integration of the single *characteristic equation*

$$\frac{dy}{dx} = \frac{b(x, y)}{a(x, y)}, \tag{2.11}$$

while the variation of u along each of these characteristics is determined by integration of the corresponding *compatibility condition*

$$\frac{du}{dx} = \frac{f(x, y, u)}{a(x, y)}. \tag{2.12}$$

Although this form of characteristic equation and compatibility condition may appear simpler than the equivalent parametric forms in (2.7) and (2.8), it is often the case that the parametric equations are easier to integrate. As a first-order linear PDE is simply a special case of a first-order semilinear PDE, the method of solution just described is applicable to both types of PDE.

It is important to recognize that the method of characteristics can fail if the Cauchy data curve Γ becomes tangent to a characteristic. This situation is illustrated in Fig. 2.2 where the characteristic C_P through the point P on Γ becomes tangent to Γ at the point Q. The reason for the possible failure of the method in such circumstances is because integration of the compatibility condition along C_P *determines* the solution u at point Q in terms of the Cauchy condition $u = h(X(\sigma_P), Y(\sigma_P))$ at P. However, this value of u will not

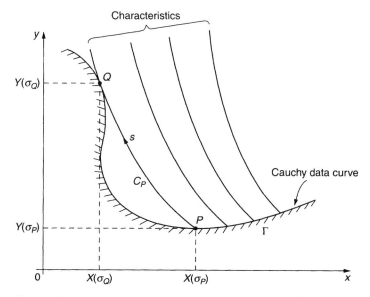

Figure 2.2 A characteristic emanating from a point on Γ becomes tangent to Γ at Q.

necessarily agree with the value $u = h(X(\sigma_Q), Y(\sigma_Q))$ assigned by the Cauchy condition at Q. Thus, in general, the Cauchy condition on the Cauchy data curve Γ is incompatible with the solution of the PDE where a characteristic becomes tangent to Γ. Cauchy problems of this type, called *characteristic Cauchy problems*, usually have no solution, although when they do exist the solution may not be unique.

Example 2.1. Use the method of characteristics to solve the initial value problem

$$u_t + cu_x = 0 \quad \text{subject to the initial condition } u(x, 0) = f(x),$$

which was considered first in Chapter 1 where the general solution $u(x, t) = f(x - ct)$ was found by inspection.

Solution: This simple initial value problem was considered first in Chapter 1 when considering the flow of vehicles moving along a road at a constant speed c. The PDE is a homogeneous linear constant coefficient PDE with the independent variables x and t and an initial line Γ coinciding with the x axis. We will solve this PDE using Eqs. (2.11) and (2.12) with x and y replaced by t and x, respectively, when from (2.11) the characteristics are seen to be determined by the characteristic equation $dx/dt = c$. Integration shows the equation of the characteristic through an arbitrary point $(\xi, 0)$ on the x axis to be $x = ct + \xi$, so the family of characteristics forms a family of parallel straight lines in the (x, t) plane, each determined by a different value of ξ. In this case the Cauchy data curve Γ (the x axis) is never tangent to a characteristic curve, so a characteristic Cauchy problem cannot arise.

The compatibility condition (2.12) becomes $du/dt = 0$, so $u(x, t) = $ constant along each member of the family of characteristics (straight lines) where, in general, the constant value is different along different characteristics. The constant value of u along the characteristic C_ξ through the point $(\xi, 0)$ must be the value of u assigned by the Cauchy condition at that point, so $u(\xi, 0) = f(\xi)$ along the line $x = ct + \xi$. Eliminating ξ between these two results shows the required solution of the Cauchy problem to be

$$u(x, t) = f(x - ct).$$

This is the solution that was found by inspection in Chapter 1. Figure 2.3 shows the pattern of characteristics in the (x, t) plane. ∎

Example 2.2. Solve the PDE

$$u_x + 2xu_y = 2xu,$$

subject to the conditions (a) $u(x, 0) = x^2$ for all x, (b) $u(0, y) = y^2$ for all y, and (c) $u(x, 0) = x^2$ for $x \geq 0$ and $u(0, y) = y^2$ for $y \geq 0$.

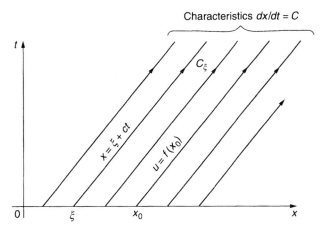

Figure 2.3 The method of characteristics applied to $u_t + cu_x = 0$, with $u(x, 0) = f(x)$.

Solution: This is a homogeneous linear variable coefficient PDE which we will again solve by using Eqs. (2.11) and (2.12), although this time without a change of independent variables. However, whereas conditions (a) and (b) are specified on the entire x and y axes, respectively, condition (c) is more complicated because the solution is only required in the first quadrant of the (x, y) plane so Cauchy conditions are specified only on the positive x and y axes.

(a) It follows from (2.11) that the characteristic equation is $dy/dx = 2x$, so as the Cauchy condition is imposed along the x axis, integration of this result shows that the characteristic C_ξ through the point $(\xi, 0)$ has the equation $y = x^2 - \xi^2$, where $(\xi, 0)$ is an arbitrary point on the x axis, as shown in Fig. 2.4a. Here, as in Example 2.1, no characteristic curve is tangent to the Cauchy data curve so no characteristic initial value problem will arise.

The compatibility condition (2.12) along each characteristic becomes $du/dx = 2xu$, and after integration we find that $\ln u - \ln k(\xi) = x^2$, where $k(\xi)$ is the integration constant associated with the solution along the characteristic C_ξ through the point $(\xi, 0)$. The parameter ξ is shown in the integration constant $k(\xi)$ to indicate that although $k(\xi)$ is constant along the characteristic C_ξ through $(\xi, 0)$, it will, in general, assume different constant values along different characteristics, each of which will depend on the value of ξ. Thus the solution along the characteristic C_ξ is seen to be given by

$$u(x, y) = k(\xi) \exp(x^2).$$

To complete the solution we must now determine the functional form of $k(\xi)$, and this is accomplished by using the Cauchy condition at the point $(\xi, 0)$ where $u(\xi, 0) = \xi^2$. Now, when $x = \xi$ and $y = 0$ the function $u(x, y)$

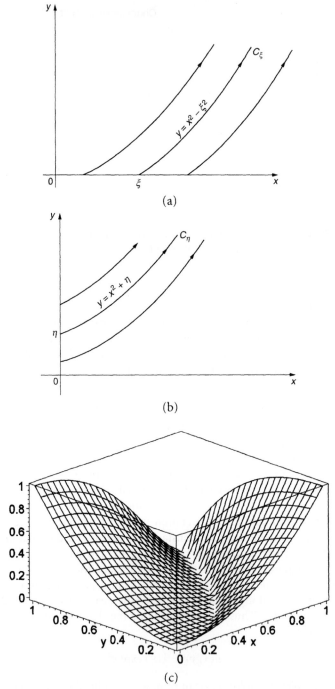

Figure 2.4   (a) The characteristics C_ξ. (b) The characteristics C_η. (c) The generalized solution in the first quadrant of the (x, y) plane.

reduces to $\xi^2 = k(\xi)\exp(\xi^2)$, so $k(\xi) = \xi^2\exp(-\xi^2)$. Substituting for $k(\xi)$ in the solution for u gives

$$u(x, y) = \xi^2\exp(x^2 - \xi^2),$$

but $\xi^2 = x^2 - y$, so in terms of x and y the solution becomes

$$u(x, y) = (x^2 - y)e^y.$$

This solution is easily checked, because differentiation shows it satisfies the PDE, and it is also seen to satisfy the Cauchy condition on the x axis.

(b) In this case the initial data curve Γ is the y axis, so the argument must proceed a little differently. As in (a) the characteristics are determined by $dy/dx = 2x$, so if a characteristic passes through the point $(0, \eta)$ on Γ, integration shows the characteristic C_η through this point has the equation $y = x^2 + \eta$, as shown in Fig. 2.4b. The compatibility condition is again $du/dx = 2xu$, but this time it applies along each member of the family of characteristics C_η. Consequently, after integration we have $\ln u - \ln g(\eta) = x^2$, where $g(\eta)$ is the integration constant associated with the solution along the characteristic C_η through the point $(0, \eta)$. Thus

$$u(x, y) = g(\eta)\exp(x^2).$$

The initial condition shows that $u(0, \eta) = \eta^2$, so setting $x = 0$ and $y = \eta$ in the solution gives $\eta^2 = g(\eta)$, and thus $u(x, y) = \eta^2\exp(x^2)$. Finally, substituting for η we arrive at the result

$$u(x, y) = (y - x^2)\exp(x^2).$$

It is again a simple matter to verify that this is the required solution.

(c) As Cauchy data are only specified on the positive x axis, it follows that the data can only determine the solution in a region that lies to the right of the characteristic $y = x^2$ in the first quadrant. Similarly, the Cauchy data specified on the positive y axis can only determine the solution that lies above the characteristic $y = x^2$ in the first quadrant. The two solutions involved have already been found in the upper half-plane in (a) and (b), so it only remains to combine them as follows to arrive at the required solution:

$$u(x, y) = \begin{cases} (x^2 - y)e^y, & x \geq 0, \ y \leq x^2 \\ (y - x^2)^2 e^{x^2}, & x \geq 0, \ y \geq x^2. \end{cases}$$

A plot of this solution is shown in Fig. 2.4c, where the bounding characteristic through the origin can be seen to separate the two parts of the solution. In fact, this is a *generalized solution*, because although it is continuous, its first-order partial derivatives exist and are continuous everywhere in the first quadrant apart from across the bounding characteristic where they are discontinuous. ∎

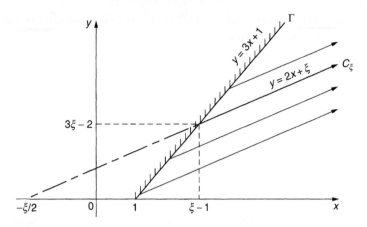

Figure 2.5   The initial line Γ and the characteristics C_ξ.

Example 2.3. Solve the PDE

$$u_x + 2u_y = 1 + u$$

for the function $u(x, y)$, subject to the Cauchy condition that $u = \sin x$ on $y = 3x + 1$.

Solution: This time, as the Cauchy condition is prescribed on an oblique straight line Γ, it will again be convenient to use Eqs. (2.11) and (2.12). The characteristic equation (2.11) becomes $dy/dx = 2$, so the equation of the characteristics is $y = 2x + \xi$, where ξ is an arbitrary integration constant that identifies a characteristic. Geometrically, the straight line characteristic C_ξ associated with the constant ξ passes through the point $(-\xi/2, 0)$ on the x axis, as shown in Fig. 2.5. The compatibility condition (2.12) becomes $du/dx = 1 + u$, so after integration we find that $\ln(1 + u) - \ln k(\xi) = x$ where, as in Example 2.1, $k(\xi)$ is an integration constant. Writing this result as $u(x, y) = k(\xi)e^x - 1$ and substituting $\xi = y - 2x$ we arrive at the result

$$u(x, y) = k(y - 2x)e^x - 1.$$

Now $u = \sin x$ on $y = 3x + 1$, so substituting for u in this last result shows the function $k(x+1)$ is such that $\sin x = k(x+1)e^x - 1$. To find $k(y-2x)$ we replace x by $y - 2x - 1$ to obtain $\sin(y - 2x - 1) = k(y - 2x)e^{y-2x-1}$. Hence we see that $k(y - 2x) = e^{2x-y+1} \sin(y - 2x - 1)$. Finally, substituting this expression for $k(y - 2x)$ into $u(x, y)$ gives

$$u(x, y) = \exp(3x - y + 1)\sin(y - 2x - 1) - 1.$$

Differentiation confirms that $u(x, y)$ satisfies the PDE, and it also satisfies the Cauchy condition on the line $y = 3x + 1$, so it is the required solution. Here no characteristic is tangent to the Cauchy data curve, which this time is an oblique straight line.   ∎

Example 2.4. Solve the PDE

$$u_x + 3u_y = u$$

for the function $u(x, y)$, subject to the Cauchy condition $u = \cos x$ on the line $y = \alpha x$. Find the value of α for which the method fails and interpret the result.

Solution: This is a linear constant coefficient PDE that could be solved either by using Eqs. (2.7) and (2.8) or by using Eqs. (2.11) and (2.12). So, to illustrate the parametric approach, we will use Eqs. (2.7) and (2.8).

The Cauchy data curve (a straight line in this case) Γ has the Cartesian equation $y = \alpha x$, and this can be expressed in terms of the parameter σ as

$$x = \sigma, \qquad y = \alpha\sigma.$$

So, in terms of σ, the Cauchy condition on Γ becomes

$$u(\Gamma) = \cos\sigma.$$

The characteristic equations (2.7) become

$$\frac{dx}{ds} = 1, \qquad \frac{dy}{ds} = 3.$$

Let C_σ be the characteristic through an arbitrary point P on Γ corresponding to $x = \sigma$, $y = \alpha\sigma$, and choose the parametrization in terms of s along the characteristic C_σ so that $s = 0$ at P. Integration of these equations then leads to the results

$$x = s + \sigma, \qquad y = 3s + \alpha\sigma,$$

from which we find that

$$\sigma = \frac{3x - y}{3 - \alpha} \quad \text{and} \quad s = \frac{y - \alpha x}{3 - \alpha}.$$

The compatibility condition (2.8) becomes $du/ds = u$. Integrating this along the characteristic C_σ through the point $(\sigma, \alpha\sigma)$ on Γ gives

$$\ln u + \ln k(\sigma) = s, \quad \text{or} \quad u(x, y) = e^s / k(\sigma),$$

where $k(\sigma)$ is the constant of integration associated with the characteristic C_σ, as shown in Fig. 2.6. The constant of integration is written $k(\sigma)$ to indicate that, in general, it is a function of σ.

As $x = s + \sigma$, $y = 3s + \alpha s$, the solution $u(x, y)$ can be written

$$u(s + \sigma, 3s + \alpha\sigma) = e^s / k(\sigma).$$

When $s = 0$ it follows from the Cauchy condition that $u(\sigma, \alpha\sigma) = \cos\sigma$, so setting $s = 0$ in this last result gives $k(\sigma) = 1/\cos\sigma$. Substituting for $k(\sigma)$ in

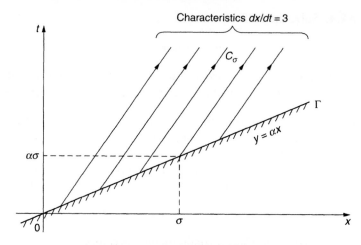

Figure 2.6   Characteristics emanating from Γ.

the expression for $u(x, y)$ we find that

$$u(x, y) = e^s \cos \sigma,$$

but $s = (y - \alpha x)/(3 - \alpha)$ and $\sigma = (3x - y)/(3 - \alpha)$, so the solution becomes

$$u(x, y) = \exp\left(\frac{y - \alpha x}{3 - \alpha}\right) \cos\left(\frac{3x - y}{3 - \alpha}\right).$$

Routine differentiation confirms that $u(x, y)$ satisfies the PDE, and $u(x, y)$ reduces to the required Cauchy condition on the Cauchy data curve (line) $y = \alpha x$, so it follows that this is the required solution.

The solution is seen to become nondifferentiable when $\alpha = 3$, due to the vanishing of the denominator in each factor. This is not surprising, because in terms of the variables x and y, all characteristics have the same slope $dy/dx = 3$, so when $\alpha = 3$ the Cauchy data line coincides with a characteristic.   ∎

As a final example we consider a Cauchy problem for a semilinear equation, because this illustrates a fundamental difference that can exist between linear and semilinear equations.

Example 2.5. Solve the PDE

$$u_x + u_y = u^2$$

for $y > 0$, subject to the Cauchy condition $u(x, 0) = \tanh x$.

Solution: This equation is semilinear due to the presence of the term u^2, and in this case the Cauchy condition is again specified on the x axis. Working with Eqs. (2.11) and (2.12) we find that the characteristic equation is $dy/dx = 1$. Integrating this result to find the characteristic C_ξ through the arbitrary point

$(\xi, 0)$ on the x axis gives $y = x - \xi$, showing that C_ξ is nowhere tangent to the Cauchy data curve $y = 0$.

The compatibility condition (2.12) becomes $du/dx = u^2$, so integrating this along the characteristic C_ξ emanating from the point $(\xi, 0)$ gives

$$-\frac{1}{u(x, y)} + k(\xi) = x,$$

where $k(\xi)$ is the integration constant associated with C_ξ. The Cauchy condition at the point $(\xi, 0)$ becomes $u(\xi, 0) = \tanh \xi$, so setting $x = \xi, y = 0$ in the above equation and using the Cauchy condition gives

$$-\frac{1}{\tanh \xi} + k(\xi) = \xi, \quad \text{so } k(\xi) = \frac{1 + \xi \tanh \xi}{\tanh \xi}.$$

Finally, using this expression for $k(\xi)$ in the equation for $u(x, y)$, and substituting $\xi = x - y$ we arrive at the solution

$$u(x, y) = \frac{\tanh(x - y)}{1 - y \tanh(x - y)}.$$

This function satisfies both the PDE and the initial condition so it is the required solution.

The example illustrates an important feature of nonlinearity in first-order PDEs, because although the Cauchy condition $u = \tanh x$ is bounded and differentiable for all x, the solution $u(x, y)$ becomes infinite on the critical curve $y \tanh(x - y) = 1$. This curve has branches in the first and third quadrants of the (x, y) plane, as shown in Fig. 2.7. As the solution is required for $y > 0$, only the branch in the first quadrant, which is seen to have the asymptotes $y = x$ and $y = 1$, is needed.

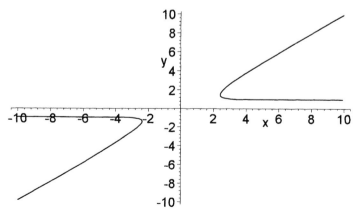

Figure 2.7   The critical curve $y \tanh(x - y) = 1$.

Inspection of Fig. 2.7 shows the solution exists and is finite for all x provided $0 < y < 1$, but that when $y > 1$ the solution will always become infinite for some x. ∎

It is appropriate that here we draw attention to a connection between first-order equations and the second-order wave equation $u_{tt} = c^2 u_{xx}$, where c is a positive constant. Note that the wave equation can be factored, and written either as

$$\left(\frac{\partial}{\partial t} + c \frac{\partial}{\partial x} \right) \left(\frac{\partial u}{\partial t} - c \frac{\partial u}{\partial x} \right) = 0 \quad \text{or as} \quad \left(\frac{\partial}{\partial t} - c \frac{\partial}{\partial x} \right) \left(\frac{\partial u}{\partial t} + c \frac{\partial u}{\partial x} \right) = 0.$$

A *degenerate* solution of the wave equation can be seen from the factored form on the left if u is a solution of the first-order PDE $u_t - cu_x = 0$. This is a *special* solution of the wave equation, because it is a solution of a first-order PDE that also happens to be a solution of a second-order equation. Another degenerate solution can be seen from the factored form on the right if u is a solution of the first-order PDE $u_t + cu_x = 0$. Each of these degenerate solutions is seen to describe the propagation of a wave of arbitrary shape (profile) moving with constant speed c, where the first wave moves to the left and the second one moves to the right along the x axis.

It will be seen in Chapter 3 that the wave equation is classified as being of *hyperbolic type*. Consequently, it is appropriate that equations of the form (2.1) that are degenerate solutions of the wave equation, and which also describe wave propagation, should be classified as being of *hyperbolic* type. Note that in PDE (2.1), one of the independent variables can be regarded as being time-like in nature, while the other behaves like a space variable.

EXERCISES 2.1

In the following Exercises solve the PDE for $u(x, y)$ subject to the given Cauchy condition on the Cauchy data curve implied by the Cauchy condition.

1. $2u_x - 5u_y = 4$, subject to the Cauchy condition $u(x, 0) = x$.

2. $u_x + 3u_y = u + 2$, subject to the Cauchy condition $u(0, y) = y$.

3. $xu_x + 2yu_y = 3u$, subject to the Cauchy condition $u(1, y) = \cos y$ for $y > 1$.

4. $u_x + u_y = 1 - u$, subject to the Cauchy condition that $u = x$ on $y = 2x$.

5. $u_x - 2u_y = u - 1$, subject to the Cauchy condition $u = 2y$ on $x = ky$, giving reasons for any restriction that must be placed on k.

6. $xu_x + yu_y = \alpha u$, subject to the Cauchy condition $u = x^2$ on $y = 1$.

7. $u_x + 2u_y = 1 + u$, subject to the Cauchy condition $u = \sin x$ on $y = 3x + 1$.

8. $yu_x + x^3 u_y = x^3 y$, subject to the Cauchy condition $u = x^4$ on $y = x^2$.

9. $u_x + u_y = u^2$, subject to the Cauchy condition $u(0, y) = \tanh y$. Comment on the relationship between the solution of this PDE and the solution of Example 2.5.

10. $u_x - 3u_y = \text{sech } u$, subject to the Cauchy condition $u(x, 0) = x$.

11. Solve $u_x + 2u_y = u$, given that $u(x, 0) = 3x, x \geq 0$ and $u(0, y) = \sin y$, $y \geq 0$.

12. Solve $u_x + xu_y = 3xu$, given that $u(x, 0) = x^2, x \geq 0$ and $u(0, y) = 1 - \cos y, y \geq 0$.

13. Consider the initial value problem $u_x + cu_y = 0$ with $c = $ constant, and the initial line Γ with the equation $y = x - 1$ on which $u = x$. Find the solution (a) by the method of characteristics and (b) by direct appeal to the general solution $u(x, y) = f(y - cx)$, where the arbitrary function f is to be determined from the initial conditions. State any condition that must be imposed on the solution in order for it to be well posed.

14. Consider the initial value problem $u_x + cu_y = 1$ with $c = $ constant, and the initial line Γ with the equation $y = 2x + 1$ on which $u = \tanh x$. Verify that the general solution of the PDE is $u(x, y) = f(y - cx) + x$ where f is an arbitrary function of its argument. Solve the initial value problem (a) by finding the function f that makes u satisfy the initial condition on Γ and (b) by using the method of characteristics. State any condition that must be imposed on the solution in order for it to be well posed.

15. Solve by the method of characteristics $u_x + 2xu_y = 2xu$, given that $u = x^2$ on the initial curve Γ with the equation $y = \frac{1}{2} x^2$.

16. Consider the initial value problem $u_x + u_y = 2$ where the initial line Γ has the equation $y = 2x$ on which $u = x^2$. Verify that the general solution of the PDE is $u(x, y) = f(y - x) + x + y$, where f is an arbitrary function of its argument. Solve the initial value problem (a) by finding the function f that makes u satisfy the initial condition on Γ and (b) by using the method of characteristics.

2.2 Quasi-Linear Equations in Two Independent Variables

First-order quasi-linear PDEs occur throughout science and engineering, and for example, they are central to the study of gas dynamics, continuum mechanics, traffic flow models, nonlinear acoustics, and groundwater flows, to mention only a few of their applications. This section will be concerned with quasi-linear PDEs in one space dimension and time, because equations of this type are directly applicable to the study of one-dimensional nonlinear waves. Thus, when considering solutions, the two independent variables involved will be taken to be a space variable x and the time t. The most general first-order quasi-linear PDE for the function $u(x, t)$ then becomes

$$a(x, t, u)u_t + b(x, t, u)u_x = f(x, t, u), \tag{2.13}$$

where $a(x, t, u)$, $b(x, t, u)$, and $f(x, t, u)$ will be assumed to be continuous functions of their arguments.

For the reason given in Section 2.1 when considering linear and semilinear PDEs, Eq. (2.13) is classified as being of *hyperbolic* type. There is, however, a fundamental difference between the properties of linear and quasi-linear PDEs that will become apparent later, although something of the difference has already been seen in Example 2.5.

In general terms, a *Cauchy problem* for (2.13) is the determination of the solution $u(x, t)$ of the PDE in some domain of the (x, t) plane containing a *Cauchy data curve* Γ defined in terms of a parameter σ by

$$x = X(\sigma) \quad \text{and} \quad t = T(\sigma),$$

on which the solution u is required to satisfy the *Cauchy condition*

$$u(\Gamma) = h(X(\sigma), T(\sigma)),$$

where $h(x, y)$ is a continuous and differentiable function of its arguments.

In wave propagation problems the Cauchy data curve Γ is usually the x axis, so then a Cauchy condition of the form $u(x, 0) = h(x)$ becomes an *initial condition* for the solution $u(x, t)$ of (2.13).

As with Eqs. (2.1) and (2.2), the method of characteristics can be used to replace PDE (2.13) by two ODEs, one of which determines the characteristic curves of the PDE while the other is a compatibility condition for $u(x, t)$ in the form of an ODE that is valid along characteristics. To see how the method of characteristics can be applied to (2.13) we again make use of the concept of a directional derivative along a curve C in the (x, t) plane parametrized in terms of s by the equations

$$x = x(s), \qquad t = T(s). \tag{2.14}$$

Then, as in Section 2.1, along C a differentiable function $u(x, t)$ becomes $u = u(x(s), y(s))$, so that

$$\frac{du}{ds} = \frac{\partial u}{\partial t}\frac{dt}{ds} + \frac{\partial u}{\partial x}\frac{dx}{ds} \tag{2.15}$$

is seen to be the directional derivative of u along C. As in the linear and semilinear cases, a comparison of the terms on the left of (2.13) with those on the right of (2.15) allows the PDE in (2.13) to be replaced by the pair of ODEs

$$\frac{dt}{ds} = a(x, t, u), \quad \text{and} \quad \frac{dx}{ds} = b(x, t, u), \tag{2.16}$$

defining the family C of *characteristic curves* of (2.13) and the *compatibility condition* for $u(x, t)$ along each characteristic C

$$\frac{du}{ds} = f(x, t, u), \tag{2.17}$$

which is again an ODE with independent variable s.

Equations (2.16) are the parametric form of the *characteristic equations* for the quasi-linear PDE in (2.13), and they should be compared with the equations in (2.7) and (2.8), and again in (2.11) and (2.12). This comparison makes clear a fundamental difference between the linear and semilinear case considered in Section 2.1 and the quasi-linear case considered here, because whereas the characteristics of linear and semilinear PDEs can be determined independently of the solution $u(x, t)$, in the quasi-linear case the characteristics depend on x, t and also on the solution $u(x, t)$. Thus, in general, it is necessary to know the solution of a quasi-linear PDE in order to construct its characteristic curves. This is why for most Cauchy problems the nonlinear system of simultaneous equations (2.16) and (2.17) comprising the equations for the characteristic and the compatibility condition must be solved numerically. Fortunately, however, we will see that a number of important special cases can be solved analytically, and these will be our concern here.

Dividing Eqs. (2.16) and (2.17) by dt/ds eliminates the parameter s and reduces the equations to the single ODE determining the family of characteristics

$$\frac{dx}{dt} = \frac{b(x, t, u)}{a(x, t, u)}, \tag{2.18}$$

when the compatibility condition along the characteristics becomes

$$\frac{du}{dt} = \frac{f(x, t, u)}{a(x, t, u)}. \tag{2.19}$$

These equations are, of course, the analogue of equations (2.11) and (2.12) for the semilinear equation (2.1).

The fundamental difference between the solutions of linear and quasi-linear PDE can be illustrated by considering a Cauchy problem for the simplest quasi-linear PDE

$$u_t + u u_x = 0, \tag{2.20}$$

which is sometimes called the **dissipationless Burgers' equation**. Let us use Eqs. (2.18) and (2.19) from the method of characteristics to find the solution of (2.20) subject to the general initial condition $u(x, 0) = h(x)$.

Equation (2.18) for the characteristic curves becomes

$$\frac{dx}{dt} = u, \tag{2.21}$$

and the compatibility condition (2.19) becomes

$$\frac{du}{dt} = 0. \tag{2.22}$$

The compatibility condition (2.22) shows that $u =$ constant along a characteristic, and it then follows from (2.21) that $dx/dt =$ constant along that characteristic, so the characteristic must be a straight line.

If we consider the characteristic C_ξ through the arbitrary point $(\xi, 0)$ on the initial line (the x axis), it follows from the initial condition that at this point $u(\xi, 0) = h(\xi)$, but $u(x, t) =$ constant along C_ξ, so we must have $u(x, t) = h(\xi)$ along C_ξ. The equation for C_ξ now follows by integrating $dx/dt = h(\xi)$ subject to the condition that the characteristic passes through the point $(\xi, 0)$. This gives the equation of the characteristic C_ξ in the parametric form

$$x = h(\xi)t + \xi. \tag{2.23}$$

Combining results shows the solution of (2.20) subject to the Cauchy (initial) condition $u(x, 0) = h(x)$ is given by

$$u(x, t) = h(\xi) \quad \text{along the lines } x = h(\xi)t + \xi. \tag{2.24}$$

Assuming the function h^{-1} inverse to h exists, the first of these equations shows that $\xi = h^{-1}(u)$, so using this to eliminate ξ from the second equation we arrive at the solution of the Cauchy problem for (2.20) in the **implicit** form

$$u(x, t) = h(x - tu). \tag{2.25}$$

This result already contains important information about the solution, because implicitly defined functions are *not* necessarily unique, so it is to be expected that in certain circumstances the solution of (2.20) may become nonunique at some critical time $t = t_c$.

To understand how this may happen, consider the situation where $h(x)$ is a strictly increasing function of x, so if $x_1 < x_2$, then $h(x_1) < h(x_2)$. In this case the characteristics through the points $(x_1, 0)$ and $(x_2, 0)$ shown in Fig. 2.8 must diverge (fan out). Note that Fig. 2.8 uses the standard convention for time-dependent problems in which the x axis is horizontal and the t axis is vertical. The result shown in Fig. 2.8 is true for any pair of values x_1 and x_2, so the solution $u(x, t)$ will be defined and unique throughout the upper half-plane $t > 0$.

Now consider the situation where $h(x)$ is a strictly decreasing function of x, so if $x_1 < x_2$, then $h(x_1) > h(x_2)$. In this case the characteristics through

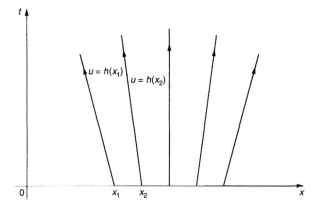

Figure 2.8   Diverging characteristics.

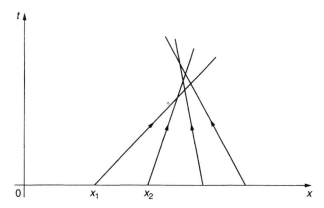

Figure 2.9   Converging characteristics.

the points $(x_1, 0)$ and $(x_2, 0)$ converge, as shown in Fig. 2.9. As the solution $u(x, t)$ has a different constant value along each characteristic, their intersection implies the nonuniqueness of the solution at some finite time $t > 0$. We show later that in this case the characteristics emanating from the initial line as t increases may form an envelope, so if the envelope starts to form when $t = t_c$, a unique solution can only exist for $0 < t < t_c$. Note that when $h(x_1) > h(x_2)$, the nonuniqueness of the solution depends only on the fact that $h(x)$ is a decreasing function of x, and not on its differentiability, so nonuniqueness can even occur when $h(x)$ is infinitely differentiable.

It is instructive to compare solution (2.25) of the quasi-linear PDE in (2.20) with the solution $u(x, t) = h(x - ct)$ of the constant coefficient equation $u_t + cu_x = 0$, subject to the same initial condition $u(x, 0) = h(x)$. It is because in the linear constant coefficient case the slope of the characteristics is $c = $ constant that the solution represents a steady translation of the initial condition along the x axis with speed c, and without change of shape or scale. In the (x, t) plane, where the solution represents a propagating wave, the function $u(x, t)$ is said to define the **wave profile** at time t. In the quasi-linear case the speed of translation of the wave depends on u, so different parts of the wave will move with different speeds, causing it to distort as it propagates. It is this distortion that can lead to the nonuniqueness of solutions in the quasi-linear case. A physical example of this phenomenon is found in the theory of shallow water waves, where the speed of propagation of a surface element of the water is proportional to the square root of the depth. This has the effect that in shallow water the crest of a wave moves faster than the trough, leading to wave breaking close to the shore line.

The related concepts of a *point of dependence* on the initial line and a *line of determinacy* formed by the characteristic through the point of dependence introduced in Section 2.1 also apply to first-order quasi-linear PDEs. If Γ_0 is the segment $x_1 < x < x_2$ of the initial line Γ, the specification of Cauchy data $u(x, 0) = h(x)$ on Γ_0 can only determine the solution along characteristics originating from points on this segment. If $h(x)$ is strictly increasing for $x_1 < x < x_2$ the characteristics will fan out (diverge) and define a solution at all points in the upper half-plane between the characteristics originating from the points x_1 and x_2 on Γ_0. However, if this is not so, the solution will become nonunique at some point between these two bounding characteristics.

An example will help to illustrate both the distortion of the wave as it propagates and the formation of the envelope of characteristics, the start of which corresponds to the time t_c when the solution first becomes nonunique. Let the initial condition for (2.20) be $u(x, 0) = \sin x$, then from (2.25) the implicit solution is seen to be

$$u(x, t) = \sin(x - tu). \tag{2.26}$$

The development of the wave profile as t increases is shown in Figs. 2.10a–2.10c. In Fig. 2.10a at time $t = 0$ the wave profile is a pure sinusoid. In Fig. 2.10b at time $t = 1$ the wave profile has steepened to the point where the tangent to the wave profile has become vertical at the point where the solution crosses the x axis, while at time $t = 1.5$ Fig. 2.10c shows that the wave is no longer single valued in periodic intervals of x. A three-dimensional plot of the solution $u(x, t)$ as a function of time for $0 \le x \le 2\pi$ and $0 \le t \le 2.0$ is shown in Fig. 2.11. This illustrates how, as the time increases, the wave profile distorts as it propagates eventually, after time $t = 1$, becoming a many-valued function with respect to x, thereby leading to the breakdown of the differentiability of the solution.

The characteristic through the arbitrary point $(\xi, 0)$ on the x axis has the equation

$$x = t \sin \xi + \xi.$$

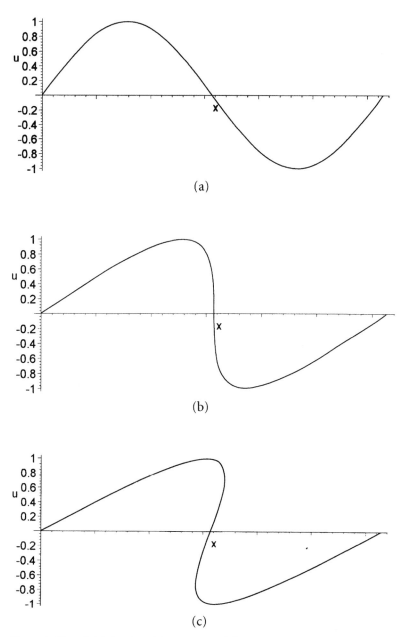

Figure 2.10 (a) Initial profile at $t = 0$. (b) Critical profile at $t = 1$. (c) Nonunique profile at $t = 1.5$.

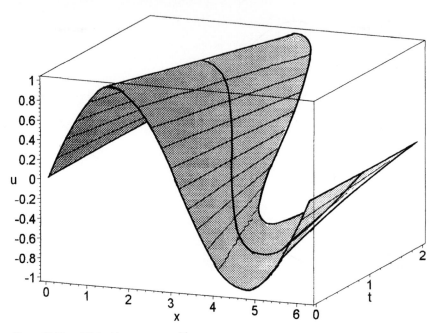

Figure 2.11   Distorting wave profile.

By defining $\varphi(x, t, \xi) = x - t \sin \xi - \xi$, the equation of this characteristic can be written in the parametric form $\varphi(x, t, \xi) = 0$, with ξ serving as the parameter. From elementary calculus it is known that, when it exists, the envelope formed by a family of curves $\varphi(x, t, \xi) = 0$ with ξ as the parameter is found by eliminating ξ between the equations $\varphi(x, t, \xi) = 0$ and $\partial \varphi / \partial \xi = 0$. A simple calculation shows that in this case the envelope has the equation $x = (t^2 - 1)^{1/2} + \arccos(-1/t)$, so using the principal branch of the inverse cosine function this becomes

$$x = (t^2 - 1)^{1/2} + \pi - \arccos(1/t).$$

As the term $(t^2 - 1)^{1/2}$ is only real valued for $t > 1$, this result confirms the uniqueness of the solution for $0 < t < 1$, because no envelope can form during this time interval. Defining the critical time as $t_c = 1$ it follows that a unique solution exists for $0 < t < t_c$, while the solution becomes nonunique for $t > t_c$. By taking different branches of the inverse cosine function, a sequence of equispaced and identical envelopes of characteristics can be constructed for all x, in each of which a cusp forms at the time $t = 1$, this being the first time at which characteristics intersect. A typical example of an envelope obtained by using the principal branch of the inverse cosine function is shown in Fig. 2.12.

The development of an infinite slope of a tangent to the wave profile at the critical time t_c can be seen by differentiating the implicit solution $u = h(x - tu)$

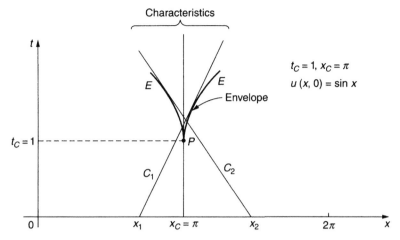

Figure 2.12   Cusp at P and characteristics C_1 and C_2 tangent to envelope E.

partially with respect to x to obtain the result

$$\frac{\partial u}{\partial x} = \frac{h'(x - tu)}{1 + th'(x - tu)}, \tag{2.27}$$

where the prime indicates differentiation of h with respect to its argument $x - tu$. Result (2.27) shows u_x will become infinite for any t such that $1 + th'$ $(x - tu) = 0$, because the denominator can only vanish if $h'(x - tu) < 0$, while at this time the numerator is nonzero. The occurrence of nonuniqueness in solutions of quasi-linear PDEs will be examined more closely when *conservation laws* and special types of discontinuous solutions called *shocks* are considered.

Example 2.6. Solve the PDE

$$u_t + uu_x = u,$$

subject to the Cauchy condition $u(x, 0) = 2x$ for $1 \leq x \leq 2$. Determine the domain in the upper-half of the (x, t) plane where the solution is defined by this Cauchy data.

Solution: The characteristic equation determined by (2.18) is $dx/dt = u$, and the compatibility condition determined by (2.19) is $du/dt = u$. Integrating the compatibility condition along the characteristic that passes through the point $(\xi, 0)$, which still must be determined, gives

$$\int_{\tau=0}^{t} \frac{du}{u} = \int_0^t d\tau,$$

and so $\ln u(x, t) - \ln u(\xi, 0) = t$. Solving this for $u(x, t)$ gives $u(x, t) = u(\xi, 0)e^t$. However, from the Cauchy condition (initial condition) we have

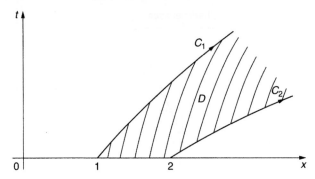

Figure 2.13   Domain of determinacy D bounded by characteristics C_1 and C_2.

$u(\xi, 0) = 2\xi$ for $1 \le \xi \le 2$, so along these characteristics $u(x, t) = 2\xi e^t$. Using this result in the characteristic equation gives $dx/dt = 2\xi e^t$. Integration of this last result from $\tau = 0$ to $\tau = t$ shows the characteristic through the point $(\xi, 0)$ has the equation $x = \xi(2e^t - 1)$. Finally, the elimination of ξ between this result and $u(x, t) = 2\xi e^t$ gives

$$u(x, t) = \frac{2xe^t}{2e^t - 1}, \quad \text{for } 1 \le x \le 2.$$

It is easily verified that this is the required solution, because differentiation confirms that $u(x, t)$ satisfies the PDE, while it also satisfies the initial condition $u(x, 0) = 2x$ when $t = 0$.

The characteristics have the equation $x = \xi(2e^t - 1)$ for $1 \le \xi \le 2$, and the solution is required in the upper-half of the (x, t) plane. The domain of determinacy D is shown in Fig. 2.13, where it lies between the two bounding characteristic curves C_1 and C_2 (lines of determinacy) originating from the ends of the interval $1 \le x \le 2$.   ∎

EXERCISES 2.2

1. Solve the PDE

$$u_t + f(u)u_x = 0,$$

subject to the Cauchy condition $u(x, 0) = g(x)$, where $f(u)$ is a continuous and differentiable function of u and $g(x)$ is a continuous function of x. Use the result to find an expression for $u_x(x, t)$ and hence find a condition that must be satisfied if $u_x(x, t)$ is to exist for $t > 0$.

2. Solve the PDE

$$u_t + 2uu_x = 0,$$

subject to the Cauchy condition $u(x, 0) = \tanh x$.

3. Solve the PDE

$$u_t - uu_x = e^t,$$

subject to the Cauchy condition $u(x, 0) = -x$.

4. Solve the PDE

$$uu_t + u_x = 0,$$

subject to the Cauchy condition $u(x, 1) = 1/x$ for $x \geq 1$.

5. Solve the PDE

$$u_t + uu_x = t,$$

subject to the Cauchy condition $u(x, 0) = -2x$. Find the domain in the upper-half of the (x, t) plane where the solution is valid.

6. Solve $u_t + uu_x = 0$ when (a) $u(x, 0) = x$, with $1 \leq x \leq 2$ and (b) $u(x, 0) = -x$, with $1 \leq x \leq 2$. In each case state the domain D in the half-plane $t \geq 0$ where the solution is valid.

7. Solve for u and v the system $u_t + 2uu_x = v - x$, with $u(x, 0) = x$ and $v_t - cv_x = 0$, with $v(x, 0) = x$.

8. Solve for u and v the system $u_t + uu_x = e^{-x}v$, with $u(x, 0) = x$ and $v_t + cv_x = 0$, with $v(x, 0) = e^x$.

2.3 Propagation of Weak Discontinuities by First-Order Equations

We again consider a scalar solution $u(x, t)$ of a first-order PDE, where x is a length and t is the time, but this time we will relax the condition that $u(x, t)$ is required to be differentiable everywhere throughout the domain D in which the solution is defined. The slightly weaker condition we now impose on $u(x, t)$ is that it is *continuous* throughout D, but differentiable everywhere in D with the exception of a curve C (there may be more than one) across which a first-order derivative of the solution may be *discontinuous*. A solution of this type will be said to have a *weak discontinuity* across C to distinguish it from the situation to be considered later, called a *shock solution*, or a *strong discontinuity*, where the solution itself is discontinuous across some curve Γ in D.

In what follows we will consider the special quasi-linear case in which the PDE is of the simpler form

$$u_t + f(u)u_x = g(u), \tag{2.28}$$

where f and g are continuous and differentiable functions of u throughout D. The argument can be extended to the general quasi-linear PDE considered in

Section 2.2, although as the general conclusion is the same, but its justification is more complicated, we will confine attention to the PDE in (2.28).

To discover the nature of the curve C in the (x, t) plane across which $u(x, t)$ is continuous, but a first-order derivative may be discontinuous, we change the independent variables in (2.28) to the new variables

$$\xi = \xi(x, t), \quad t' = t, \tag{2.29}$$

where at this stage the new curvilinear coordinate ξ must still be chosen. In the change of variables in (2.29) the time remains unchanged, although in the new coordinate system t' is now measured along the curves $\xi(x, t) = \text{constant}$. Provided the Jacobian J of the transformation is nonvanishing, that is, as long as

$$J = \frac{\partial(\xi, t')}{\partial(x, t)} = \begin{vmatrix} \xi_x & \xi_t \\ t'_x & t'_t \end{vmatrix} = \xi_x \neq 0,$$

the transformation from the (x, t) plane to the (ξ, t') plane is one-to-one, and in operator form

$$\frac{\partial}{\partial t} \equiv \frac{\partial \xi}{\partial t}\frac{\partial}{\partial \xi} + \frac{\partial t'}{\partial t}\frac{\partial}{\partial t'} \equiv \frac{\partial \xi}{\partial t}\frac{\partial}{\partial \xi} + \frac{\partial}{\partial t'}$$

$$\frac{\partial}{\partial x} \equiv \frac{\partial \xi}{\partial x}\frac{\partial}{\partial \xi} + \frac{\partial t'}{\partial x}\frac{\partial}{\partial t'} \equiv \frac{\partial \xi}{\partial x}\frac{\partial}{\partial \xi}. \tag{2.30}$$

Using results (2.30) to replace derivatives with respect to t and x in (2.28) by derivatives with respect to t' and ξ changes the PDE to

$$u_{t'} + (\xi_t + f(u)\xi_x)u_\xi = \tilde{g}(u), \tag{2.31}$$

where g becomes $\tilde{g}$ after the change of variable.

In (2.31) u_ξ is now the derivative of u normal to the curves $\xi = \text{constant}$, and $u_{t'}$ is the derivative of u along the curves $\xi = \text{constant}$. Let us now choose ξ by setting $\xi = \varphi$ where, if possible, the curve $\varphi(x, t) = 0$ is the member of the family of curves $\varphi = \text{constant}$ across which u and $u_{t'}$ are continuous, but u_φ is discontinuous. Then, along $\varphi(x, t) = 0$, we have

$$\varphi_t + \varphi_x dx/dt = 0,$$

so

$$\frac{dx}{dt} = -\frac{\partial \varphi}{\partial t} \bigg/ \frac{\partial \varphi}{\partial x}. \tag{2.32}$$

Replacing ξ by φ in (2.31), and using the fact that $\varphi_x \neq 0$, because it is the Jacobian of the transformation, allows the PDE to be rewritten as

$$u_{t'} + \varphi_x \left(\frac{\varphi_t}{\varphi_x} + f(u) \right) u_\varphi = \tilde{g}(u),$$

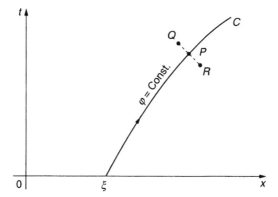

Figure 2.14 Differencing across the characteristic C at P.

and from (2.32) this becomes

$$u_{t'} + \varphi_x \left(-\frac{dx}{dt} + f(u) \right) u_\varphi = \tilde{g}(u). \qquad (2.33)$$

If the curve C is now identified with a characteristic curve of (2.28), the bracketed term in (2.33) vanishes, because $dx/dt = f(u)$ is the characteristic equation of (2.28). As a result (2.33) reduces to

$$u_{t'} = \tilde{g}(u). \qquad (2.34)$$

Differentiating Eq. (2.34) with respect to φ, and then interchanging the order of differentiation in the term on the left, gives

$$(u_\varphi)_{t'} = b(u)u_\varphi, \qquad (2.35)$$

where $b(u) = \partial \tilde{g}/\partial u$.

Now let P be an arbitrary point on C, with points Q and R close to P but on opposite sides of C on a line through P perpendicular to C, as shown in Fig. 2.14.

Differentiating (2.35) with respect to φ, and then differencing the result across C at P, allowing Q and R to tend to P, and using the fact that $b(u)$ is continuous across C leads to the ODE

$$\frac{d}{dt'}[[u_\varphi]] = b(u)[[u_\varphi]], \qquad (2.36)$$

where $[[u_\varphi]]$ denotes the jump in the derivative of u with respect to φ across C and, because this ODE holds along C, we have replaced $\partial/\partial t'$ by d/dt'.

As P was an arbitrary point on the curve C, result (2.36) shows that a weak discontinuity in a solution of (2.28) can be permitted, in which case the jump in the directional derivative of u normal to C propagates along a characteristic of the PDE. The strength of the jump $[[u_\varphi]]$ along C is seen to obey the ODE in (2.36), with the initial value determined by the jump $[[u_\varphi]]$ across C on the initial line.

When no jump in u_φ occurs across a characteristic the solution is continuously differentiable across the characteristic, and although result (2.36) remains true it becomes degenerate because then $[[u_\varphi]] \equiv 0$. Combining these observations with the results of Section 2.2 establishes the following important property of first-order quasi-linear PDEs in the two independent variables x and t.

Propagation of weak discontinuities

Let $u(x, 0) = \tilde{u}(x)$ be an initial condition (Cauchy condition) for (2.28) defined over some interval of the x axis, where $\tilde{u}(x)$ is continuous but $d\tilde{u}/dx$ has at most a finite number of jump discontinuities at $x = x_1, x_2, \ldots, x_n$. Then each element of the initial condition associated with a point $(x, 0)$ on the initial line, including those at the points $(x_1, 0), (x_2, 0), \ldots, (x_n, 0)$, will propagate along the characteristic curve through the corresponding point on the initial line. As linear and semilinear PDEs are special cases of quasi-linear PDEs, this conclusion also applies to them in general, although the preceding argument has only established this property for linear and semilinear PDEs of the type shown in (2.28).

The example that follows illustrates the propagation of weak discontinuities along characteristics in the case of a rather simple quasi-linear PDE with an initial condition that is continuous for all x but with a discontinuity in u_x at two different places on the x axis.

Example 2.7. Solve the PDE

$$u_t + uu_x = 0,$$

subject to the Cauchy conditions:

(a)

$$u(x, 0) = \begin{cases} -1, & -\infty < x \le -a \\ x/a, & -a < x < a \\ 1, & a \le x < \infty \end{cases} \quad \text{and}$$

(b)

$$u(x, 0) = \begin{cases} 1, & -\infty < x \le -a \\ -x/a, & -a < x < a \\ -1, & a \le x < \infty. \end{cases}$$

Solution: Cauchy condition (a): The characteristics of the PDE are determined by the ODE $dx/dt = u$, and the compatibility condition is $du/dt = 0$.

Reasoning as in Section 2.2, the compatibility condition shows $u = $ constant along a characteristic, so it follows from the characteristic equation that the characteristics are straight lines. Any characteristic issuing out from a point $(\xi, 0)$ on the x axis in the interval $-\infty < x \leq -a$ will have slope $dx/dt = -1$, so these are straight lines $x = \xi - t$, along which $u(x, t) = -1$. Similarly, any characteristic issuing out from the point $(\xi, 0)$ in the interval $a \leq x < \infty$ as t increases will have slope $dx/dt = 1$, and so are the straight lines $x = \xi + t$, along which $u(x, t) = 1$. Thus for $t > 0$, and to the left of the characteristic $x = -(a + t)$, the PDE will have the constant solution $u(x, t) = -1$, while for $t > 0$, and to the right of the characteristic $x = a + t$, it will have the constant solution $u(x, t) = 1$.

It follows from the initial condition that the solution at the point $(\xi, 0)$ in the interval $-a < x < a$ is $u(\xi, 0) = \xi/a$, so along the characteristic through this point $u(x, t) = \xi/a$. Thus the characteristic through this point is found by integrating $dx/dt = \xi/a$, showing that $x = (a + t)\xi/a$. However, along this characteristic we have seen that $u(x, t) = \xi/a$, so after the elimination of ξ the solution in the truncated wedge-shaped region covered by characteristics issuing out from any point in the interval $-a < x < a$ as t increases is

$$u(x, t) = \frac{x}{a + t}.$$

The solution in this truncated wedge-shaped region is called a *simple wave solution*. The solution of this Cauchy problem involves piecing together the simple wave and the two adjacent constant solutions to obtain a solution that is continuous for $t > 0$. A result such as this is often called a *global solution*, because it is defined for all x and any time $t > 0$ after the imposition of the initial condition. A graph of the characteristics is shown in Fig. 2.15, and a plot of the solution when $a = 1$ is given in Fig. 2.16. It can be seen from Fig. 2.16 that the discontinuities in the initial condition at $x = \pm 1$ on the initial line propagate as weak discontinuities along the characteristics through these two points.

Cauchy condition (b): Reasoning as in (a) above, the characteristic through the point $(\xi, 0)$ in the interval $-\infty < x \leq -a$, along which $u(x, t) = 1$, is the

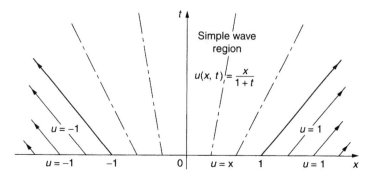

Figure 2.15   Constant states bounding a simple wave region.

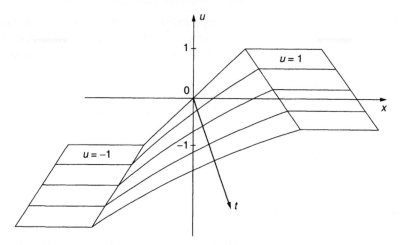

Figure 2.16   Evolution of the solution with time.

straight line $x = \xi + t$, and the characteristic through the point $(\xi, 0)$ in the interval $a \leq x < \infty$, along which $u(x, t) = -1$, is the straight line $x = \xi - t$. Similarly, as the initial condition at the point $(\xi, 0)$ in the interval $-a < x < a$ on the initial line is $u(\xi, 0) = -\xi/a$, it follows that $u(x, t) = -\xi/a$ along this characteristic, the equation of which is found by integrating $dx/dt = -\xi/a$ to obtain $x = (a - t)\xi/a$. Eliminating ξ between this equation and $u(x, t) = -\xi/a$ shows that now the solution is given by

$$u(x, t) = \frac{x}{t - a}$$

and this solution is seen to become nondifferentiable when $t = a$.

Consequently, a unique continuous solution that is piecewise differentiable only exists in the strip $-\infty < x < \infty$, $0 < t < a$, and so it is not a global solution. Note that the breakdown of the solution when $t = a$ depends only on the values of the initial condition at $x = \pm a$ on the initial line, and that it is *not* caused by the nondifferentiability of the initial data at these points.

In case (a), which gave rise to a global solution, the initial condition was strictly *increasing* in the interval $-a < x < a$, whereas in case (b) it was strictly *decreasing*. It was this feature of the initial condition that caused the characteristics in case (a) to *diverge* giving rise to a global solution, while in case (b) they *converged* leading to nonuniqueness and so to a loss of differentiability of the solution when $t = a$.   ∎

As a final example we examine the above problem in the limit as $a \to 0$ when the Cauchy condition becomes discontinuous at the origin, rather than continuous and piecewise differentiable, as was the case in Example 2.7.

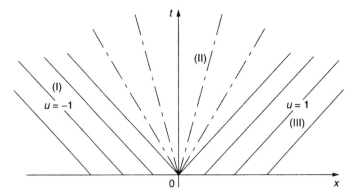

Figure 2.17   Simple wave centered on the origin.

In general, a first-order quasi-linear PDE for $u(x, t)$ subject to a piecewise constant initial condition is called a **Riemann problem**.

Example 2.8. Consider the following Riemann problems for the PDE

$$u_t + uu_x = 0,$$

subject to the piecewise constant Cauchy conditions:

$$\text{(a) } u(x, 0) = \begin{cases} -1, & -\infty < x < 0 \\ 1, & 0 < x < \infty, \end{cases} \quad \text{and} \quad \text{(b) } u(x, 0) = \begin{cases} 1, & -\infty < x < 0 \\ -1, & 0 < x < \infty. \end{cases}$$

Solution: Cauchy condition (a): In this Riemann problem the piecewise constant initial condition *increases* as x increases. It follows from Example 2.7 that in the region $t > 0$ to the left of the characteristic $x = -t$ drawn from the origin the solution is constant and given by $u(x, t) = -1$, while in the region $t > 0$ to the right of the characteristic $x = t$ drawn from the origin the solution is again constant and given by $u(x, t) = 1$. Thus, in the upper-half plane $t > 0$, the solution is known in regions I and III shown in Fig. 2.17, but it still must be determined in the wedge-shaped region II, where the equation of the bounding characteristic to the left of the region can be written $x/t = -1$ and the equation of the bounding characteristic to the right becomes $x/t = 1$.

We know from the characteristic equation and the compatibility condition for this PDE that when a differentiable solution exists the characteristics must be straight lines along which the solution is constant. This implies that for a differentiable solution to exist in region II, the characteristics in this region must be the straight lines (or *rays*) $x/t = \zeta$ drawn from the origin with $-1 \leq \zeta \leq 1$. Accordingly, let us try to find a solution of the form $u(x, t) = U(\zeta)$ in region II. We have

$$u_t = U'(\zeta)\frac{\partial \zeta}{\partial t} = -\frac{1}{x}\zeta^2 U'(\zeta) \quad \text{and} \quad u_x = U'(\zeta)\frac{\partial \zeta}{\partial x} = \frac{1}{x}\zeta U'(\zeta),$$

where a prime indicates differentiation with respect to ζ. Substituting these results into the PDE $u_t + u u_x = 0$ gives

$$-\frac{1}{x}\zeta^2 U'(\zeta) + \frac{1}{x}\zeta U(\zeta) U'(\zeta) = 0.$$

Away from the singularity at the origin in region II, the variables x, ζ, and $U'(\zeta)$ are all nonzero, so after cancelation of the factor $\zeta\, U'(\zeta)/x$ the equation reduces to $U(\zeta) = \zeta$. Replacing ζ by x/t shows that in region II the solution is given by

$$u(x, t) = x/t, \quad \text{for } -1 \le x/t \le 1.$$

The solution in the upper-half plane $t > 0$ is now seen to be

$$u(x, t) = \begin{cases} -1, & -\infty < x/t \le -1 \\ x/t, & -1 < x/t < 1 \\ 1, & 1 \le x/t < \infty. \end{cases}$$

This specially simple solution is seen not to depend on x and t independently, but on the quotient x/t, and it is because of this that the original PDE reduced to an ODE for $U(x/t)$, and hence to the solution $u(x, t)$. Solutions of this type are called **similarity solutions**, because although the equation is a PDE, its solution $u(x, t)$ is the same (similar) everywhere the single variable $\zeta = x/t$ is constant (see Section 7.4). In this case this means that the solution is constant along the characteristics $x/t = \zeta$ with $-1 \le \zeta \le 1$. A plot of the solution is shown in Fig. 2.18, from which it can be seen that the discontinuity in the initial conditions at the origin is resolved immediately, after which the solution develops as a continuous piecewise differentiable function.

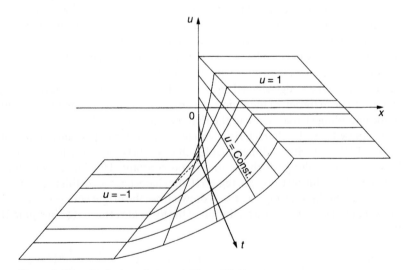

Figure 2.18   Evolution of the solution with time.

Note that the solution subject to initial condition (a) can be obtained from the solution of Example 2.7 in the limit $a \to 0$. So, in this case, the solution of a Riemann problem can be obtained as the limit of an initial condition that is continuous and piecewise differentiable.

When this solution is interpreted as a wave, the solution in region II is called a **centered simple wave solution** with its center at the origin, because all characteristics traversing this region pass through the origin. The solution shows how, provided values of the piecewise constant parts of the initial condition *increase* as x increases, a centered simple wave solution provides a *continuous transition* between adjacent constant solutions. This property is of considerable importance in applications where the initial condition increases, but is discontinuous at a finite number of points, since it allows solutions to be pieced together.

Again using the terminology of wave propagation, where $u(x, t)$ describes a physical quantity, constant solutions are usually referred to as **constant states**, because they correspond to a situation in which a physical quantity remains constant with respect to changes in position x and time t. In this Riemann problem the constant state in region I is joined continuously by the centered simple wave solution in region II to the constant state in region III. In gas dynamics, where a similar problem describes the expansion of a compressible gas from a constant state at one pressure to a constant state at a lower pressure, region II is called an **expansion region**, and the characteristics in region II are then said to form an **expansion fan**.

Cauchy condition (b): This Riemann problem is totally different from the one in case (a), because here the piecewise constant initial condition *decreases* as x increases. The pattern of characteristics for this problem is shown in Fig. 2.19, from which it can be seen that the characteristics intersect immediately, so a continuous piecewise differentiable solution cannot exist for $t \geq 0$. This situation will be examined later, when it will be shown that the discontinuity in the initial condition can be resolved by the introduction of a propagating nondifferentiable discontinuous solution called a *shock solution*, which in general does *not* move along a characteristic. ∎

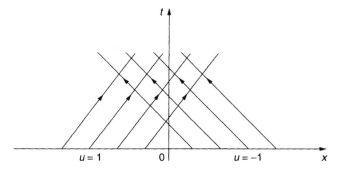

Figure 2.19 Intersecting characteristics.

EXERCISES 2.3

1. Solve the PDE $u_t - uu_x = 0$ subject to the Cauchy condition

$$u(x,0) = \begin{cases} -1, & -\infty < x < -1 \\ x, & -1 \le x \le 1 \\ 1, & 1 \le x < \infty. \end{cases}$$

2. Solve the PDE $u_t + uu_x = 0$ subject to the Cauchy condition

$$u(x,0) = \begin{cases} 0, & -\infty < x < -1 \\ 1 - |x|, & -1 \le x \le 1 \\ 0, & 1 < x < \infty. \end{cases}$$

3. Solve the PDE $u_t + u_x = u^2$ subject to the Cauchy condition

$$u(x,0) = \begin{cases} -a, & -\infty < x < -a \\ x, & -a \le x \le a \\ a, & a < x < \infty. \end{cases}$$

4. Solve the PDE $u_t + u^2 u_x = 0$ subject to the Cauchy condition

$$u(x,0) = \begin{cases} -1, & -\infty < x < -1 \\ x, & -1 \le x \le 3 \\ 3, & 3 < x < \infty. \end{cases}$$

5. Solve the PDE $u_t + 3u^3 u_x = 0$ subject to the Cauchy condition

$$u(x,0) = \begin{cases} -1, & -\infty < x < -1 \\ x, & -1 \le x \le 4 \\ 4, & 4 < x < \infty. \end{cases}$$

6. Find the implicit solution of the PDE $u_t + uu_x = 0$ subject to the Cauchy condition

$$u(x,0) = \begin{cases} 1/e, & -\infty < x < -1 \\ e^{-|x|}, & -1 \le x \le 1 \\ 1/e, & 1 < x < \infty. \end{cases}$$

By considering the behavior of the characteristics originating from the interval $-1 \le x \le 1$ on the initial line, show that a unique solution only exists in the time interval $0 < t < e/(e-1)$.

7. Solve the Riemann problem for the PDE $u_t + u^2 u_x = 0$ subject to the piecewise constant Cauchy condition

$$u(x,0) = \begin{cases} 2, & x < 0 \\ 3, & x > 0. \end{cases}$$

8. Solve the Riemann problem for the PDE $u_t - uu_x = 0$ subject to the piecewise constant Cauchy condition

$$u(x,0) = \begin{cases} 1, & x < 0 \\ -2, & x > 0. \end{cases}$$

9. Solve the Riemann problem for the PDE $u_t + uu_x = 0$ subject to the piecewise constant Cauchy condition

$$u(x,0) = \begin{cases} 1, & -\infty < x < -1 \\ 2, & -1 < x < 1 \\ 3, & 1 < x < \infty. \end{cases}$$

2.4 Discontinuous Solutions, Conservation Laws, and Shocks

The two previous sections showed how the solution of a Cauchy problem for a first-order quasi-linear PDE for a function $u(x, t)$ with x being a distance and t the time may evolve until at some critical time $t = t_c$ it becomes nonunique. The question that must now be asked is whether, by relaxing still further our concept of a solution as a continuous and piecewise differentiable solution of a PDE, it is possible to define a more general type of solution beyond the critical time t_c.

The generalization of a classical solution of a PDE to a discontinuous solution is closely related to the concept of a conservation law. All conservation laws depend on the same fundamental idea, namely that the rate of change with respect to time of the amount of a quantity u of interest contained in a volume V bounded by a closed surface S is equal to the flux of the u across S. Here, the *flux* of u across S means the amount of u that crosses S in a unit time.

A simple physical example of a conservation law is provided by considering the time variation of the mass of a compressible fluid contained inside a fixed volume V. Clearly, if fluid is neither introduced into V by an internal source nor removed from it by an internal sink, the rate of change of the mass of the fluid in V with respect to time must be equal to the difference between the mass of fluid entering V and leaving it in a unit time through the surface S which encloses V.

Because conservation laws arise frequently, and often involve several space variables and time, we begin by defining a conservation law for a quantity $u(\mathbf{r}, t)$ defined in a volume V bounded by a surface S, where $\mathbf{r}$ is a position vector in V and t is the time. Then, by restricting this more general result to the one-dimensional case, we will relate the concept of a conservation law to a first-order quasi-linear PDE involving a variable $u(x, t)$ that is a function of a single space variable x and the time t.

Let us identify the function $u(\mathbf{r}, t)$ with the density of a quantity of interest at position vector $\mathbf{r}$ and time t in volume V. Then the amount m of the quantity u contained in V at time t is

$$m = \int_V u(\mathbf{r}, t) \, dV,$$

so the rate of change of m with respect to time is

$$\frac{dm}{dt} = \frac{d}{dt} \int_V u(\mathbf{r}, t) \, dV. \tag{2.37}$$

If $\mathbf{F}$ is a vector function defined throughout V, then by definition the *flux* of $\mathbf{F}$ through an element of area dA with unit normal $\mathbf{n}$ is $dq = \mathbf{F} \cdot \mathbf{n} \, dA$ or, equivalently, as $dq = \mathbf{F} \cdot d\mathbf{A}$, where $d\mathbf{A} = \mathbf{n} \, dA$. Identifying dA with an element of area dS belonging to the surface S enclosing volume V, and letting $\mathbf{n}$ be a unit vector directed out of V, the flux Q of $\mathbf{F}$ leaving V through its bounding surface S is given by integrating dq over S to obtain

$$Q = \int_S \mathbf{F} \cdot d\mathbf{S}. \tag{2.38}$$

Equating the time rate of change of the amount of u in V to the loss of an amount Q of u through the surface leads to the balance condition $dm/dt = -Q$, where the negative sign is necessary because Q represents the loss of u through S. In terms of (2.37) and (2.38), this condition becomes

$$\frac{d}{dt} \int_V u(\mathbf{r}, t) \, dV = - \int_S \mathbf{F} \cdot d\mathbf{S}. \tag{2.39}$$

For obvious reasons, since this is a conservation law expressed in terms of integrals, it is called an *integral conservation law* for the quantity $u(\mathbf{r}, t)$. A simple example of an integral conservation law was encountered in Section 1.2 when considering traffic flow along a road.

In the cases that will concern us here $\mathbf{F} = \mathbf{F}(u(\mathbf{r}, t))$, where $\mathbf{F}$ is a known function of $u(\mathbf{r}, t)$, so (2.39) becomes

$$\frac{d}{dt} \int_V u(\mathbf{r}, t) \, dV + \int_S \mathbf{F}(u) \cdot d\mathbf{S} = 0. \tag{2.40}$$

Applying the Gauss divergence theorem to the second term in (2.40) gives

$$\frac{d}{dt} \int_V u(\mathbf{r}, t)\, dV + \int_V \operatorname{div} \mathbf{F}(u)\, dV, \tag{2.41}$$

so, provided u is differentiable with respect to t, these integrals can be combined to give the result

$$\int_V (u_t + \operatorname{div} \mathbf{F}(u))\, dS = 0. \tag{2.42}$$

An integral such as this, with a bounded integrand, is well defined even if the integrand is discontinuous across a finite number of surfaces inside V, so (2.42) remains valid when u is allowed to become discontinuous. However, provided u and $\mathbf{F}$ are continuous and differentiable, the arbitrary nature of the volume V used when deriving this integral conservation law implies that (2.42) can only be true if its integrand vanishes throughout V. When this occurs, (2.42) is seen to be equivalent to the PDE

$$u_t + \operatorname{div} \mathbf{F}(u) = 0. \tag{2.43}$$

In a one-dimension result (2.43) reduces to

$$u_t + (F(u))_x = 0. \tag{2.44}$$

After performing the indicated differentiation in the second term on the left of (2.44) we obtain

$$u_t + f(u)u_x = 0, \tag{2.45}$$

where $f(u) = dF/du$. This is now seen to be a quasi-linear PDE of the type considered in (2.28), but without the term $g(u)$, which, depending on its sign, can be interpreted as either a source or a sink. This shows it is appropriate to call (2.45) a conservation law in differential form or, more simply, a PDE *conservation law*.

It is important to recognize that the PDE conservation law in (2.45) only follows from the integral conservation law in (2.41) when u and $F(u)$ are continuous and differentiable functions, so (2.45) is not applicable at any point where $u(x, t)$ is discontinous.

To examine what happens when $u(x, t)$ becomes discontinuous we must return to the one-dimensional form of the integral conservation law in (2.41) and consider it to be defined over some interval $a \le x \le b$, when the result simplifies to

$$\frac{d}{dt} \int_a^b u(x, t)\, dx + \int_a^b (F(u))_x\, dx = 0. \tag{2.46}$$

If $F(u)$ is a continuous and differentiable function, it follows from the fundamental theorem of calculus that (2.46) can be written

$$\frac{d}{dt} \int_a^b u(x, t) \, dx + F(u(b, t)) - F(u(a, t)) = 0. \tag{2.47}$$

Now let us suppose that $u(x, t)$ is discontinuous across some moving point $x = s(t)$ strictly inside the interval $[a, b]$, so that $a < s(t) < b$. After dividing the interval $[a, b]$ into two parts separated by the point $x = s(t)$, (2.47) becomes

$$\frac{d}{dt} \int_a^{s(t)_L} u \, dx + \frac{d}{dt} \int_{s(t)_R}^b u \, dx + F(u(s(t)_L, t)) - F(u(s(t)_R, t)) = 0, \tag{2.48}$$

where $s(t)_L$ is the limit of x as $x \to s(t)$ from the left of $x = s(t)$ and $s(t)_R$ is the corresponding limit of x as $x \to s(t)$ from the right. After applying Leibniz' theorem for the differentiation of an integral with respect to a parameter to the first two terms in (2.48) we obtain

$$\int_a^{s(t)_L} u_t \, dx + \frac{ds}{dt} u(s(t)_L, t) + \int_{s(t)_R}^b u_t \, dx$$

$$- \frac{ds}{dt} u(s(t)_R, t) + F(u(s(t)_L)) - F(u(s(t)_R)) = 0. \tag{2.49}$$

Finally, letting $a \to s(t)_L$ and $b \to s(t)_R$, the bounded nature of u_t causes the first and third integrals on the left of (2.49) to vanish, leading to the result that across the discontinuity

$$(u_L - u_R) \cdot U = F(u)_L - F(u)_R, \tag{2.50}$$

where $F(u)_L = F(u(s(t)_L)), F(u)_R = F(u(s(t)_R))$ and $U = ds/dt$ is the speed with which the discontinuity moves.

Result (2.50) is called the *jump condition* that must be satisfied by a discontinuous solution of the PDE in (2.45), where $F(u) = \int f(u) \, du$. Denoting the jumps in $u(x, t)$ and $F(u)$ by $[[u]]$ and $[[F(u)]]$, respectively, the jump condition in (2.50) becomes

$$U \cdot [[u]] = [[F(u)]]. \tag{2.51}$$

It can be seen from (2.50) that the jump condition involves the three quantities u_L, u_R, and U, so if u_L and u_R are specified the propagation speed of the discontinuous solution is determined by $U = [[F(u)]]/[[u]]$. Result (2.50) suggests that specifying any two of these three quantities will determine the third uniquely. However, when $F(u)$ is nonlinear, the jump condition (2.50), equivalently (2.51), is itself nonlinear, so it does *not* necessarily follow that specifying any two of these quantities will determine the third one uniquely. The jump condition must always be used with care, in order to ensure that when

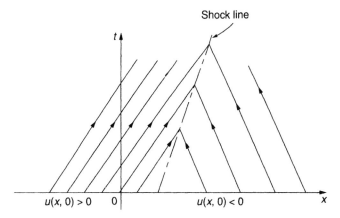

Figure 2.20 Shock line caused by intersecting characteristics.

working with a physical problem a discontinuity is produced in a physically realizable manner caused by the intersection of characteristics.

Definition 2.2. (Shocks)
A discontinuous solution will be called a *shock* if it satisfies the jump condition in (2.50), and the characteristics of the PDE in (2.45) impinge on both sides of the path followed by the discontinuous solution in the (x, t) plane.

The path followed by a shock in the (x, t) plane is called a *shock line*, and (2.50) shows that when u_L and u_R are constant the shock line will be a straight line. A typical pattern of straight line characteristics in relation to a shock line is shown in Fig. 2.20.

Example 2.9. Where possible, find a shock solution for the PDE

$$u_t + u u_x = 0,$$

subject to the piecewise constant Cauchy conditions

(a) $u(x, t) = \begin{cases} 1, & -\infty < x < 0 \\ -1, & 0 < x < \infty, \end{cases}$ (b) $u(x, t) = \begin{cases} -1, & -\infty < x < 0 \\ 1, & 0 < x < \infty. \end{cases}$

Solution: Cauchy condition (a): This is the problem considered in Example 2.8, but with the Cauchy conditions interchanged. In Example 2.8 we were unable to solve the problem in (a) above by means of the method of characteristics due to the immediate intersection of the characteristics for $t > 0$ (see Fig. 2.19). Let us now apply jump condition (2.50) to be satisfied by a discontinuous solution.

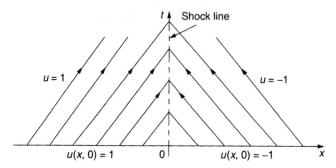

Figure 2.21   Stationary shock caused by intersecting characteristics.

In the given PDE the function $f(u) = u$ and $F(u) = \int u\,du = \frac{1}{2}u^2$, so condition (2.51) becomes $U[[u]] = [[\frac{1}{2}u^2]]$. As $u_L = 1$, $u_R = -1$, $[[u]] = 2$ and $[[\frac{1}{2}u^2]] = 0$, the jump condition reduces to $2U = 0$, showing that $U = 0$. In this case the path followed by the discontinuity is stationary and coincides with the t axis. This discontinuous solution is indeed a shock, because the characteristics of the PDE impinge on both sides of its path (the shock line) as shown in Fig. 2.21.

Cauchy condition (b): The solution can be found in that of Example 2.8(a) to be

$$u(x, t) = \begin{cases} -1, & -\infty < x/t \le -1 \\ x/t, & -1 < x/t < 1 \\ 1, & 1 \le x/t < \infty. \end{cases}$$

No shock occurs here because the initial discontinuity is resolved immediately by a centered simple wave.

Let us examine the consequence of applying the jump condition to this shock-free problem. Once again $f(u) = u$ and $F(u) = \frac{1}{2}u^2$, but now $u_L = -1$ and $u_R = 1$, so this time $[[u]] = -2$ while again $[[F(u)]] = 0$, leading to the jump condition $-2U = 0$, and so to $U = 0$. The path followed by this discontinuous solution again coincides with the t axis, but we know from the solution found earlier that this path lies in the center of a simple wave region (see Fig. 2.14), and so cannnot have characteristics impinging on it from each side. Consequently this discontinuous solution is a mathematically possible discontinuous solution, but it is *not* a shock since it is not caused by the intersection of characteristics, and so it must be discarded as being a nonphysical solution. ∎

In general it is possible for more than one shock to occur, after which the shocks can interact with one another, while a shock can also interact with a smooth solution.

However, as the analysis of these important interactions is complicated it will not be considered here. The final example illustrates how the interaction of a simple wave and a shock can arise, although the analysis will only be developed up to the time when the interaction is about to take place.

Example 2.10. Consider the PDE

$$u_t + uu_x = 0,$$

subject to the piecewise constant Cauchy condition

$$u(x,0) = \begin{cases} -1, & -\infty < x < 0 \\ -\frac{1}{2}, & 0 < x < 1 \\ -1, & 1 < x < \infty. \end{cases}$$

Find the solution up to the time when the different types of wave interact.

Solution: Using the method of Example 2.8(a) there is seen to be a centered simple wave region with its center at the origin in which the solution is given by $u(x, t) = x/t$ for $-1 \le x/t \le -\frac{1}{2}$. The characteristics originating from the interval $0 < x < 1$ on the initial line and the characteristics originating from the interval $1 < x < \infty$ on the initial line converge for $t > 0$, so a shock originates from the point $(1, 0)$ on the initial line. To the immediate left of this point $u_L = -\frac{1}{2}$ and to the right $u_R = -1$, so as $f(u) = u$ and $F(u) = \frac{1}{2}u^2$ the jump condition across the shock is $U((-\frac{1}{2}) - (-1)) = \frac{1}{2}((-\frac{1}{2})^2 - (-1)^2)$, showing the shock speed to be $U = dx/dt = -\frac{3}{4}$. The shock speed is constant and the shock line must pass through the point $(1, 0)$, so the equation of the shock line is $t = \frac{4}{3}(1 - x)$.

The solution corresponding to the initial condition $u(x, 0) = -1$ in the interval $-\infty < x < 0$, for $t > 0$ and to the left of the characteristic $x/t = -1$ is $u(x, t) = -1$. The solution between the characteristic $x/t = -\frac{1}{2}$ and the shock line $t = \frac{4}{3}(1 - x)$ for $t > 0$ is $u(x, t) = -\frac{1}{2}$, and the solution to the right of the shock for $t > 0$ is $u(x, t) = -1$.

The characteristic $x/t = -\frac{1}{2}$ bounding the simple wave region to the right and the shock line $t = \frac{4}{3}(1 - x)$ intersect at the point $x = -2$ and $t = 4$, so as the interaction between the shock and the centered simple wave starts when $t = 4$ the solution described above is only valid for $0 < t < 4$. The pattern of characteristics in the constant state regions and the centered simple wave region is shown in Fig. 2.22 together with the shock line. ∎

The consequence of generalizing a continuous piecewise smooth differentiable solution to allow discontinuous solutions was seen to be that mathematical solutions can arise that are not physically realizable. The choice of physically realizable solutions of conservation laws was resolved here by the selection principle given in Definition 2.2, which admits shocks, and also all classical (differentiable) solutions as special cases.

In physical problems conservation laws arise naturally in an unambiguous manner, and so lead to appropriate jump conditions to be satisfied by any

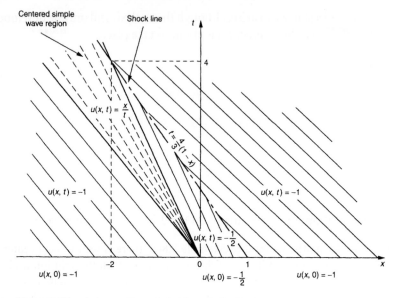

Figure 2.22   Intersection of shock and simple wave.

shocks that may occur. However, a change of variable in a conservation law can lead to a different conservation law, so it is necessary to examine the relationship between the shock solution of a physical conservation law and the shock solution of any new conservation law that may be derived from it.

To illustrate what happens, we again consider the PDE $u_t + uu_x = 0$, which can be written in the conservation form $u_t + \frac{\partial}{\partial x}(\frac{1}{2}u^2)u_x = 0$ already used in Examples 2.9 and 2.10. Multiplying the PDE by nu^{n-1}, with n being a positive integer, leads to the new system of infinitely many PDE conservation laws

$$\frac{\partial}{\partial t}(u^n) + \frac{\partial}{\partial x}\left(\frac{n}{n+1}u^{n+1}\right) = 0, \quad n = 1, 2, \ldots, \tag{2.52}$$

where now the variable is u^n. The jump condition for this family of conservation laws derived from (2.50), which involves the new variable u^n, is

$$U[[u^n]] = \frac{n}{n+1}[[u^{n+1}]], \quad \text{for } n = 1, 2, \ldots, \tag{2.53}$$

so the propagation speed of a discontinuous solution for this system is seen to be

$$U = \frac{n[[u^{n+1}]]}{(n+1)[[u^n]]}. \tag{2.54}$$

Result (2.54) shows that the propagation speed U of a discontinuous solution of (2.52) depends on n, and so is not the same as the propagation speed of a discontinuous solution of the original PDE $u_t + uu_x = 0$. To illustrate the different values of U that can arise we consider the initial condition that gave rise to the shock in Example 2.10. Setting $n = 1$ reduces (2.52) to the conservation law used in Example 2.10 where $u_L = -\frac{1}{2}$, $u_R = -1$ and (2.54) then shows, as expected, that $U = -\frac{3}{4}$. However, setting $n = 2$ in (2.54) with the same values of u_L and u_R shows that the propagation speed for a discontinuous solution of the PDE conservation law

$$\frac{\partial}{\partial t}(u^2) + \frac{\partial}{\partial x}\left(\frac{2}{3}u^3\right) = 0 \tag{2.55}$$

is $U = -\frac{7}{9}$.

This simple example demonstrates the importance of the fact that, when working with physical problems, the variable occurring in a physically derived conservation law should not be changed.

EXERCISES 2.4

1. Solve the PDE $u_t + u^2 u_x = 0$ subject to the piecewise constant Cauchy conditions

$$\text{(a)} \quad u(x, t) = \begin{cases} 2, & x < 0 \\ 3, & x > 0 \end{cases} \quad \text{and} \quad \text{(b)} \quad u(x, t) = \begin{cases} 3, & x < 0 \\ 2, & x > 0. \end{cases}$$

In each case comment on the nature of the solution obtained.

2. State the restriction on $f(u)$, u_L, and u_R that will ensure the existence of a shock solution for the PDE $u_t + f(u)u_x = 0$ subject to the piecewise constant Cauchy condition

$$u(x, 0) = \begin{cases} u_L, & x < a \\ u_R, & x > a. \end{cases}$$

Find the shock speed and the equation of the shock line when (a) $f(u) = -2u$ and (b) $f(u) = 2u$, stating in each case the relationship that must exist between u_L and u_R.

3. Consider the solution of the PDE $u_t - 2uu_x = 0$, subject to the piecewise constant Cauchy condition

$$u(x, 0) = \begin{cases} \frac{1}{2}, & x < 0 \\ 1, & 0 < x < 1 \\ \frac{1}{2}, & x > 1. \end{cases}$$

Find the solution up to the time at which different types of wave interact.

4. Show that the discontinuous solution of the linear first-order constant coefficient PDE $u_t + cu_x = 0$ subject to the piecewise constant Cauchy condition

$$u(x,0) = \begin{cases} u_L, & x < a \\ u_R, & x > a \end{cases}$$

propagates along the characteristic through the point $(a,0)$, irrespective of whether $u_L < u_R$ or $u_L > u_R$. Give reasons why a discontinuity in an initial condition for a general linear or semilinear first-order PDE must propagate along the characteristic through the point on the initial line where the discontinuity in the Cauchy data is located.

5. Solve the linear first-order PDE $u_t + 2tu_x + 2tu = 0$ subject to the discontinuous Cauchy condition

$$u(x,0) = \begin{cases} 2x, & x < 1 \\ x, & x > 1 \end{cases}$$

for the cases (a) when $x < 1$ and (b) when $x > 1$. If u_L is the solution at time t to the immediate left of the characteristic through the point $(1,0)$, and u_R is the corresponding solution to the immediate right, find $[[u]] = u_L - u_R$ as a function of t and hence show that $[[u]] \to 0$ as $t \to \infty$.

6. Write the PDE $u_t + 2u^n u_x = 0$ in conservation form and find the jump condition that must be satisfied by a discontinuous solution. Multiply the PDE by nu^{n-1} and use the result to find a family of conservation laws that depend on the positive integer n. Derive the jump condition that must be satisfied by a discontinuous solution of these conservation laws, and hence show that its propagation speed depends on n.

Classification of Equations and Reduction to Standard Form

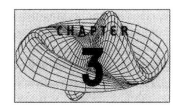

3.1 Classification of PDEs and Their Reduction to Standard Form

Several second-order PDE governing the solution of different physical problems were derived in Section 1.2, and in each case the problem was seen to have associated with it a different set of auxiliary conditions. In the case of a string or membrane whose oscillations are governed by the wave equation, to obtain a unique solution it was seen that both the initial shape and speed had to be specified at the start of the problem, corresponding to the imposition of Cauchy conditions. The situation was different in the case of the heat equation because there, although the solution could be required to satisfy Dirichlet, Neumann, or Robin conditions on the boundary, and the initial temperature distribution could be specified, it was not possible to specify the initial rate of change of the temperature. Finally, appropriate conditions for the Laplace equation were seen to be different again, because the equation has no timelike variable. So, although no initial conditions can be imposed on solutions of the Laplace equation, Dirichlet, Neumann, or Robin conditions can be specified on the physical boundary of the region in which a solution is required.

A typical physical example involving the Laplace equation subject to Dirichlet conditions is the determination of the electrostatic potential distribution throughout a vacuum in a rectangular cavity, the electrically conducting walls of which are maintained at different fixed electric potentials. Equivalently, when interpreted in terms of temperatures, these same boundary conditions imposed on the Laplace equation will lead to the steady-state temperature distribution in an isotropic rectangular solid block of heat-conducting material when its sides are maintained at different fixed temperatures.

We now show that each of the three second-order PDEs derived in Section 1.2 belongs to a fundamentally different class, known respectively, as equations of *hyperbolic, parabolic,* and *elliptic type*. It is because each of these PDEs belongs to a different class that the auxiliary conditions are different in each case.

Each class of PDE has associated with it a standard form of the governing equation in which the second-order terms have been reduced to their simplest form. These are called the *standard forms of the PDEs,* or their *canonical forms,* and for second-order linear PDEs they are as follows:

The first standard form of a second-order hyperbolic equation is

$$u_{xy} = \Phi(x, y, u, u_x, u_y).$$

The second standard form of a second-order hyperbolic equation is

$$u_{xx} - u_{yy} = \Phi(x, y, u, u_x, u_y).$$

The standard form of a second-order parabolic equation is

$$u_x - u_{yy} = \Phi(x, y, u, u_x, u_y).$$

The standard form of a second-order elliptic equation is

$$u_{xx} + u_{yy} = \Phi(x, y, u, u_x, u_y).$$

In each of the above cases the form of the function $\Phi(x, y, u, u_x, u_y)$ involving the lower-order terms is unimportant since it does not enter into the classification of the equation.

A general second-order linear PDE involving the independent variables x and y and the independent variable $u(x, y)$ can be written

$$A(x, y)u_{xx} + 2B(x, y)u_{xy} + C(x, y)u_{yy} = \Phi(x, y, u, u_x, u_y), \quad (3.1)$$

where $u, u_x,$ and u_y occur linearly in the function Φ. Once this PDE has been classified as being of hyperbolic, parabolic, or elliptic type we will see how the PDE can be reduced to its standard form. In terms of this classification the wave equation will be seen to be of *hyperbolic type*, the heat equation of *parabolic type*, and the Laplace equation of *elliptic type*. The standard forms of these second-order equations are both important and useful, because while preserving the inherent mathematical properties of the original PDE they simplify its structure and so make easier the task of finding a solution. In what follows it will be assumed that the coefficients $A(x, y)$, $B(x, y)$, and $C(x, y)$ in (3.1) do not vanish simultaneously, because in that case the second-order PDE degenerates to one of first order. The term *constant coefficient second-order* linear PDE is used to refer to an equation of the form (3.1) in which the coefficients of u and all its partial derivatives are constants as, for example, in the equation

$$u_{xx} + 3u_{xy} - 2u_{yy} = 4u_x + u_y + 5u + \sin(x + y).$$

It was seen in Section 1.2 that the geometry of a physical problem influenced the algebraic structure of the PDE governing the solution. This suggests that some other choice of independent variables might simplify the structure of the PDE, and so lead to a form of the equation that is easier to study. Accordingly, we start by transforming the independent variables x and y to the new independent variables ξ and η through the change of variables

$$\xi = \xi(x, y), \qquad \eta = \eta(x, y), \tag{3.2}$$

where ξ and η are both continuous and twice differentiable functions with respect to x and y, and such that their Jacobian J is nonvanishing, where

$$J = \frac{\partial(\xi, \eta)}{\partial(x, y)} = \begin{vmatrix} \xi_x & \xi_y \\ \eta_x & \eta_y \end{vmatrix} \neq 0. \tag{3.3}$$

The nonvanishing of the Jacobian of the transformation ensures that a one-to-one transformation exists between the original independent variables (x, y) and the new variables (ξ, η). This simply means that the new independent variables can serve as new coordinate variables without ambiguity.

Applying the chain rule shows the first- and second-order partial derivatives of u with respect to x and y can be expressed in terms of ξ and η as follows:

$$
\begin{aligned}
u_x &= \xi_x u_\xi + \eta_x u_\eta, \qquad u_y = \xi_y u_\xi + \eta_y u_\eta \\
u_{xx} &= \xi_x^2 u_{\xi\xi} + 2\xi_x \eta_x u_{\xi\eta} + \eta_x^2 u_{\eta\eta} + \xi_{xx} u_\xi + \eta_{xx} u_\eta \\
u_{yy} &= \xi_y^2 u_{\xi\xi} + 2\xi_y \eta_y u_{\xi\eta} + \eta_y^2 u_{\eta\eta} + \xi_{yy} u_\xi + \eta_{yy} u_\eta \\
u_{xy} &= \xi_x \xi_y u_{\xi\xi} + (\xi_x \eta_y + \xi_y \eta_x) u_{\xi\eta} + \eta_x \eta_y u_{\eta\eta} + \xi_{xy} u_\xi + \eta_{xy} u_\eta.
\end{aligned}
\tag{3.4}
$$

Substituting these results into (3.1) produces the equivalent PDE

$$a(\xi, \eta) u_{\xi\xi} + 2b(\xi, \eta) u_{\xi\eta} + c(\xi, \eta) u_{\eta\eta} = \hat{\Phi}(\xi, \eta, u, u_\xi, u_\eta), \tag{3.5}$$

where Φ becomes $\hat{\Phi}$ after the transformation, and

$$
\begin{aligned}
a(\xi, \eta) &= A\xi_x^2 + 2B\xi_x\xi_y + C\xi_y^2 \\
b(\xi, \eta) &= A\xi_x\eta_x + B(\xi_x\eta_y + \xi_y\eta_x) + C\xi_y\eta_y \\
c(\xi, \eta) &= A\eta_x^2 + 2B\eta_x\eta_y + C\eta_y^2.
\end{aligned}
\tag{3.6}
$$

At this stage the form of PDE (3.5) is no simpler than that of the original equation (3.1), but this is to be expected because so far the choice of the new variables ξ and η has been arbitrary. However, before showing how to choose the new coordinate variables so (3.1) is simplified, routine algebraic manipulation of results (3.6) shows that

$$b^2 - ac = (B^2 - AC)J^2. \tag{3.7}$$

This has established the important fact that a change of variable for which $J \neq 0$ leaves unchanged the signs of the expressions $b^2 - ac$ and $B^2 - AC$, each of which has been derived only from the coefficients of the second-order terms belonging to the corresponding PDE.

The quantity $d = B^2 - AC$ is called the *discriminant* of PDE (3.1), and because the sign of d is an invariant of the PDE with respect to all coordinate transformations for which $J \neq 0$, it is used to classify PDE (3.1) according to type as follows:

- the PDE is said to be of *hyperbolic* type in any region where

$$d = B^2 - AC > 0;$$

- the PDE is said to be of *parabolic* type in any region where

$$d = B^2 - AC = 0;$$

- the PDE is said to be of *elliptic* type in any region where

$$d = B^2 - AC < 0.$$

The terms *hyperbolic, parabolic,* and *elliptic* used to identify the three different types of PDE represented by (3.1) according to the behavior of its discriminant are long established. They are simply three convenient names, and these should not be taken to imply any geometrical properties of the solutions of the associated PDE. The names come from a formal comparison between the coefficients A, B, and C of PDE (3.1) and the corresponding coefficients of the *quadratic terms* in the general second-degree algebraic equation

$$Ax^2 + Bxy + Cy^2 + Dx + Ey + F = 0.$$

The form of plane geometrical curve represented by this algebraic equation is determined by its discriminant $d = B^2 - AC$. It is a hyperbola if $d > 0$, a parabola if $d = 0$, and an ellipse if $d < 0$, and it is because of this that these same names are used to classify the PDE in (3.1) depending on the nature of its discriminant.

In (3.1) the coefficients A, B, and C depend on x and y, so the classification of a second-order linear PDE with variable coefficients is seen to depend on its second-order terms and also on the choice of point (x, y) where the PDE is classified. Consequently, when variable coefficient equations are involved, it is possible for a PDE to change its classification from one part of a region to another. The matching of the solution of an equation of one type to the solution of an equation of a different type across a line dividing adjacent parts of a region is difficult, so when seeking solutions we will only consider equations that do not change type in the region under consideration. Clearly, if the coefficients A, B, and C in (3.1) are constants, the classification of the PDE cannot change throughout the region where the PDE is defined.

Before leaving the topic of equations that change type it is appropriate to mention the important practical problem associated with the mathematics of transonic air flow, where a change takes place from subsonic to supersonic flow. In this case an unavoidable change of type occurs, because steady subsonic flow is governed by an elliptic equation, whereas supersonic flow is governed by an equation of hyperbolic type. The prototype equation involved here is the *Tricomi equation*

$$u_{xx} + x u_{yy} = 0, \tag{3.8}$$

for which the discriminant is seen to be $d = -x$. Consequently the Tricomi equation is hyperbolic when $x < 0$, elliptic when $x > 0$, and degenerately parabolic along the y axis. Returning now to the transformed PDE in (3.5), and recalling the structure of the standard forms of the three classes of PDE, we see that (3.5) can be reduced to a standard form if the coordinate transformation $\xi = \xi(x, y), \eta = \eta(x, y)$ can be chosen such that:

(a) $a = c = 0$, corresponding to the first hyperbolic standard form;

(b) $a = b = 0$, corresponding to the parabolic standard form;

(c) $a = c, b = 0$, corresponding to the elliptic standard form.

The hyperbolic case, $d = B^2 - AC > 0$

If the PDE is such that $d = B^2 - AC > 0$ and $A = C = 0$, then division of (3.1) by $2B$ reduces it to an equation of the form

$$u_{xy} = \Psi(x, y, u_x, u_y), \tag{3.9}$$

where $\Psi = \Phi/2B(x, y)$. Expressed in this form the PDE is in the *first standard form (canonical form) of a hyperbolic equation*, so in this special case no change of variables is needed to simplify its structure.

The linear change of variables $X = x + y$, $Y = x - y$ converts (3.9) into

$$u_{XX} - u_{YY} = \tilde{\Psi}(X, Y, u_X, u_Y), \tag{3.10}$$

which is the *second standard form (canonical form) of a hyperbolic equation*, where the function Ψ becomes the function $\tilde{\Psi}$ after the change of variables from x and y to X and Y.

If PDE (3.1) is of hyperbolic type, but it is not already expressed in either of its two standard forms, the change of variables involving ξ and η in (3.2) must be chosen to bring about such a simplification. Let us see how a hyperbolic equation of type (3.1) can be reduced to the first standard form. Let us suppose $A \neq 0$ (a similar argument applies if $C \neq 0$), then to ensure that $a = c = 0$ in (3.5) it follows from (3.6) that ξ and η must be solutions of

$$A\xi_x^2 + 2B\xi_x\xi_y + C\xi_y^2 = 0 \quad \text{and} \quad A\eta_x^2 + 2B\eta_x\eta_y + C\eta_y^2 = 0. \quad (3.11)$$

The first of Eqs. (3.11) can be factored to give

$$(\xi_x + \lambda_1\xi_y)(\xi_x + \lambda_2\xi_y) = 0, \quad (3.12)$$

where

$$\lambda_1(x, y) = \frac{B - \sqrt{d}}{A} \quad \text{and} \quad \lambda_2(x, y) = \frac{B + \sqrt{d}}{A}. \quad (3.13)$$

Note that as $d > 0$ the λ_1 and λ_2 in (3.13) are real and distinct.

The second of the equations in (3.11) can be factored in an identical manner, so as only two distinct factorizations are possible between the two equations these are seen to lead to the two equations

$$\xi_x + \lambda_1(x, y)\xi_y = 0 \quad \text{and} \quad \eta_x + \lambda_2(x, y)\eta_y = 0. \quad (3.14)$$

These are the equations that define the new coordinate variables ξ and η that are necessary to make $a = c = 0$ in (3.5).

As the total derivative of ξ along the coordinate line $\xi(x, y) = $ constant is $d\xi = 0$, it follows that $\xi_x dx + \xi_y dy = 0$, or $\xi_x/\xi_y = -dy/dx$, with a corresponding result along the coordinate line $\eta(x, y) = $ constant. Using these results in (3.14) shows that the required variables ξ and η are determined by the respective solutions of the two ordinary differential equations

$$\frac{dy}{dx} = \lambda_1(x, y) \quad \text{and} \quad \frac{dy}{dx} = \lambda_2(x, y), \quad (3.15)$$

called the *characteristic equations* of the PDE in (3.1).

Integration of the first of these ordinary differential equations leads to the family of curvilinear coordinates $\xi(x, y) = c_1$, while integration of the second one gives another family of curvilinear coordinates $\eta(x, y) = c_2$, where c_1 and c_2 are arbitrary constants of integration. These two families of curvilinear coordinates $\xi(x, y) = c_1$ and $\eta(x, y) = c_2$, where c_1 and c_2 can be regarded as parameters, are called the *characteristic curves* of the hyperbolic equation in (3.1) or, more simply, the *characteristics* of the equation. The fact that $d > 0$ means that the two families of characteristic curves are two families of *real* curves in the (x, y) plane.

The effect of changing the independent variables in (3.1) from x and y to ξ and η followed by division by $b(\xi, \eta)$ is to reduce (3.5) to the *first standard form for a hyperbolic equation*

$$u_{\xi\eta} = \Psi(\xi, \eta, u, u_\xi, u_\eta), \tag{3.16}$$

which was encountered in (3.9). The change of variables $\xi = X+Y, \eta = X-Y$ transforms (3.16) from the first standard form into the *second standard form for a hyperbolic equation,*

$$u_{XX} - u_{YY} = \tilde{\Psi}(X, Y, u_X, u_Y), \tag{3.17}$$

given in (3.10).

When the coefficients A, B, and C in (3.1) are constants, the right-hand sides of the ordinary differential equations in (3.15) become constants, showing that then the two families of characteristic curves associated with the PDE reduce to two different families of parallel straight lines.

Example 3.1. Show that the homogeneous wave equation

$$u_{tt} - c^2 u_{xx} = 0$$

is unconditionally hyperbolic. Reduce it to the first standard form for a hyperbolic equation and hence find its general solution.

Solution: To interpret the results for (3.1) that involve the independent variables x and y in terms of the wave equation $u_{tt} - c^2 u_{xx} = 0$, where the independent variables are t and x, it will be necessary to replace x and y in (3.1) and (3.4) by t and x. The wave equation is a constant coefficient equation with $A = 1$, $B = 0$, and $C = -c^2$, so the discriminant $d = c^2 > 0$, showing the equation to be unconditionally hyperbolic.

The characteristic equations (3.15) become

$$\frac{dx}{dt} = c \quad \text{and} \quad \frac{dx}{dt} = -c,$$

and after integration to determine the characteristic curves of the wave equation we find that

$$x - ct = \xi \quad \text{and} \quad x + ct = \eta,$$

where ξ and η are constants of integration. As the wave equation is a constant coefficient PDE, its two families of characteristic "curves" reduce to the two different families of parallel straight lines $x - ct = $ constant and $x + ct = $ constant.

Substituting these expressions for ξ and η into (3.4) gives

$$u_{tt} = c^2 u_{\xi\xi} - 2c^2 u_{\xi\eta} + c^2 u_{\eta\eta}, \qquad u_{xx} = u_{\xi\xi} + 2u_{\xi\eta} + u_{\eta\eta},$$

so in terms of the characteristic coordinates ξ and η the wave equation $u_{tt} - c^2 u_{xx} = 0$ reduces to

$$u_{\xi\eta} = 0.$$

The result of reducing the homogeneous wave equation to its first standard form has been to produce a particularly simple equation for which the general solution is easily seen to be

$$u = f(\xi) + g(\eta),$$

where f and g are arbitrary twice differentiable functions of their respective arguments. The form of the general solution of the wave equation in terms of the original variables x and t follows by substituting for ξ and η, when we obtain

$$u(x, t) = f(x - ct) + g(x + ct).$$

This is an important result that will be needed later, and it is one of the few cases where a general solution of a PDE can be found.   ■

Example 3.2. Find a condition that ensures the equation

$$x^2 u_{xx} - y^2 u_{yy} = u_x + 2y^2 + 1$$

is hyperbolic, and reduce it to the first standard form for a hyperbolic equation.

Solution: This PDE involves the same independent variables as (3.1) so the general results derived for that equation may be used without relabeling the independent variables. In the given equation $A = x^2$, $B = 0$, and $C = -y^2$, so the discriminant $d = x^2 y^2$. The equation will be hyperbolic when $d > 0$, and this is true throughout the (x, y) plane with the sole exception of the coordinate axes $x = y = 0$.

The characteristic equations (3.15) become

$$\frac{dy}{dx} = -\frac{y}{x} \quad \text{and} \quad \frac{dy}{dx} = \frac{y}{x},$$

and when integrated to find the equations of the characteristic curves they lead to the results

$$xy = \xi \quad \text{and} \quad y/x = \eta,$$

where ξ and η are constants of integration.

From the results in (3.4) we have

$$u_x = yu_\xi - (y/x^2)u_\eta,$$

$$u_{xx} = y^2 u_{\xi\xi} - 2(y^2/x^2)u_{\xi\eta} + (y^2/x^4)u_{\eta\eta} + 2(y/x^3)u_\eta,$$

$$u_{yy} = x^2 u_{\xi\xi} + 2u_{\xi\eta} + (1/x^2)u_{\eta\eta}.$$

Substituting for u_x, u_{xx}, and u_{yy} in the differential equation and using $y = \sqrt{\xi\eta}$ and $x = \sqrt{\xi/\eta}$ reduces the PDE to the required first standard form for a hyperbolic equation

$$\frac{\partial^2 u}{\partial \xi \partial \eta} = \frac{1}{2\xi}\left(1 + \frac{1}{2}\sqrt{\frac{\eta}{\xi}}\right)u_\eta - \frac{1}{4\sqrt{\xi\eta}}u_\xi - \frac{1}{4\xi\eta} - \frac{1}{2}.$$

This is true everywhere except the coordinate axes where the discriminant vanishes.   ∎

The parabolic case, $d = B^2 - AC = 0$

If PDE (3.1) is parabolic, its discriminant $d = B^2 - AC = 0$. In the special case that $A = B = 0$ division by C (that cannot vanish because then the PDE ceases to be of second order) reduces it to the form

$$u_{yy} = \Psi(x, y, u, u_x, u_y), \tag{3.18}$$

where $\Psi = \Phi/C(x, y)$. This is the *standard form* (*canonical form*) *for a parabolic equation*, and in the special case that $A = C = 0$ the PDE is already in its standard form so no change of coordinates is necessary.

We now show how a general parabolic equation can be reduced to the standard form in (3.18) by means of a suitable change of independent variables. As the discriminant $d = B^2 - AC = 0$, the two characteristic equations in (3.15) reduce to the single equation

$$\frac{dy}{dx} = \frac{B}{A}, \tag{3.19}$$

and after integration this gives rise to a single family of real curvilinear characteristic curves $\xi(x, y) = $ constant. Equations (3.15) do not now determine the second family of characteristic curves $\eta(x, y) = $ constant, so this family can be chosen arbitrarily, subject only to the conditions that $\eta(x, y)$ is continuous and twice differentiable with respect to x and y, and the Jacobian J in (3.3) does not vanish. In applications η is chosen to simplify the calculations as much as possible, so it is often taken to be either $\eta = x$ or $\eta = y$.

As the PDE is parabolic $B^2 = AC$, the term $b(\xi, \eta)$ in (3.6) can be rewritten as

$$b(\xi, \eta) = A\left(\xi_x \eta_x + \frac{B}{A}(\xi_x \eta_y + \xi_y \eta_x) + \frac{B^2}{A^2}\xi_y \eta_y\right)$$
$$= A\left(\xi_x + \frac{B}{A}\xi_y\right)\left(\eta_x + \frac{B}{A}\eta_y\right). \tag{3.20}$$

Comparing Eq. (3.19), which defines the characteristic curves $\xi(x, y) = $ constant, with Eqs. (3.15) and (3.14) shows the first bracketed factor on the right of (3.20) is zero, so $b(\xi, \eta) = 0$. Similarly, we can show that $a(\xi, \eta) = 0$.

Thus the change to these curvilinear coordinates reduces the PDE to the form

$$c(\xi, \eta) u_{\eta\eta} = \hat{\Phi}(\xi, \eta, u, u_\xi, u_\eta).$$

Division by $c(\xi, \eta)$ then allows a general parabolic equation of the form (3.1) to be rewritten as

$$u_{\eta\eta} = \Psi(\xi, \eta, u, u_\xi, u_\eta), \tag{3.21}$$

where $\Psi = \hat{\Phi}/c(\xi, \eta)$. Result (3.21) is now in the required *standard form for a parabolic equation*. As the choice of η is arbitrary, the form taken by $\Psi(\xi, \eta, u, u_\xi, u_\eta)$ will depend on the choice of η.

Example 3.3. Reduce the following PDE to standard form

$$x^2 y^2 u_{xx} + 2xy u_{xy} + u_{yy} = 0.$$

Solution: Identifying the coefficients of this PDE with those of (3.1) gives $A = x^2 y^2$, $B = xy$, and $C = 1$, so the discriminant $d = B^2 - AC = 0$ showing the equation to be parabolic. Equation (3.19), which determines the family of characteristic curves $\xi(x, y) = $ constant, becomes $dy/dx = 1/xy$, so after integration we find that

$$\frac{1}{2} y^2 = \ln x + \ln \xi,$$

where ξ is an integration constant. The family of curvilinear coordinates determined by this single characteristic equation is thus

$$\xi = \frac{1}{x} \exp\left(\frac{1}{2} y^2\right).$$

The other characteristic coordinate variable η can be chosen arbitrarily, so for simplicity we set $\eta = x$. Thus a coordinate transformation that will reduce the PDE to its standard form is $\xi = \frac{1}{x} \exp\left(\frac{1}{2} y^2\right)$ and $\eta = x$.

Using these results in (3.4) gives

$$u_{xx} = \frac{1}{x^4} \exp(y^2) u_{\xi\xi} - \frac{2}{x^2} \exp\left(\frac{1}{2} y^2\right) u_{\xi\eta} + u_{\eta\eta} + \frac{2}{x^2} \exp\left(\frac{1}{2} y^2\right) u_\xi$$

$$u_{xy} = -\frac{y}{x^3} \exp(y^2) u_{\xi\xi} + \frac{y}{x} \exp\left(\frac{1}{2} y^2\right) u_{\xi\eta} - \frac{y}{x^2} \exp\left(\frac{1}{2} y^2\right) u_\xi$$

$$u_{yy} = \frac{y^2}{x^2} \exp(y^2) u_{\xi\xi} + \frac{y^2}{x^2} \exp\left(\frac{1}{2} y^2\right) u_\xi.$$

Substituting these results into the original PDE gives

$$u_{\eta\eta} + \frac{\exp\left(\frac{1}{2} y^2\right)}{x^3} u_\xi = 0.$$

Returning to the original variables x and y to ξ and η this becomes

$$u_{\eta\eta} = -\frac{\xi}{\eta^2} u_\xi.$$

■

Elliptic equations, $d = B^2 - AC < 0$

The PDE in (3.1) will be of *elliptic type* if its discriminant $d = B^2 - AC < 0$. In the specially simple case that $A = C$ and $B = 0$ (3.1) becomes

$$Au_{xx} + Au_{yy} = \Phi(x, y, u, u_x, u_y),$$

and division by A reduces it to the *standard form for an elliptic equation*

$$u_{xx} + u_{yy} = \hat{\Phi}(x, y, u, u_x, u_y), \tag{3.22}$$

where $\hat{\Phi} = \Phi/A(x, y)$. So, in this case, no transformation is needed to bring the PDE to its standard form.

When the PDE in (3.1) is of elliptic type, but not in standard form, the new coordinate variables in (3.2) must be chosen so that in the transformed equation (3.5) the coefficients $a(\xi, \eta) = c(\xi, \eta)$ and $b(\xi, \eta) = 0$. Imposing the condition $a(\xi, \eta) = c(\xi, \eta)$ on the coefficients in (3.5) shows that interchanging ξ and η leaves these coefficients invariant. So, to determine the new coordinates ξ and η, we first replace them in the condition $b(\xi, \eta) = 0$ by $\varphi(x, y)$ and then, after solving for φ_x/φ_y, use the two results that are obtained to determine ξ and η. Replacing ξ and η in $b(\xi, \eta) = 0$ by $\varphi(x, y)$ leads to the single equation

$$A\varphi_x^2 + 2B\varphi_x\varphi_y + C\varphi_y^2 = 0,$$

so as $A \neq 0$ this can be written

$$\left(\frac{\varphi_x}{\varphi_y}\right)^2 + 2\frac{B}{A}\left(\frac{\varphi_x}{\varphi_y}\right) + \frac{C}{A} = 0. \tag{3.23}$$

If $\varphi(x, y) = $ constant is a coordinate line, the total derivative of φ along this line is $d\varphi = 0$, so $\varphi_x dx + \varphi_y dy = 0$ or

$$\frac{dy}{dx} = -\frac{\varphi_x}{\varphi_y}. \tag{3.24}$$

Setting $-\varphi_x/\varphi_y = \lambda$ in (3.23) gives

$$\lambda^2 - 2\frac{B}{A}\lambda + \frac{C}{A} = 0,$$

and solving for λ we arrive at the two expressions

$$\lambda_\pm = \frac{B \pm \sqrt{B^2 - AC}}{A}.$$

However, $d = B^2 - AC < 0$, so the $\lambda_\pm$ are complex and (3.24) leads to the two complex characteristic equations

$$\frac{dy}{dx} = \frac{B - i\sqrt{AC - B^2}}{A} \quad \text{and} \quad \frac{dy}{dx} = \frac{B + i\sqrt{AC - B^2}}{A}. \quad (3.25)$$

As the expressions on the right of (3.25) are complex, after integrating the first of these it is necessary to add a complex integration constant $\xi + i\eta$. However, x and y are *real* variables, so after integrating the second expression it is necessary to add the complex integration constant $\xi - i\eta$, which is the complex conjugate of the first integration constant. Solving these two equations for ξ and η gives the required change of variables $\xi = \xi(x, y)$ and $\eta = \eta(x, y)$.

With this new choice of independent variables (3.5) simplifies to

$$a(\xi, \eta) u_{\xi\xi} + a(\xi, \eta) u_{\eta\eta} = \hat{\Phi}(\xi, \eta, u, u_\xi, u_\eta),$$

and division by $a(\xi, \eta)$ reduces it to the *standard form for an elliptic equation*

$$u_{\xi\xi} + u_{\eta\eta} = \Psi(\xi, \eta, u, u_\xi, u_\eta), \quad (3.26)$$

where $\Psi = \hat{\Phi}/a(\xi, \eta)$.

Example 3.4. Classify and reduce to standard form the PDE

$$u_{xx} + y u_{yy} + \frac{1}{2} u_y + 4y u_x = 0.$$

Solution: Comparison with (3.1) gives $A = 1$, $B = 0$, and $C = y$, so the discriminant $d = B^2 - AC = -y$. Thus the PDE is elliptic when $y > 0$, hyperbolic when $y < 0$ and degenerately parabolic when $y = 0$. ∎

Elliptic case, $y > 0$

Equations (3.25) become

$$\frac{dy}{dx} = -\sqrt{-y} \text{ or } \frac{dy}{dx} = -i\sqrt{y} \quad \text{and} \quad \frac{dy}{dx} = \sqrt{-y} \text{ or } \frac{dy}{dx} = i\sqrt{y}.$$

Integrating these complex characteristic equations gives

$$2\sqrt{y} = -ix + \xi - i\eta \quad \text{and} \quad 2\sqrt{y} = ix + \xi + i\eta,$$

and solving for ξ and η gives

$$\xi = 2\sqrt{y} \quad \text{and} \quad \eta = -x.$$

Substituting these expressions into Eqs. (3.4) gives

$$u_x = -u_\eta, \quad u_y = \frac{1}{\sqrt{y}} u_\xi, \quad u_{xx} = u_{\eta\eta}, \quad u_{yy} = \frac{1}{y} u_{\xi\xi} - \frac{1}{2 y^{3/2}} u_\xi.$$

Inserting these results in the original PDE reduces it to the standard form for an elliptic equation

$$u_{\xi\xi} + u_{\eta\eta} = \xi^2 u_\eta.$$

Hyperbolic case, $y < 0$

The characteristic equations become

$$\frac{dy}{dx} = -\sqrt{-y} \quad \text{and} \quad \frac{dy}{dx} = \sqrt{-y},$$

with the respective solutions

$$-2\sqrt{-y} = -x + \xi \quad \text{and} \quad -2\sqrt{-y} = x + \eta,$$

so

$$\xi = x - 2\sqrt{-y} \quad \text{and} \quad \eta = -x - 2\sqrt{-y}.$$

Substitution into Eqs. (3.4) then gives

$$u_x = u_\xi - u_\eta, \quad u_y = (1/\sqrt{-y})(u_\xi + u_\eta), \quad u_{xx} = u_{\xi\xi} - 2u_{\xi\eta} + u_{\eta\eta}$$

$$u_{yy} = -(1/y)u_{\xi\xi} - (2/y)u_{\xi\eta} - (1/y)u_{\eta\eta} + \frac{1}{2}(-y)^{-3/2}(u_\xi + u_\eta).$$

Inserting these results in the original PDE reduces it to the standard form for a hyperbolic equation

$$u_{\xi\eta} = \frac{1}{16}(\xi + \eta)^2(u_\eta - u_\xi).$$

The method of classification of second-order PDE with coefficients $A(x, y)$, $B(x, y)$, and $C(x, y)$ also applies to second-order quasi-linear PDE, although in such cases the classification is likely to be local. This is because, in general, the coefficients are then $A(x, y, u, u_x, u_y)$, $B(x, y, u, u_x, u_y)$, and $C(x, y, u, u_x, u_y)$, so the value of the discriminant depends on the point (x^0, y^0) that is chosen, and the values of u, u_x, and u_y at (x^0, y^0).

Timelike and spacelike arcs

It has been shown that the characteristic curves associated with a second-order hyperbolic equation provide a natural coordinate system, which when used simplifies the structure of the equation. Another property of characteristics is the way they distinguish between two different types of curve on which auxiliary conditions are specified. A curve, or arc Γ, such as that shown in Fig. 3.1a, from which two characteristics radiate out from points P on the *same* side of Γ as time increases, are called **spacelike arcs**, and they represent curves on which *initial conditions* can be specified. A curve, or arc Γ, like that in Fig. 3.1b, such

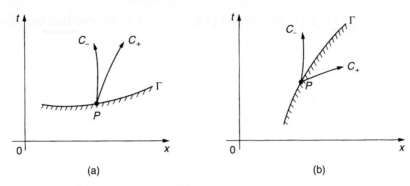

Figure 3.1   (a) A spacelike arc. (b) A timelike arc.

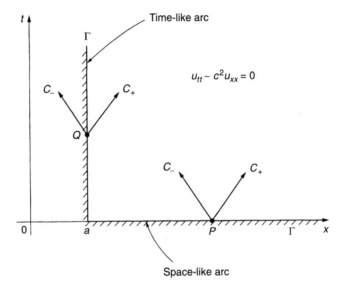

Figure 3.2   A mixed initial boundary value problem for the wave equation.

that the characteristic curves radiating out from points P as time increases are *separated* by the arc Γ, are called **timelike arcs**, and they represent curves on which boundary conditions can be specified. For general hyperbolic equations the characteristics are *curves* as shown in Figs. 3.1a and 3.1b, although when an equation like the wave equation with constant coefficients is involved the characteristics become families of straight lines.

In the *mixed initial boundary value problems* to be considered in Section. 4.3, both initial and boundary conditions are imposed, so both types of arc occur. Figure 3.2 shows a typical situation for the wave equation with initial data imposed on the x axis for $x > a$, which is spacelike, and boundary values imposed on the line $x = a$, $t > 0$, which is timelike. This particular situation will arise later when studying the wave equation and the reflection of waves by a fixed boundary. Note that in Fig. 3.2, because the wave equation is involved, the characteristics are shown as straight lines.

EXERCISES 3.1

1. Use the chain rule to show how the change of variable $\xi = \xi(x, y)$, $\eta = \eta(x, y)$ applied to a continuous twice differentiable function $u(x, y)$ leads to the following results, where ξ and η are continuous and twice differentiable with respect to x and y and their Jacobian is nonvanishing:

$$u_x = \xi_x u_\xi + \eta_x u_\eta, \qquad u_y = \xi_y u_\xi + \eta_y u_\eta$$

$$u_{xx} = \xi_x^2 u_{\xi\xi} + 2\xi_x \eta_x u_{\xi\eta} + \eta_x^2 u_{\eta\eta} + \xi_{xx} u_\xi + \eta_{xx} u_\eta$$

$$u_{xy} = \xi_x \xi_y u_{\xi\xi} + (\xi_x \eta_y + \xi_y \eta_x) u_{\xi\eta} + \eta_x \eta_y u_{\eta\eta} + \xi_{xy} u_\xi + \eta_{xy} u_\eta$$

$$u_{yy} = \xi_y^2 u_{\xi\xi} + 2\xi_y \eta_y u_{\xi\eta} + \eta_y^2 u_{\eta\eta} + \xi_{yy} u_\xi + \eta_{yy} u_\eta.$$

In Exercises 2 through 15 classify the given PDE and reduce it to standard form.

2. $2u_{xx} + 6u_{xy} + 5u_{yy} + u_x = 0$.

3. $u_{xx} - 2u_{xy} + u_{yy} + 3u_x - u_y = 0$.

4. $u_{xx} + 6u_{xy} + 10u_{yy} + (x + y)u_x + u = 0$.

5. $u_{xx} + 6u_{xy} + 9u_{yy} + 3yu_y = 0$.

6. $u_{xx} + 2u_{xy} - 3u_{yy} + 4xu_x + 4yu_y - 2u = 0$.

7. $u_{xx} - 2\cos x\, u_{xy} + (2 - \sin^2 x)u_{yy} + u = 0$.

8. $u_{xx} + 2\cosh x\, u_{xy} + (\sinh^2 x - 8)u_{yy} + u = 0$.

9. $yu_{xx} - xu_{yy} = 0$ for $x > 0$, $y > 0$.

10. $x^2 u_{xx} + 4xyu_{xy} + 4y^2 u_{yy} + 2u_y = 0$.

11. $yu_{xx} - xu_{yy} = 0$ for $x < 0$, $y > 0$.

12. $y^2 u_{xx} + x^2 u_{yy} = 0$.

13. $y^2 u_{xx} + 2xyu_{xy} + 2x^2 u_{yy} + xu_x = 0$.

14. $y^2 u_{xx} + u_{yy} = 0$.

15. $u_{xx} + 2u_{xy} + \cos^2 x\, u_{yy} = 0$.

3.2 Classification of Second-Order PDE in Many Independent Variables

An equivalent way of classifying a second-order PDE of the type in (3.1) involves considering the pattern of the signs of the second-order terms once the PDE has been reduced to a standard form free from mixed partial derivatives. To see how this works, we recall that the second standard form for a hyperbolic equation was of the form

$$u_{xx} - u_{yy} = \Phi(x, y, u, u_x, u_y),$$

while the standard form for a parabolic equation was

$$u_{yy} = \Phi(x, y, u, u_x, u_y),$$

and the standard form for an elliptic equation was

$$u_{xx} + u_{yy} = \Phi(x, y, u, u_x, u_y).$$

When expressed in words these results show that PDE (3.1) is:

- hyperbolic if when expressed in standard form with no mixed derivatives the coefficients of the two second-order derivatives are nonvanishing and opposite in sign;

- parabolic if when expressed in standard form with no mixed derivatives only one second-order derivative term is present;

- elliptic if when expressed in standard form with no mixed derivatives the coefficients of both second-order derivatives are nonvanishing and of the same sign.

This alternative classification allows the linear algebra of quadratic forms to be used to classify PDEs involving many independent variables. With these ideas in mind, we now consider the classification of the more general quasi-linear second-order PDE involving n independent variables $x_1, x_2, \ldots, x_n$ of the form

$$\sum_{i,j=1}^{n} A_{ij}(x_1, \ldots, x_n, u, u_{x_1}, \ldots, u_{x_n}) u_{x_i x_j} + \sum_{i=1}^{n} B_i(x_1, \ldots, x_n, u) u_{x_i}$$

$$= \Phi(x_1, \ldots, x_n). \tag{3.27}$$

To proceed further we will consider this PDE at a specific point $(x_1^0, x_2^0, \ldots, x_n^0)$ where the solution is u^0, and denote the values of A_{ij} at this point by a_{ij}. Thus, locally, (3.27) can be approximated by a constant coefficient equation in which the second-order terms are given by

$$\sum_{i,j=1}^{n} a_{ij} u_{x_i x_j}. \tag{3.28}$$

We now assume the equality of mixed derivatives, which is always permissible, and set $a_{ij} = a_{ji}$.

Denote the $n \times n$ matrix of coefficients of the terms $u_{x_i x_j}$ by $\mathbf{A} = \mathbf{A}[a_{ij}]$, and call this the *coefficient matrix* of (3.28), while remembering that it only contains the coefficients of the second-order derivatives. Then, as $a_{ij} = a_{ji}$, the coefficient matrix $\mathbf{A}$ is seen to be symmetric.

Following the form of argument used in Section 3.1 we now change to a new set of independent variables $\xi_1, \xi_2, \ldots, \xi_n$ by introducing the matrix column vector $\mathbf{x}$ with its elements being the original independent variables $x_1, \ldots, x_n$, and defining the new independent variables in the matrix column vector $\boldsymbol{\xi}$ with elements $\xi_1, \ldots, \xi_n$ through the condition $\boldsymbol{\xi} = \mathbf{Qx}$, where for the time being $\mathbf{Q} = \mathbf{Q}[q_{ij}]$ is an arbitrary $n \times n$ matrix with constant coefficients q_{ij}.

Repeated application of the chain rule in the form

$$\frac{\partial}{\partial x_i} \equiv \sum_{k=1}^{n} \frac{\partial \xi_k}{\partial x_i} \frac{\partial}{\partial \xi_k}$$

to the second-order derivatives of $u(x_1, \ldots, x_n)$ in (3.28) with respect to $x_1, \ldots, x_n$ transforms them into derivatives of $u(\xi_1, \ldots, \xi_n)$ with respect to $\xi_1, \ldots, \xi_n$, allowing (3.28) to be reexpressed as

$$\sum_{i,j=1}^{n} a_{ij} u_{x_i x_j} = \sum_{k,l=1}^{n} \left(\sum_{i,j=1}^{n} c_{ki} a_{ij} c_{lj} \right) u_{\xi_k \xi_l}. \tag{3.29}$$

The coefficient matrix of the terms $u_{\xi_k \xi_l}$ in this transformed expression is now seen to be equal to $\mathbf{QAQ}^{\mathrm{T}}$, where the superscript T indicates the matrix transpose operation and $\mathbf{Q}$ is a diagonalizing matrix for $\mathbf{A}$.

It is a standard result from linear algebra that any real symmetric matrix can always be diagonalized by means of a suitable orthogonal matrix. So, if the matrix $\mathbf{Q}$ is taken to be a diagonalizing matrix for $\mathbf{A}$, it follows that

$$\mathbf{QAQ}^{\mathrm{T}} = \Lambda = \begin{bmatrix} \lambda_1 & & & & \\ & \lambda_2 & & \mathbf{0} & \\ & & \cdot & & \\ & & & \cdot & \\ & \mathbf{0} & & & \cdot \\ & & & & & \lambda_n \end{bmatrix}, \tag{3.30}$$

where Λ is a diagonal matrix with the elements $\lambda_1, \lambda_2, \ldots, \lambda_n$ on its leading diagonal being the eigenvalues of $\mathbf{A}$. The numbers $\lambda_1, \lambda_2, \ldots, \lambda_n$ are all real, because a symmetric $n \times n$ matrix always has n real eigenvalues.

To construct $\mathbf{Q}$, suppose the eigenvalues $\lambda_1, \lambda_2, \ldots, \lambda_n$ and the corresponding eigenvectors $\mathbf{x}_1, \mathbf{x}_2, \ldots, \mathbf{x}_n$ of $\mathbf{A}$ are known. The eigenvectors are then scaled to produce a set of normalized eigenvectors $\hat{\mathbf{x}}_1, \hat{\mathbf{x}}_2, \ldots, \hat{\mathbf{x}}_n$ such that $\hat{\mathbf{x}}_i \hat{\mathbf{x}}_i^T = 1$, where again the superscript T represents the transpose operation and $i = 1, 2, \ldots, n$. The matrix $\mathbf{Q}$ is then constructed by using the normalized eigenvectors as its columns. As a result, the diagonal matrix Λ then has the eigenvalues of $\mathbf{A}$ arranged along its leading diagonal in the same order as the corresponding normalized eigenvectors form the columns of $\mathbf{Q}$.

Equation (3.27) is classified as follows, and when the equation is quasi-linear the classification will usually depend on both the point $(x_1^0, x_2^0, \ldots, x_n^0)$ and the solution u^0 at this point:

- PDE (3.27) is called *hyperbolic* at a point $(x_1^0, x_2^0, \ldots, x_n^0)$, where the solution is u^0, if none of the eigenvalues λ_i of its coefficient matrix $\mathbf{A}$ vanish, and one eigenvalue has a sign opposite to that of all the others.

- PDE (3.27) is called *parabolic* at a point $(x_1^0, x_2^0, \ldots, x_n^0)$, where the solution is u^0, if only one of the eigenvalues λ_i of its coefficient matrix $\mathbf{A}$ vanishes and all the remaining eigenvalues have the same sign.

- PDE (3.27) is called *elliptic* at a point $(x_1^0, x_2^0, \ldots, x_n^0)$, where the solution is u^0, if none of the eigenvalues λ_i of its coefficient matrix $\mathbf{A}$ vanish, and all have the same sign.

When more than two independent variables are involved, other intermediate types of PDE exist, depending on the number of eigenvalues that vanish and the pattern of signs of the remaining eigenvalues. As such equations are of lesser practical importance than equations of hyperbolic, parabolic, and elliptic type they will not be considered further.

Example 3.5. Show how the wave equation, heat equation, and Laplace equation in three space dimensions are, respectively, unconditionally hyperbolic, parabolic, and elliptic.

Solution: The wave equation in three space dimensions and time involves the four independent variables x, y, z, and t, and the four partial derivatives u_{xx}, u_{yy}, u_{zz}, and u_{tt} are present in the PDE $u_{tt} = c^2(u_{xx} + u_{yy} + u_{zz})$, so its coefficient matrix is

$$\mathbf{A} = \begin{bmatrix} 1 & 0 & 0 & 0 \\ 0 & -c^2 & 0 & 0 \\ 0 & 0 & -c^2 & 0 \\ 0 & 0 & 0 & -c^2 \end{bmatrix}.$$

As $\mathbf{A}$ is already in diagonalized form it can be seen immediately that it has four nonzero eigenvalues, three of which are of one sign and one of opposite sign.

So the wave equation is hyperbolic, and this is true unconditionally because the eigenvalues never change sign.

To examine the heat equation in three space dimensions and time we need to consider a PDE of the form $u_t = k(u_{xx} + u_{yy} + u_{zz})$. There are four independent variables x, y, z, and t, but only the three second-order partial derivatives u_{xx}, u_{yy}, and u_{zz} are present in the equation as u_{tt} is missing. Writing the heat equation as $u_t = k(0 \cdot u_{tt} + u_{xx} + u_{yy} + u_{zz})$, the coefficient matrix $\mathbf{A}$ is seen to be

$$\mathbf{A} = \begin{bmatrix} 0 & 0 & 0 & 0 \\ 0 & k & 0 & 0 \\ 0 & 0 & k & 0 \\ 0 & 0 & 0 & k \end{bmatrix}.$$

Matrix $\mathbf{A}$ has three positive eigenvalues and one zero eigenvalue, so the three-dimensional unsteady heat equation is parabolic. The heat equation is unconditionally parabolic because the pattern of the eigenvalues never changes.

The Laplace equation in three space dimensions is $u_{xx} + u_{yy} + u_{zz} = 0$. Only the three independent variables x, y, and z are involved and all three second-order derivatives u_{xx}, u_{yy}, and u_{zz} are present in the equation. The coefficient matrix is $\mathbf{A} = \mathbf{I}$, where $\mathbf{I}$ is the 3×3 unit matrix, so as this has three eigenvalues, each equal to 1, the Laplace equation is unconditionally elliptic. ■

Example 3.6. Classify the quasi-linear equation

$$u_{tt} + u_{xx} + \left(1 + u_x^2\right)u_{yy} + xuu_{zz} = 0.$$

Solution: The coefficient matrix is

$$\mathbf{A} = \begin{bmatrix} 1 & 0 & 0 & 0 \\ 0 & 1 & 0 & 0 \\ 0 & 0 & \left(1 + u_x^2\right) & 0 \\ 0 & 0 & 0 & xu \end{bmatrix},$$

so as this is already in diagonal form inspection shows it has the four eigenvalues $1, 1, 1 + u_x^2$, and xu. The first three eigenvalues are unconditionally positive, while the fourth eigenvalue is positive if $xu > 0$, 0 on the y axis, and negative if $xu < 0$. Thus the equation is elliptic when $xu > 0$, degenerately parabolic on the y axis, and hyperbolic when $xu < 0$. ■

It is useful to recognize that the diagonalization of the coefficient matrix $\mathbf{A}$ of PDE (3.27), with its coefficients evaluated at the point $(x_1^0, x_2^0, \ldots, x_n^0)$, where the solution is u^0, is closely related to the algebraic problem of reducing to a sum of squares a quadratic form with the same coefficient matrix $\mathbf{A}$. The reduction of a quadratic form to a sum of squares is not unique, although the pattern of signs of the coefficients of squares, and the presence of zero coefficients where a variable is absent, remains the same. This means that the same result is also true of the quadratic form associated with the coefficient matrix **A** of a PDE. In linear algebra, this property of a quadratic form is often called *Sylvester's law of inertia*.

In terms of the quadratic form $\mathbf{x}^T\mathbf{A}\mathbf{x}$ associated with the coefficient matrix **A** of PDE (3.27), the equation will be hyperbolic if, when it is reduced to the sum of squares, $n-1$ of the coefficients multiplying the squares are of one sign and one is of opposite sign. The PDE will be parabolic if one of the n coefficients is zero and the remaining $n-1$ are nonzero and all have the same sign. Finally, the PDE will be elliptic if none of the n coefficients vanish and all are of the same sign (positive or negative).

When n is small, because of the difficulty of finding the eigenvalues and eigenvectors of **A**, it is usually easier to classify the PDE (3.27) by examining the associated quadratic form $\mathbf{x}^T\mathbf{A}\mathbf{x}$.

Example 3.7. Use the associated quadratic form to classify the PDE

$$3u_{xx} + 2u_{yy} + 6u_{xy} - 2u_{xz} - 4u_{yz} = 0.$$

Solution: The quadratic form associated with this PDE is $Q(x, y, z) = \mathbf{x}^T\mathbf{A}\mathbf{x}$, where

$$\mathbf{A} = \begin{bmatrix} 3 & 3 & -1 \\ 3 & 2 & -2 \\ -1 & -2 & 0 \end{bmatrix} \quad \text{and} \quad \mathbf{x} = \begin{bmatrix} x \\ y \\ x \end{bmatrix},$$

so

$$Q(x, y, z) = 3x^2 + 2y^2 + 6xy - 2xz - 4yz.$$

Comparison of the PDE with Q shows the quadratic form can be obtained directly from the PDE without using the matrix **A**, simply by replacing u_{xx} by x^2, u_{yy} by y^2, u_{xy} by xy, u_{xz} by xz, and u_{yz} by yz.

To express Q as a sum of squares we use a simple method due to Lagrange. This involves first separating out all terms containing the same coefficient, say x, completing the square, and compensating for any additional terms that arise by subtracting squares and products of terms that do not contain x (that is to say, terms involving y and z). This process is then repeated using the remaining terms, until Q is represented as a sum of squares. Examination of this form of Q then determines the classification of the PDE.

Grouping terms involving the factor x in Q gives

$$Q = 3\left(x^2 + 2xy - \frac{2}{3}xz\right) + 2y^2 - 4yz.$$

Now

$$3\left(x + y - \frac{1}{3}z\right)^2 = 3x^2 + 3y^2 + \frac{1}{3}z^2 + 6xy - 2xz - 2yz,$$

so

$$3\left(x^2 + 2xy - \frac{2}{3}xz\right) = 3\left(x + y - \frac{1}{3}z\right)^2 - 3y^2 - \frac{1}{3}z^2 + 2yz.$$

Substituting this in Q we have

$$Q = 3\left(x + y - \frac{1}{3}z\right)^2 - 3y^2 - \frac{1}{3}z^2 + 2yz + 2y^2 - 4yz,$$

or

$$Q = 3\left(x + y - \frac{1}{3}z\right)^2 - y^2 - 2yz - \frac{1}{3}z^2.$$

Repeating the process with the remaining three terms, and this time grouping terms containing the factor y, gives

$$Q = 3\left(x + y - \frac{1}{3}z\right)^2 - (y^2 + 2yz) - \frac{1}{3}z^2.$$

This can now be written

$$Q = 3\left(x + y - \frac{1}{3}z\right)^2 - (y + z)^2 + z^2 - \frac{1}{3}z^2,$$

and so, finally,

$$Q = 3\left(x + y - \frac{1}{3}z\right)^2 - (y + z)^2 + \frac{2}{3}z^2.$$

This is the required decomposition of Q into a sum of squares. As the three variables x, y, and z are involved, and the sum involves three squares, two of which are of one sign while the third is of opposite sign, so the PDE must be hyperbolic. ∎

As already remarked, different decompositions of Q involving the sum of squares are possible. However, Sylvester's law of inertia ensures that the pattern of the coefficients involved is always the same, so the precise form of a decomposition is irrelevant when it is used for the classification of a PDE.

In Example 3.7, the classification of the PDE by means of its associated quadratic form involves far less effort than basing the classification on the eigenvalues of $\mathbf{A}$ if these must be found by hand computation. This can be seen from the fact that in this case the eigenvalues of $\mathbf{A}$ are 0.22086, 6.23217, and -1.45303, and these values can only be found by numerical computation.

EXERCISES 3.2

In Exercises 1 through 4 use the associated quadratic form to classify the given PDE.

1. $4u_{xx} + 9u_{yy} + 3u_{zz} - 8u_{xy} + 4u_{xz} + 2u_{yz} + u_x + u = 0$.

2. $u_{xx} + 7u_{yy} + 12u_{zz} + 4u_{xy} - 12u_{yz} + 3xu_y + yu_z = 0$.

3. $7u_{yy} + 11u_{zz} + 4u_{xy} - 2u_{xz} - 12u_{yz} + (1 + x)\sin u = 0$.

4. $2u_{xx} + 21u_{yy} + 12u_{zz} - 12u_{xy} + 12u_{yz} + 3u_y + 1 = 0$.

5. Perform the calculations leading to result (3.29).

3.3 Well-Posed Problems for Hyperbolic, Parabolic, and Elliptic Partial Differential Equations

In Section 1.5 attention was drawn to the fact that inappropriate auxiliary conditions imposed on a PDE, or an attempt to find a solution in the wrong type of region, such as one that is *open* instead of *closed*, can lead to an ill-posed boundary value problem. The example considered there concerned the Laplace equation, which was shown to be improperly posed when Cauchy conditions are imposed on the finite edge of a semi-infinite strip, with Dirichlet conditions being imposed on the two sides of the strip which extended to infinity (making it an *open* region).

Summarized in Table 3.1, according to the type of PDE involved, are the most frequently occurring forms of boundary condition and region that lead to properly posed problems for linear hyperbolic, parabolic, and elliptic equations.

It will be recalled that a region is **closed** when every point on its boundary belongs to the region, and it is **open** if some or all of the points on its boundary do not belong to the region. In the context of PDEs, the term *region* refers to a region in space when no time is involved, or to a region in space–time when a time-dependent problem is involved. Thus, if $u(x, y)$ satisfies a PDE in a region D of the (x, y) plane, such that $a \le x \le b, c \le y \le d$, then D is *closed*, because every boundary point belongs to D, while if D is such that $a \le x \le b, c \le y < d$ then D is *open*, because the points on $y = d$, with

Type of PDE	Type of Boundary Condition	Type of Region
Hyperbolic	Cauchy conditions	Open
Parabolic	Dirichlet, Neumann, or mixed	Open
Elliptic	Dirichlet, Neumann, or mixed	Closed

Table 3.1

$a \leq x \leq b$ do not belong to D. Similarly, if $u(x, t)$ satisfies a PDE in the region D of space–time such that $a \leq x \leq b$, $t > 0$, then D is *open*, because t is unrestricted as $t \to \infty$.

It can be seen from this that the example in Section 1.5 was ill posed because *Cauchy conditions* were imposed on part of the boundary of an *open region*, as a result of which neither the boundary conditions nor the region were appropriate for an *elliptic* equation.

Linear Wave Propagation in One or More Space Dimensions

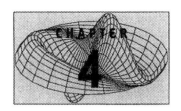

4.1 Linear Waves and the Wave Equation

The wave equation describes physical phenomena as diverse as sound waves, ripples on the surface of water, vibrations of strings and plates, and electromagnetic waves. Two features characteristic of waves are that they travel with finite speed through the medium in which they occur, and they transmit energy through the medium without causing it to suffer any permanent change. In what follows we will see how the mathematics of wave propagation governed by the linear wave equation interprets these properties, and how solutions of the one-dimensional wave equation can be found by D'Alembert's formula, which also provides insight into how the initial conditions influence the wave.

Without first specifying the physical meaning of a propagating disturbance ϕ, which might for example be the amplitude of a ripple on the surface of water, we will first consider the general situation where a disturbance propagates in the positive direction along the x axis with a constant speed c. The disturbance ϕ will depend on x and t, and when $t = 0$ the effect of the disturbance can be described by a function $f(x)$ called the initial **wave profile**. This represents the instantaneous disturbance at time $t = 0$ as a function of x, as would be recorded in a "flash photograph." In the event that the disturbance propagates without a change of shape, a subsequent "flash photograph" at a time t would show a new wave profile that could be obtained from the original profile at $t = 0$ by translating it along the positive x axis through a distance ct. Taking a new origin at the point $x = ct$, and denoting distances measured from this origin by X, we can write $X = x - ct$. The wave profile relative to this new origin will be $\phi = f(X)$, so in terms of the original variables we can write

$$\phi(x, t) = f(x - ct). \tag{4.1}$$

This represents the most general equation for a **progressive** one-dimensional wave moving in the positive direction along the x axis with constant speed c, without change of shape or scale. Reversing the sign of c gives

$$\phi(x, t) = f(x + ct), \tag{4.2}$$

and this describes a progressive wave of the same form, which now moves in the negative direction along the x axis.

The simplest example of a wave of this type, and one which is of fundamental importance in the study of linear wave propagation, is the **harmonic wave** with a profile at $t = 0$ given by $\phi(x, 0) = a \cos mx$. If such a wave moves to the right with speed c, the wave profile at time t will be given by

$$\phi(x, t) = a \cos[m(x - ct)]. \tag{4.3}$$

The maximum magnitude of the disturbance represented by (4.3) is a, and this is called the **amplitude** of the wave. As (4.3) is a **periodic** function of its argument $x - ct$, the wave profile will be repeated regularly at distances $\lambda = 2\pi/m$, and this distance λ is called the **wavelength** of the wave. When expressed in terms of λ, result (4.3) becomes

$$\phi(x, t) = a \cos\left(\frac{2\pi}{\lambda}(x - ct)\right). \tag{4.4}$$

The time taken for one wavelength of the wave to pass any fixed point is denoted by τ and called the **period** of the periodic wave. Result (4.4) shows that as t increases by one period τ, so one complete cycle of values of ϕ will evolve, showing that $2\pi c\tau/\lambda = 2\pi$, from which it follows that

$$\tau = \lambda/c. \tag{4.5}$$

Two other definitions of importance in the study of linear periodic waves are the **frequency** of the wave, which will be denoted by f, and the number of waves in a unit length, which will be denoted by n. The frequency f is the number of wavelengths that pass a fixed point in unit time, so

$$f = 1/\tau, \tag{4.6}$$

from which it follows that

$$c = \lambda f, \tag{4.7}$$

so the number of waves in a unit length will be

$$n = 1/\lambda, \tag{4.8}$$

so in terms of f and n, the wave profile in (4.4) becomes

$$\phi(x, t) = a \cos[2\pi(nx - ft)]. \tag{4.9}$$

It is convenient to introduce two new quantities called the **wave number** and the **angular frequency** (or **radian frequency**) of the periodic wave in (4.3), and to denote them by k and ω, respectively, where

$$k = 2\pi n \quad \text{and} \quad \omega = 2\pi f, \tag{4.10}$$

as a result of which (4.9) becomes

$$\phi(x, t) = a \cos(kx - \omega t). \tag{4.11}$$

It often happens that two similar harmonic waves $\phi_1(x, t)$ and $\phi_2(x, t)$ occur with one displaced along the x axis relative to the other, so we can write

$$\phi_1(x, t) = a \cos(kx - \omega t) \quad \text{and} \quad \phi_2(x, t) = a \cos[(kx - \omega t) + \varepsilon].$$

When this happens, the displacement of wave profile $\phi_2(x, t)$ along the x axis relative to $\phi_1(x, t)$ is ε/k. The quantity ε is called the **phase** of $\phi_2(x, t)$ relative to $\phi_1(x, t)$, and the waves are said to be **in phase** if $\varepsilon = 2r\pi$, for $r = 1, 2, \ldots$, and to be exactly **out of phase** if $\varepsilon = (2r - 1)\pi$, for $r = 1, 2, \ldots$. This means that when waves $\phi_1(x, t)$ and $\phi_2(x, t)$ are in phase their profiles coincide, but when they are exactly out of phase their arguments differ by π, so such waves are often said to be $180°$ out of phase.

In wave representations like (4.11), the quantity

$$v_p = \omega/k \tag{4.12}$$

is called the **phase speed** of the wave in the positive x direction, because it is the speed with which a point of constant phase moves along the x axis.

An important special case arises when two harmonic waves of the type in (4.11) have equal amplitudes but propagate in opposite directions, because then the result of adding these waves gives

$$\phi(x, t) = a \cos(kx - \omega t) + a \cos(kx + \omega t)$$
$$= 2a \cos(kx) \cos(\omega t). \tag{4.13}$$

This form of wave is called a **stationary wave**, to distinguish it from the **progressive waves** considered earlier. It has the important property that the wave profile no longer propagates, and although the fundamental sinusoidal shape remains unchanged, the shape is modulated (multiplied) by the factor $\cos(\omega t)$. This has the effect that the sign and magnitude of the displacement changes periodically as the time t increases. At the points where $x = \pm\pi/(2k), \pm 3\pi/(2k), \pm 5\pi/(2k), \ldots$, the displacement of the stationary wave remains zero for all time, and these points are called the **nodes** of the wave.

The points midway between the nodes, where the stationary wave experiences maximum displacement in opposite directions periodically with time, are called the **antinodes** of the wave.

Result (4.1) generalizes immediately to the case of a **plane wave** in space, because it can be interpreted to mean that the disturbance ϕ at each point on a plane normal to the x axis, and moving with speed c in the positive x direction, is the same. In what follows, the term **wavefront** will be used to refer to an identifiable leading edge of a propagating localized disturbance.

Result (4.1) can be generalized to the case of a plane wave moving with constant speed c in the direction of an arbitrary unit normal v in space. Let a line parallel to v pass through the origin of an orthogonal Cartesian coordinate system $O(x, y, z)$, and take the position of the plane to be at a distance ξ from the origin along this line. Then from (4.1) we see that the wave profile of a wave moving in the positive ξ direction can be written

$$\phi(\xi, t) = f(\xi - ct).$$

However, if the direction cosines of the ξ axis relative to the Cartesian system are (n_1, n_2, n_3), so that $n_1^2 + n_2^2 + n_3^2 = 1$, we can write $\xi = n_1 x + n_2 y + n_3 z$, and then in terms of the $O(x, y, z)$ coordinate system the plane wave with normal v becomes

$$\phi(x, y, z, t) = f(n_1 x + n_2 y + n_3 z - ct). \tag{4.14}$$

Clearly, a wave of similar type moving in the negative ξ direction will be of the form

$$\phi(x, y, z, t) = g(n_1 x + n_2 y + n_3 z + ct). \tag{4.15}$$

If the functions f and g in (4.14) and (4.15) are twice differentiable with respect to their arguments, routine differentiation shows that both equations satisfy the three-dimensional wave equation

$$\frac{\partial^2 \phi}{\partial x^2} + \frac{\partial^2 \phi}{\partial y^2} + \frac{\partial^2 \phi}{\partial z^2} = \frac{1}{c^2} \frac{\partial^2 \phi}{\partial t^2}. \tag{4.16}$$

The one-dimensional form of (4.16), in terms of the x space coordinate, is

$$\frac{\partial^2 \phi}{\partial x^2} = \frac{1}{c^2} \frac{\partial^2 \phi}{\partial t^2}. \tag{4.17}$$

The change of variables $u = x - ct, v = x + ct$ reduces (4.17) to

$$\frac{\partial^2 \phi}{\partial u \, \partial v} = 0,$$

and integration then gives the general solution

$$\phi(x, t) = f(x - ct) + g(x + ct), \qquad (4.18)$$

where f and g are arbitrary twice differentiable functions of their arguments. This shows that the general solution of the one-dimensional wave equation (4.17) can be interpreted as the sum of two arbitrary waves; the wave f moving to the right with constant speed c, and the wave g moving to the left with constant speed c.

The argument used to establish (4.14) and (4.15) shows the general plane wave solution of the three-dimensional wave equation (4.16) for a wave with unit normal ν is

$$\phi(x, y, z, t) = f(n_1 x + n_2 y + n_3 z - ct) + g(n_1 x + n_2 y + n_3 z + ct) \quad (4.19)$$

where, again, $n_1^2 + n_2^2 + n_3^2 = 1$.

Plane waves are not the only progressive waves that are possible and if, for example, we consider the wave equation with spherical symmetry, where the solution ϕ only depends on the radius r and the time t, the equation reduces to

$$\frac{\partial^2 \phi}{\partial r^2} + \frac{2}{r} \frac{\partial \phi}{\partial r} = \frac{1}{c^2} \frac{\partial^2 \phi}{\partial t^2}. \qquad (4.20)$$

When this is written in the form

$$\frac{\partial^2}{\partial r^2} (r\phi) = \frac{1}{c^2} \frac{\partial^2}{\partial t^2} (r\phi),$$

the general solution is seen to be

$$r\phi(r, t) = f(r - ct) + g(r + ct),$$

and hence we arrive at a progressive wave of the form

$$\phi(r, t) = \frac{1}{r} [f(r - ct) + g(r + ct)], \qquad (4.21)$$

where again f and g are arbitrary twice differentiable functions of their arguments.

The wave equation is linear and homogeneous, so if ϕ_1 and ϕ_2 are solutions, then because the operation of differentiation is linear, $a_1 \phi_1 + a_2 \phi_2$ will also be a solution. This is a particular example of what is called the **linear superposition** of solutions, and it has already been encountered in (4.13), (4.18), and (4.21). Linear superposition is of fundamental importance in the study of linear PDEs, and later it will be used extensively when solving linear PDEs by the method of separation of variables.

A progressive plane wave with a constant phase ε has the form

$$\phi(x, t) = a \cos(kx - \omega t + \varepsilon),$$

and it can be written in the complex exponential form

$$\phi(x, t) = \text{Re}\{A \exp[i(kx - \omega t + \varepsilon)]\}, \tag{4.22}$$

where Re denotes the *real part* of a complex quantity. Apart from a change of phase, a harmonic wave can be represented by (4.22) with Re replaced by Im, signifying the *imaginary part* of the complex quantity. In this representation the complex constant A is called the **complex amplitude** of the wave and we set $|A| = a$. By convention the notation in (4.22) is usually abbreviated by omitting the symbol Re and writing

$$\phi(x, t) = A \exp[i(kx - \omega t + \varepsilon)], \tag{4.23}$$

with the understanding that $\phi(x, t)$ may be either the real or the imaginary part of (4.23). The complex form has the advantage that the single notation represents two linearly independent solutions of the wave equation.

EXERCISES 4.1

1. Explain why

$$\phi = a \cos(Ax + By + Cz - ct)$$

represents a plane wave that propagates with speed $c/\sqrt{A^2 + B^2 + C^2}$. Find a unit normal $\hat{n}$ to the plane of the wave and hence find its wavelength.

2. Verify that

$$\phi = A \cos mx \cos nt + B \sin mx \sin nt$$

and

$$\phi = C \exp[ni(x + ct)] + D \exp[-ni(x + ct)]$$

are both stationary wave solutions of a one-dimensional wave equation and find the wave speeds.

3. By considering the two forms of stationary wave solutions in Exercise 2, find a solution of the one-dimensional wave equation that is never infinite for real values of x and t, and is such that $\phi = 0$ when $x = 0$, and when $t = 0$.

4. Verify that

$$\phi = A \cos mx \cos ny \cos rct + B \sin mx \sin ny \sin rct,$$
$$\text{with } m^2 + n^2 = r^2,$$

$$\phi = C \cos mx \exp(ny) \cos rct + D \sin mx \exp(-ny) \sin rct,$$
$$\text{with } m^2 - n^2 = r^2$$

are both solutions of the two-dimensional wave equation

$$\frac{1}{c^2}\frac{\partial^2\phi}{\partial t^2} = \frac{\partial^2\phi}{\partial x^2} + \frac{\partial^2\phi}{\partial y^2}.$$

5. Show that three harmonic waves, each of equal amplitude but with a phase difference of $2\pi/3$ between the first and second wave, and the same phase difference between the second and third wave, have a zero sum.

6. Show that $\phi = r^{-1/2}\cos(\theta/2)f(r \pm ct)$ is a progressive wave in two space dimensions, where r and θ are plane polar coordinates and f is an arbitrary twice differentiable function of its argument. Explain why θ must be restricted if ϕ is to be a single-valued function.

4.2 The D'Alembert Solution and the Telegraph Equation

It is unusual to be able to find a general solution of a PDE, but an exception arises when considering the one-dimensional wave equation

$$\frac{\partial^2\phi}{\partial t^2} = c^2\frac{\partial^2\phi}{\partial x^2}, \tag{4.24}$$

subject to the Cauchy conditions

$$\phi(x,0) = h(x) \quad \text{and} \quad \phi_t(x,0) = k(x), \tag{4.25}$$

where h and k are arbitrarily assigned differentiable functions for $-\infty < x < \infty$. The subsequent argument will show that h must be at least twice differentiable and k at least once differentiable.

The solution of this **pure initial value problem** that will be found, called the D'Alembert solution (or formula), will show precisely how the functions h and k enter into the solution. We will begin our derivation by using the general solution of (4.24)

$$\phi(x,t) = f(x - ct) + g(x + ct) \tag{4.26}$$

found in (4.18). Setting $t = 0$ and using the first initial condition in (4.25) shows that

$$h(x) = f(x) + g(x). \tag{4.27}$$

Partial differentiation of (4.26) with respect to t, followed by setting $t = 0$ and using the second initial condition in (4.25), gives

$$k(x) = -cf'(x) + cg'(x). \tag{4.28}$$

Integrating (4.28) with respect to x from an arbitrary point a to x gives

$$\frac{1}{c} \int_a^x k(s)ds = f(a) - f(x) + g(x) - g(a).$$

Using (4.27) then allows this to be written as

$$f(x) = \frac{1}{2}h(x) - \frac{1}{2c} \int_a^x k(s)ds - \frac{1}{2}[g(a) - f(a)], \qquad (4.29)$$

so as $g(x) = h(x) - f(x)$ we also have

$$g(x) = \frac{1}{2}h(x) + \frac{1}{2c} \int_a^x k(s)ds + \frac{1}{2}[g(a) - f(a)]. \qquad (4.30)$$

Replacing x by $x - ct$ in (4.29) and by $x + ct$ in (4.30), using the general solution (4.26), and adding the results we find that

$$\phi(x, t) = \frac{1}{2}\left\{ h(x - ct) + h(x + ct) - \frac{1}{c} \int_a^{x-ct} k(s)ds + \frac{1}{c} \int_a^{x+ct} k(s)ds \right\}.$$

Reversing the limits in the first integral, and compensating by changing the sign of the integral, allows the two integrals to be combined to give

$$\phi(x, t) = \frac{1}{2}[h(x - ct) + h(x + ct)] + \frac{1}{2c} \int_{x-ct}^{x+ct} k(s)ds. \qquad (4.31)$$

This result is the **D'Alembert solution** of the homogeneous wave equation (4.24) subject to the pure initial conditions (4.25).

Examination of (4.31) provides fundamental information about the way the solution is influenced by the Cauchy conditions in (4.25), as can be seen by considering Figs. 4.1a and 4.1b. Figure 4.1a shows the region D in the upper part of the (x, t) plane bounded by the two straight lines with slopes $\pm c$ drawn through the general point (x_0, t_0) at Q. These lines intercept the x axis at point A located at $x_0 - ct_0$ and point B located at $x_0 + ct_0$. It will be recognized from Chapter 3.1 that the lines QA and QB are the characteristic curves of the wave equation traced backward from Q until they intercept the initial line at A and B. As x is a space coordinate and t is the time in the wave equation, the slopes of these characteristics are $dx/dt = \pm c$ so that c has the dimensions of a speed.

Examination of the D'Alembert solution in (4.31) shows the initial condition h only occurs in the first bracketed term and, furthermore, only its values at the ends of the interval I on the initial line determined by $x_0 - ct_0 \le x \le x_0 + ct_0$ are involved. The initial condition k also only influences the solution through its values on this same interval I, although its effect occurs as an integral of k over the complete interval I. It is because of these results that the interval

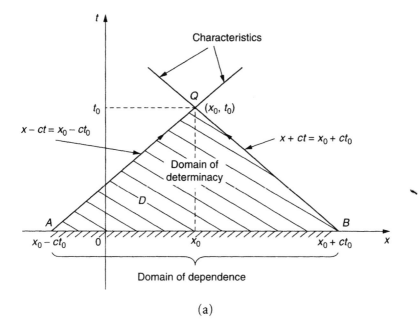

(a)

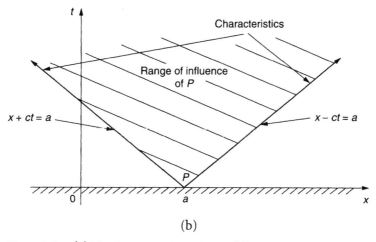

(b)

Figure 4.1 (a) The domain of dependence AB and determinacy D. (b) Range of influence of P.

I is called the **domain of dependence** of the solution at (x_0, t_0) on the initial data specified on I. Result (4.31) shows that the specification of the initial data on interval I determines the solution of the wave equation throughout the triangular region D in Fig. 4.1a, and that initial data outside this interval have no influence on the solution at (x_0, t_0).

Region D is called the *domain of determinacy* of the solution corresponding to initial data specified on the domain of dependence I, with the understanding that the initial data specified on I determines the solution at each point inside D. This is a direct generalization of the line of determinacy introduced when considering first-order PDEs.

Figure 4.1b shows the two characteristics that emerge into the upper half of the (x, t) plane from a point P on the initial line as t increases. This shows the initial line is *spacelike*. The shaded region R in Fig. 4.1b represents all points in the upper half of the (x, t) plane that can be influenced by initial data at P. Because of this the region R is called the **range of influence** of P.

The *uniqueness* of the pure initial value problem (4.24) and (4.25) for the wave equation follows directly from the explicit form of the D'Alembert solution. The effect of the initial conditions on the D'Alembert solution is illustrated in the computer plot of the analytical solution of the pure initial value problem

$$\frac{\partial^2 \phi}{\partial t^2} = \frac{\partial^2 \phi}{\partial x^2},$$

subject to the Cauchy conditions

$$h(x) = \begin{cases} 0, & -\infty < x < -\pi \\ \sin x, & -\pi \le x \le \pi \\ 0, & \pi < x < \infty \end{cases} \quad \text{and} \quad k(x) = \begin{cases} 0, & -\infty < x < -\pi \\ \cos x/2, & -\pi \le x \le \pi, \\ 0, & \pi < x < \infty \end{cases}$$

which are only nonzero in a finite interval that, in this case, is $-\pi < x < \pi$. Localized initial conditions of this type are said to have **compact support**, where the **support** of the data is the finite interval over which the data are nonzero.

In this problem, for convenience, the wave propagation speed has been chosen to be $c = 1$, and a 3d-plot of the solution has been made for $-9 \le x \le 9$ and $0 \le t \le 8$. The D'Alembert solution is

$$\phi(x, t) = \frac{1}{2}[h(x - t) + h(x + t)] + \frac{1}{2}\int_{x-t}^{x+t} k(s)\,ds,$$

but when evaluating this formula it is necessary to remember the piecewise nature of the functions h and k. In order to take account of the piecewise definitions of h and k we will need to make use of the Heaviside unit step function

$$H(x - x_0) = \begin{cases} 0, & x < x_0 \\ 1, & x > x_0. \end{cases} \tag{4.32}$$

The first group of terms in the D'Alembert solution can now be written

$$\frac{1}{2}[h(x-t)+h(x+t)] = \frac{1}{2}[H(x-t+\pi)-H(x-t-\pi)]\sin(x-t)$$
$$+\frac{1}{2}[H(x+t+\pi)-H(x+t-\pi)]\sin(x+t),$$
$$|x| < \infty, t \geq 0,$$

and after integration the integral term in the D'Alembert solution becomes

$$\frac{1}{2}\int_{x-t}^{x+t} k(s)ds = H(x+t+\pi)\sin\left[\frac{1}{2}(x+t)\right] + H(x+t+\pi)$$
$$- H(x+t-\pi)\sin\left[\frac{1}{2}(x+t)\right] + H(x+t-\pi)$$
$$- H(x-t+\pi)\sin\left[\frac{1}{2}(x-t)\right] - H(x-t+\pi)$$
$$+ H(x-t-\pi)\sin\left[\frac{1}{2}(x-t)\right] - H(x-t-\pi),$$

$$|x| < \infty, t \geq 0.$$

Figure 4.2a shows a plot of $\phi_1(x,t) = \frac{1}{2}[h(x-t)+h(x+t)]$, which can be interpreted as the solution of the initial value problem when $k(s) \equiv 0$. The plot shows how in this case the initial sinusoid decomposes into two propagating waves, one of which moves to the left and the other to the right, each with unit speed $c = 1$. The waves are only nonzero in an interval of length 2π (they are *localized in space*), and before separating they interact linearly (additively), but after separation when $t = \pi$, each becomes a replica of the initial disturbance $h(x)$, though with half of its amplitude.

Figure 4.2b shows a plot of $\phi_2(x,t) = \frac{1}{2}\int_{x-t}^{x+t} k(s)ds$, and so can be interpreted as a solution of the initial value problem when $h(s) \equiv 0$. The plot shows how, in this case, the solution starts with a zero displacement at time $t = 0$, and subsequently grows to a constant amplitude of magnitude 2 that spreads equally to the left and right with speed $c = 1$ as t increases. Finally, Fig. 4.2c shows a plot of the solution of the full initial value problem $\phi(x,t) = \frac{1}{2}[h(x-t)-h(x+t)] + \frac{1}{2}\int_{x-t}^{x+t} k(s)ds$.

An important extension of the D'Alembert solution occurs when the non-homogeneous wave equation

$$\frac{\partial^2 \phi}{\partial t^2} - c^2 \frac{\partial^2 \phi}{\partial x^2} = f(x,t) \tag{4.33}$$

is involved and is required to satisfy the initial conditions in (4.25). One way of solving this nonhomogeneous pure initial value problem is by noticing

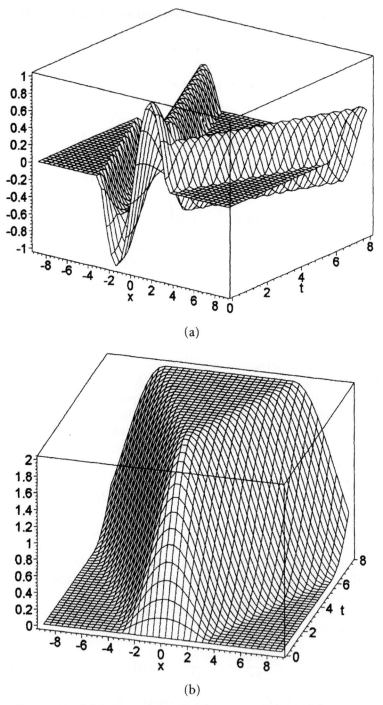

(a)

(b)

Figure 4.2   (a) A plot of $\phi_1(x, t)$. (b) A plot of $\phi_2(x, t)$. (c) A plot of $\phi(x, t)$.

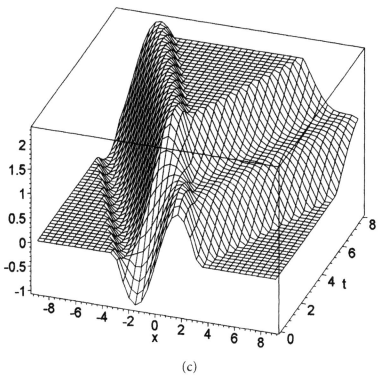

(c)

Figure 4.2 (*Continued*).

that if we set $\phi = \phi_1 + \phi_2$, where ϕ_1 is a solution of the homogeneous wave equation (4.24) subject to the initial conditions in (4.25), then ϕ_2 will be a solution of

$$\frac{\partial^2 \phi_2}{\partial t^2} - c^2 \frac{\partial^2 \phi_2}{\partial x^2} = f(x, t) \tag{4.34}$$

subject to the *homogeneous* initial conditions

$$\phi_2(x, 0) = 0 \quad \text{and} \quad \phi_{2t}(x, 0) = 0. \tag{4.35}$$

As the solution ϕ_1 is given by the D'Alembert solution, we need only solve the simplified initial value problem in (4.34) and (4.35), and then use linear superposition to combine the two results.

Integrating (4.34) over the region D in Fig. 4.1a gives

$$\iint_D \left(\frac{\partial^2 \phi_2}{\partial t^2} - c^2 \frac{\partial^2 \phi_2}{\partial x^2} \right) dx \, dt = \iint_D f(x, t) dx \, dt.$$

An application of Green's theorem to the integral on the left allows it to be replaced by a line integral around the triangular boundary ∂D of D, so

$$-\int_{\partial D} \frac{\partial \phi_2}{\partial t}\,dx + c^2 \frac{\partial \phi_2}{\partial x}\,dt = \iint_D f(x,t)\,dx\,dt. \tag{4.36}$$

The boundary ∂D comprises the three directed straight line segments BQ, QA, and AB, and along BQ $dx/dt = -c$, while along QA $dx/dt = c$. Using these results in (4.36) gives

$$\int_{BQ} c\left(\frac{\partial \phi_2}{\partial t}\,dt + \frac{\partial \phi_2}{\partial x}\,dx\right) - \int_{QA} c\left(\frac{\partial \phi_2}{\partial t}\,dt + \frac{\partial \phi_2}{\partial x}\,dx\right)$$
$$-\int_{AB}\left(\frac{\partial \phi_2}{\partial t}\,dx + c^2 \frac{\partial \phi_2}{\partial x}\,dt\right) = \iint_D f(x,t)\,dx\,dt. \tag{4.37}$$

Examination of (4.37) shows that the bracketed terms in the first two integrals are simply the total differential $d\phi_2$, while the first term in the third integral vanishes because of the initial condition $\phi_{2t}(x,0) = 0$, while the second term vanishes because $dt = 0$ on AB.

Consequently, result (4.37) simplifies to

$$\int_{BQ} c\,d\phi_2 - \int_{QA} c\,d\phi_2 = \iint_D f(x,t)\,dx\,dt.$$

Evaluating the two directed integrals on the left we arrive at the result

$$c\phi_2(Q) - c\phi_2(B) + c\phi_2(Q) - c\phi_2(A) = \iint_D f(x,t)\,dx\,dt.$$

However, because of the first of the initial conditions in (4.35), $\phi_2(A) = \phi_2(B) = 0$, so this last result reduces to

$$\phi_2(Q) = \frac{1}{2c} \iint_D f(x,t)\,dx\,dt. \tag{4.38}$$

When (4.38) is expressed in terms of the geometry of the triangle ABQ in Fig. 4.1a it becomes

$$\phi_2(x_0, t_0) = \frac{1}{2c} \int_0^{t_0} \int_{x_0-c(t_0-\tau)}^{x_0+c(t_0-\tau)} f(\xi, \tau)\,d\xi\,d\tau. \tag{4.39}$$

Omitting the suffix zero from this last result, and using the fact that $\phi = \phi_1 + \phi_2$, we see that the solution of the nonhomogeneous wave equation (4.34) subject to the pure initial conditions (4.35) is

$$\phi(x, t) = \frac{1}{2}[h(x - ct) + h(x + ct)] + \frac{1}{2c} \int_{x-ct}^{x+ct} k(s)\,ds$$
$$+ \frac{1}{2c} \int_0^t \int_{x-c(t-\tau)}^{x+c(t-\tau)} f(\xi, \tau)\,d\xi\,d\tau. \tag{4.40}$$

This, then, is the generalization of the D'Alembert solution to the case when the wave equation contains a nonhomogeneous term $f(x, t)$.

Examination of (4.40) shows that now the solution at (x, t) depends not only on the initial conditions on the interval $x - ct \leq x \leq x + ct$ on the initial line, but also on the integral of $f(x, t)$ over the triangle with its vertices at the points $(x, t), (x - ct)$, and $(x + ct)$. Thus, when there is a nonhomogeneous term, the domain of dependence of the solution at (x, t) becomes the *entire* triangular region D, and not simply the line AB.

The influence of the nonhomogeneous term can be seen by examination of the computer plot of the solution of the equation with wave speed $c = 1$

$$\frac{\partial^2 \phi}{\partial t^2} - \frac{\partial^2 \phi}{\partial x^2} = \frac{1 + x}{1 + t^2}$$

subject to the localized Cauchy conditions

$$h(x) = \begin{cases} 0, & -\infty < x < -3\pi/2 \\ \cos x, & -3\pi/2 \leq x \leq 3\pi/2, \\ 0, & 3\pi/2 < x < \infty \end{cases} \quad k(x) = \begin{cases} 0, & -\infty < x < -\pi \\ -\sin x, & -\pi < x < \pi/2 \\ 0, & \pi/2 < x < \infty. \end{cases}$$

Setting

$$\phi_1(x, t) = \frac{1}{2}[h(x - t) + h(x + t)] \quad \text{and} \quad \phi_2(x, t) = \frac{1}{2}\int_{x-t}^{x+t} k(s)\,ds,$$

the contribution ϕ_3 made to the solution by the nonhomogeneous term is given by

$$\phi_3(x, t) = \frac{1}{2}\int_0^t \int_{x-(t-\tau)}^{x+(t-\tau)} f(\xi, \tau)\,d\xi\,d\tau$$

$$= \frac{1}{2}\int_0^t d\tau \int_{x-(t-\tau)}^{x+(t-\tau)} \frac{1 + \xi}{1 + \tau^2}\,d\xi$$

$$= -\frac{1}{2}x\ln(1 + t^2) + xt\arctan(t) - \frac{1}{2}\ln(1 + t^2) + t\arctan(t).$$

Plots of $\phi_1(x, t), \phi_2(x, t)$ are shown in Figs. 4.3a and 4.3b, while a plot of their combined effect $\phi_1(x, t) + \phi_2(x, t)$ is shown in Fig. 4.3c. A plot of the complete solution $\phi(x, t) = \phi_1(x, t) + \phi_2(x, t) + \phi_3(x, t)$ is shown in Fig. 4.3d. It is seen from this that the nonhomogeneous term has had the effect of tilting and changing the scale of the solution of the homogeneous equation shown in Fig. 4.3c.

It has been shown that the homogeneous one-dimensional constant coefficient wave equation describes the propagation of waves without change of shape, but this is a special case, since the situation becomes more complicated when wave propagation is governed by more general hyperbolic equations, like

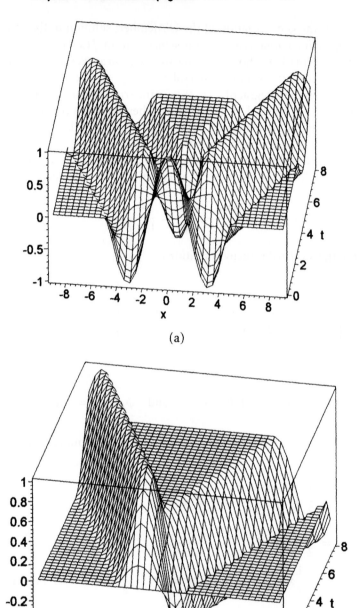

(a)

(b)

Figure 4.3   (a) A plot of $\phi_1(x, t)$. (b) A plot of $\phi_2(x, t)$. (c) A plot of $\phi_1(x, t) + \phi_2(x, t)$. (d) A plot of the complete solution $\phi(x, t) = \phi_1(x, t) + \phi_2(x, t) + \phi_3(x, t)$.

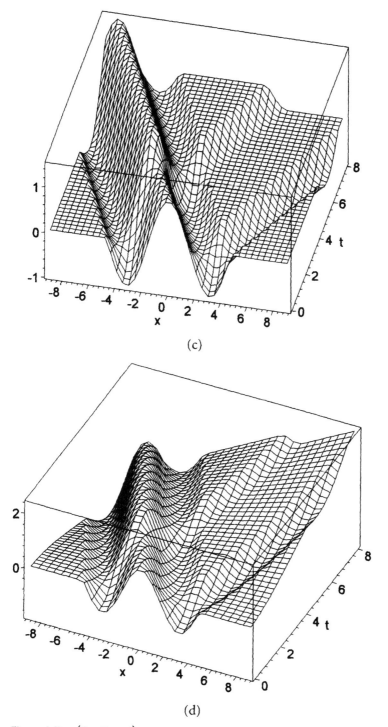

(c)

(d)

Figure 4.3   (Continued).

the **telegraph equation**

$$\frac{\partial^2 \phi}{\partial t^2} = \frac{1}{LC}\frac{\partial^2 \phi}{\partial x^2} - \left(\frac{R}{L} + \frac{G}{C}\right)\frac{\partial \phi}{\partial t} - \frac{RG}{LC}\phi, \tag{4.41}$$

which was derived in Section 1.2. For convenience, we will write a general telegraph equation in the form

$$\frac{\partial^2 \phi}{\partial t^2} = c^2\frac{\partial^2 \phi}{\partial x^2} - p\frac{\partial \phi}{\partial t} - q\phi. \tag{4.42}$$

Then, in terms of (4.41), the parameters are $c^2 = 1/(LC)$, $p = (R/L + G/C)$, and $q = RG/LC$.

If we consider a harmonic voltage wave moving to the right with speed ω/k, the exponential representation introduced in (4.23) allows us to set

$$\phi(x, t) = A\exp[i(kx - \omega t)].$$

If this harmonic wave is to satisfy (4.42), the parameters ω and k cannot be arbitrary, so to find their relationship we must substitute $\phi(x, t)$ into (4.42). The result of the substitution, after cancelation of the nonzero factor $A\exp[i(kx - \omega t)]$, is that

$$\omega^2 + ip\omega - (c^2k^2 + q) = 0. \tag{4.43}$$

Equation (4.43) is the compatibility relation between the angular frequency ω of the harmonic wave and its wavelength $\lambda = 2\pi/k$. Hence, as the propagation speed is ω/k, waves with different frequencies will propagate with different speeds, causing the initial wave profile to distort as it propagates. To examine this more closely, we resolve the wave profile at time $t = 0$ into different frequency components, each of which will propagate with a different speed. Consequently, when the frequency components are recombined to find the wave profile at a subsequent time $t > 0$, the wave profile will have changed from its initial shape. This phenomenon is common in optics, where the separation of a beam of light into its constituent colors (frequencies) after passage through a prism is called **dispersion**. This same term is used to describe the compatibility condition (4.43), which is called the **dispersion relation** for the PDE. Different PDEs describing wave propagation have different dispersion relations, and in the case of the ordinary wave equation, obtained from (4.43) by setting $p = q = 0$, we see that $\omega/k = c = $ constant. Thus, as shown previously, waves governed by this homogeneous wave equation propagate without change of shape or dissipation.

Returning to the wave propagation governed by the telegraph equation, and solving dispersion relation (4.43) for ω, gives

$$\omega = -\frac{ip}{2} \pm \frac{\sqrt{4c^2k^2 + (4q - p^2)}}{2},$$

so after substituting this result into $\phi(x,t) = A\exp[i(kx - \omega t)]$ we find that

$$\phi(x,t) = A\exp\left(-\frac{pt}{2}\right)\exp\left[i\left(kx \pm \frac{\sqrt{4c^2k^2 + (4q - p^2)}}{2}t\right)\right]. \quad (4.44)$$

In terms of wave propagation in the distributed parameter telephone line shown in Fig. 1.11 where $p = (R/L + G/C) > 0$, we see that the factor $\exp(-pt/2)$ will cause the wave to attenuate as it propagates, in addition to its being distorted due to the effect of dispersion. However, if the parameters in the line are adjusted so that $4q - p^2 = 0$, which in result (4.41) corresponds to requiring $R/L = G/C$, result (4.44) simplifies to

$$\phi(x,t) = A\exp\left(-\frac{pt}{2}\right)\exp[ik(x \pm ct)]. \quad (4.45)$$

The wave propagation is now different, because although attenuation is still present, distortion has been removed, since now all frequencies propagate at the same speed c. Waves of this type are said to be **relatively undistorted**, because although they attenuate as they propagate, they preserve their shape. This condition is used in telephone lines where the attenuation is corrected by the insertion of amplifiers at regular intervals along the line. An illustration of relatively undistorted wave propagation is shown in Fig. 4.4 where a pulse decays exponentially with time while preserving its shape.

We mention in passing that the telegraph equation arises in various other contexts as, for example, in situations that can be modeled by using a continuum

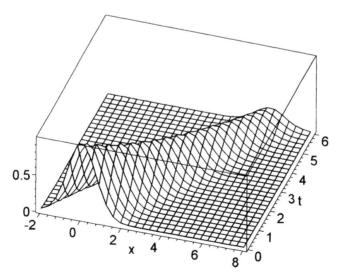

Figure 4.4   An example of relatively undistorted wave propagation.

approximation to a large system of coupled masses and springs moving in a viscous medium, and also in the theory of the diffusion of Brownian particles when the effect of particle inertia is taken into account.

EXERCISES 4.2

1. Find a solution to the nonhomogeneous one-dimensional wave equation

$$\frac{\partial^2 \phi}{\partial t^2} - c^2 \frac{\partial^2 \phi}{\partial x^2} = \sin(kx - \omega t),$$

subject to the homogeneous initial conditions $\phi(x, 0) = 0$ and $\phi_t(x, 0) = 0$. Consider case (a) $\omega / k \neq c$, and the *resonance case* (b) $\omega / k = c$. Show that in case (a) the solution comprises three sinusoidal waves that propagate with different amplitudes, and with speeds $\pm c$ and the phase speed ω / k. Show that in case (b) the solution comprises two harmonic waves that propagate with speeds $\pm c$, and another harmonic wave whose amplitude grows linearly with time.

In each of the following problems make a three-dimensional computer plot of the solution of the homogeneous one-dimensional wave equation for the stated Cauchy conditions, assuming that the wave speed $c = 1$.

2. Plot the solution for $|x| < 4, 0 < t < 3$, given that

$$h(x) = \begin{cases} 0, & -\infty < x < -1 \\ 1, & -1 < x < 1, \\ 0, & 1 < x < \infty \end{cases} \qquad g(x) = \begin{cases} 0, & -\infty < x < -1 \\ 1/2, & -1 < x < 1 \\ 0, & 1 < x < \infty. \end{cases}$$

3. Plot the solution for $|x| < 4, 0 < t < 6$, given that

$$h(x) = \begin{cases} 0, & -\infty < x < -1 \\ 1 - |x|, & -1 < x < 1, \\ 0, & 1 < x < \infty \end{cases} \qquad g(x) = \begin{cases} 0, & -\infty < x < -1 \\ \frac{1}{1+x^2} - \frac{1}{2}, & -1 < x < 1 \\ 0, & 1 < x < \infty. \end{cases}$$

4. Plot the solution for $|x| < 6, 0 < t < 6$, given that

$$h(x) = \frac{1}{1 + x^2}, \qquad g(x) = \begin{cases} 0, & -\infty < x < -1 \\ 2(1 - |x|), & -1 \le x \le 1 \\ 0, & 1 < x < \infty. \end{cases}$$

In each of the following problems make a three-dimensional computer plot of the solution of the nonhomogeneous one-dimensional wave equation for the given nonhomogeneous term $f(x, t)$ subject to the stated Cauchy conditions.

5. Plot the solution for $|x| < 5, 0 < t < 4$, given that $c = 1$,

$$h(x) = \begin{cases} 0, & -\infty < x < -1 \\ 1, & -1 < x < 1, \\ 0, & 1 < x < \infty \end{cases} \qquad k(x) \equiv 0 \text{ and } f(x, t) = \frac{1}{40} xt.$$

6. Plot the solution for $|x| < 7, 0 < t < 7$, given that $c = 2$,

$$h(x) = \begin{cases} 0, & -\infty < x < -4 \\ 1, & -4 < x < 4, \\ 0, & 4 < x < \infty \end{cases} \qquad k(x) \equiv 0 \text{ and } f(x, t) = \cos(x - 2t).$$

Comment on the result.

7. The modified telegraph equation $c^2 \phi_{xx} = \phi_{tt} + p\phi_t$ describes the situation where, provided $p > 0$, a wave propagated by the wave equation is subjected to a dissipative effect caused by the term $p\phi_t$. Show the substitution $\phi = u \exp(-pt/2)$ converts the modified telegraph equation to $c^2 u_{xx} = u_{tt} - \frac{1}{4} p^2 u$. When the term $\exp(-pt/2)$ is written $\exp(-t/t_0)$, with $t_0 = 2/p$, the quantity t_0 is called the **modulus of decay** of the wave.

Deduce that when p is sufficiently small for p^2 to be neglected, the solution of a wave moving to the right can be written as $\phi = \exp(-pt/2) f(x - ct)$, where f is any twice differentiable function of its argument. Write $f(x - ct)$ in the form $f(x - ct) = \exp[-p(x - ct)/2c]g(x - ct)$, where g is an arbitrary twice differentiable function of its argument, which is permissible because f is an arbitrary function. Substitute $f(x - ct)$ into the expression for ϕ and hence show that $\phi = \exp(-px/2c)g(x - ct)$. This expression is similar to the previous one for ϕ, although now the exponential decay factor involves x instead of t.

8. The telegraph equation can be generalized to

$$\phi_{tt} = c^2 \phi_{xx} - (a + b)\phi_t - ab\phi.$$

Show by making the substitution $\psi = \phi \exp\{\frac{1}{2}(a + b)t\}$, that ψ satisfies the equation

$$\psi_{tt} - c^2 \psi_{xx} = \frac{1}{4}(a - b)^2 \psi.$$

Use this result to deduce that relatively undistorted wave propagation is possible if $a = b$, and that a progressive wave of the form

$$u = \exp(-at) f(x \pm ct)$$

can propagate in either direction, where f is an arbitrary twice differentiable function of its argument.

4.3 Mixed Initial and Boundary Value Problems for the Wave Equation

Section 4.2 developed the D'Alembert solution $\phi(x, t)$ for the one-dimensional wave equation in an unbounded region of space $-\infty < x < \infty$, $t \geq 0$. However, this situation is restrictive, because problems arise where solutions are required in semi-infinite one-dimensional regions in space such that either $x \geq a$ or $x \leq b$, and $t \geq 0$, and also in finite regions like $a \leq x \leq b$ and $t \geq 0$, to neither of which is the D'Alembert solution directly applicable.

When problems of this type occur with the one-dimensional wave equation, in addition to specifying **initial conditions** on the part of the initial line $t = 0$ where the solution is defined, it is also necessary to specify a **boundary condition** on any space boundary involved. In the semi-infinite case $x \geq a$ and $t \geq 0$, a boundary condition must be imposed on the boundary $x = a$ together with initial conditions involving the specification of $\phi(x, 0)$ and $\phi_t(x, 0)$ on the part of the initial line $x \geq a$, $t \geq 0$. Similarly, in a finite region $a \leq x \leq b$ and $t \geq 0$, boundary conditions must be imposed on $x = a$ and $x = b$, and initial conditions involving the specification of $\phi(x, t)$ to satisfy given values of $\phi(x, 0)$ and $\phi_t(x, 0)$ on the segment of the initial line $a \leq x \leq b$, $t = 0$ must be satisfied. Problems like this are called **mixed initial and boundary value problems** for the one-dimensional wave equation, and the purpose of this section will be to determine how the D'Alembert solution for the homogeneous one-dimensional wave equation can be applied to such problems.

Two different approaches, the first of which combines the D'Alembert solution with a geometrical argument involving characteristics, while the second is a purely analytical approach, will be described. For the first method we need to consider Fig. 4.5, which shows a parallelogram-shaped region G whose sides are formed by pairs of straight line characteristics $\xi = x - ct$ and $\eta = x + ct$ of the one-dimensional wave equation

$$\frac{1}{c^2}\frac{\partial^2 \phi}{\partial t^2} = \frac{\partial^2 \phi}{\partial x^2}.$$

The center of the parallelogram is located at the point (x_0, t_0), and its corners $A, B, C,$ and D are located at the respective points $(x_0 + \xi, t_0 + \eta)$, $(x_0 + c\eta, t_0 + \xi/c)$, $(x_0 - \xi, t_0 - \eta)$, and $(x_0 - c\eta, t_0 - \xi/c)$.

Using the fact that the general solution of the one-dimensional wave equation can be written $\phi(x, t) = f(x - ct) + g(x + ct)$, and applying this result to each corner of the parallelogram followed by some elementary algebra which is left as an exercise, shows that

$$\phi(A) + \phi(C) = \phi(B) + \phi(D). \tag{4.46}$$

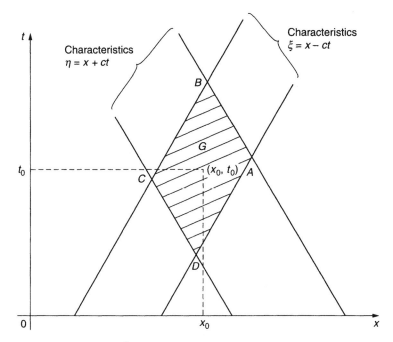

Figure 4.5   Region G bounded by characteristics of the wave equation.

This means that when the solution ϕ is known at any three corners of G, result (4.46) will determine its value at the fourth corner. To see how this enables the D'Alembert solution to determine the solution of the wave equation for $x \geq a, t \geq 0$ satisfying the boundary condition $\phi(a, t) = F(t)$ on the boundary $x = a$ we need to consider Fig. 4.6.

The D'Alembert formula will determine the solution in the region D above the initial line, and bounded at the left by the characteristic $x - ct = a$, and we will assume that on the spacelike boundary $x = a, t \geq 0$ the solution is $\phi(a, t) = F(t)$, where $F(t)$ is some known boundary condition. Next we construct the parallelogram $PQRS$ with its sides formed by characteristics such that the corner Q lies on the boundary $x = a$ at $t = \tau$, while the corners R and S lie on the boundary of D. Then the D'Alembert formula determines the solution at R and S, and the boundary condition shows the solution at Q is $F(\tau)$, so result (4.46) now determines the solution at P as

$$\phi(P) = F(\tau) + \phi(S) - \phi(R). \qquad (4.47)$$

Clearly, by choosing Q, R, and S suitably, point P can be made to lie anywhere to the right of $x = a$ and above the characteristic $x - ct = a$, so by means of this geometrical construction the solution ϕ can be found throughout $x \geq a$, $t \geq 0$, and so the mixed initial boundary value problem is solved.

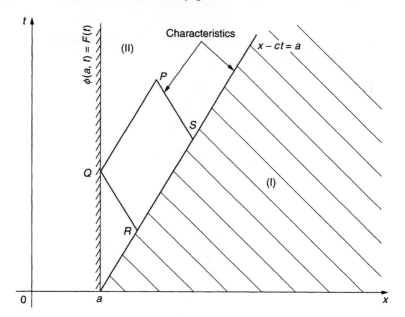

Figure 4.6   The wave equation in the semi-infinite region $x \geq a, t \geq 0$.

This approach has established that a suitable boundary condition for the homogeneous one-dimensional wave equation in a semi-infinite region is the specification of the solution ϕ on the space boundary. Other boundary conditions are also possible, as will be shown by the following considerations.

We take as our starting point the general solution of the homogeneous one-dimensional wave equation

$$\phi(x, t) = f(x - ct) + g(x + ct). \tag{4.48}$$

Our first task will be to determine the form of functions f and g for the initial boundary value problem for

$$\frac{1}{c^2} \frac{\partial^2 \phi}{\partial t^2} = \frac{\partial^2 \phi}{\partial x^2}, \tag{4.49}$$

in the semi-infinite domain $x > 0, t > 0$, subject to the initial conditions

$$\phi(x, 0) = h_1(x) \quad \text{and} \quad \phi_t(x, 0) = k_1(x), \quad \text{for } x > 0 \tag{4.50}$$

and the homogeneous boundary condition on $x = 0$,

$$\phi(0, t) = 0, \quad \text{for } t \geq 0. \tag{4.51}$$

Making the identifications $h(x) = h_1(x)$ and $k(x) = k_1(x)$, the D'Alembert formula (4.31) determines the solution $\phi(x, t)$ in region (I) of the (x, t) plane, which lies to the right of the characteristic $x - ct = 0$, such that $x > 0$,

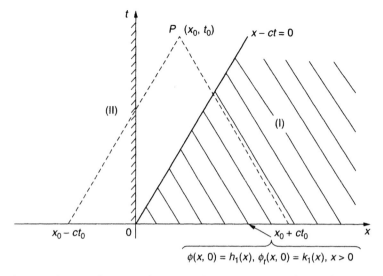

Figure 4.7   A solution in the semi-infinite region $x \geq 0$, $t \geq 0$.

$t > 0$, as shown in Fig. 4.7. To determine $\phi(x, t)$ in region (II) to the left of the characteristic $x - ct = 0$, such that $x > 0$, $t > 0$ as shown in Fig. 4.7, it is necessary to know $h(x)$ and $k(x)$ for $x < 0$. These initial data can be extended as follows. Setting $x = 0$ in the general solution $\phi(x, t) = f(x - ct) + g(x + ct)$ and using the boundary condition (4.51) we find that $0 = f(-ct) + g(ct)$. As ct is the argument of the functions f and g of a single real variable, this result simply asserts that

$$f(-x) = -g(x). \tag{4.52}$$

Expressing results (4.29) and (4.30) in terms of $h_1(x)$ and $k_1(x)$, using (4.52), and setting $a = 0$ gives

$$f(x) = \frac{1}{2}h_1(x) - \frac{1}{2c}\int_0^x k_1(s)\,ds - \frac{1}{2}[g(0) - f(0)] \tag{4.53}$$

and

$$g(x) = \frac{1}{2}h_1(x) + \frac{1}{2c}\int_0^x k_1(s)\,ds + \frac{1}{2}[g(0) + f(0)]. \tag{4.54}$$

If we now replace x by $-x$ in (4.53) we find that

$$f(-x) = \frac{1}{2}h_1(-x) - \frac{1}{2c}\int_0^{-x} k_1(s)\,ds - \frac{1}{2}[g(0) - f(0)]. \tag{4.55}$$

However, setting the dummy variable $s = -u$ in the integral, and then replacing u by s, shows that (4.50) is equivalent to

$$f(-x) = \frac{1}{2}h_1(-x) + \frac{1}{2c}\int_0^x k_1(-s)ds - \frac{1}{2}[g(0) - f(0)], \quad (4.56)$$

so expressions (4.54) and (4.56) will only satisfy (4.52) if

$$h_1(-x) = -h_1(x) \quad \text{and} \quad k_1(-x) = -k_1(x). \quad (4.57)$$

Thus, for $\phi(x, t)$ to be defined by the D'Alembert formula in region (II), it is necessary to extend the functions h_1 and k_1 for negative x by requiring that they are both *odd functions*. This has shown that the solution of the mixed initial boundary value problem (4.49), (4.50), and (4.51) for $x \geq 0$, $t = 0$ coincides with the solution provided by the D'Alembert formula subject to the *pure initial value problem*

$$\phi(x, 0) = h(x) \quad \text{and} \quad \phi_t(x, 0) = k(x), \quad (4.58)$$

in the semi-infinite region $t \geq 0$, provided

$$h(x) = \begin{cases} h_1(x), & x \geq 0 \\ -h_1(x), & x < 0 \end{cases} \quad \text{and} \quad k(x) = \begin{cases} k_1(x), & x \geq 0 \\ -k_1(x), & x < 0. \end{cases} \quad (4.59)$$

This purely mathematical solution can be interpreted physically by considering the boundary $x = 0$ to act as a **reflecting barrier**. In this interpretation the barrier has the property that when it is reached by a wave moving to the left, the wave experiences a change of sign before being reflected back and moving to the right.

The specification of ϕ on a boundary $x = $ constant is called a **fixed boundary condition**, as opposed to a **free boundary condition** where $\partial\phi/\partial x$ is specified on the boundary $x = $ constant.

The effect of a reflecting barrier is best seen by considering the wave solution for a problem with a reflecting barrier at $x = 0$ in which the wave speed $c = 1$, and the initial conditions are

$$h_1(x) = \begin{cases} 0, & 0 \leq x < 2\pi \\ \sin(x + 2\pi), & 2\pi \leq x \leq 3\pi \\ 0, & x > 3\pi \end{cases} \quad \text{and} \quad k_1(x) \equiv 0 \quad \text{for } x \geq 0,$$

so ϕ is a localized sine pulse, and $\phi_t(x, 0)$ is identically zero. A computer plot of the solution of this simple problem is given in Fig. 4.8. This shows the resolution of the initial pulse into two identically shaped pulses, each with half of the initial amplitude, one of which propagates to the right and the other to the left, both with speed $c = 1$. The reflection of the leftward moving pulse at the boundary $x = 0$ is seen to be accompanied by a reversal of sign, and after

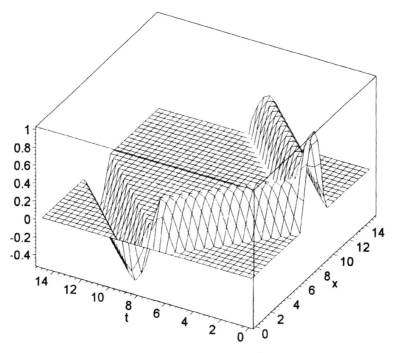

Figure 4.8   Reflection at a fixed boundary $x = 0$.

the interaction close to the boundary is completed the wave then moves to the right with speed $c = 1$.

The method just used to express the solution $\phi(x, t)$ in terms of the D'Alembert formula when a fixed boundary condition is imposed can also be used to find the solution when a free boundary condition is imposed. This leads to the solution of the mixed initial and boundary value problem

$$\frac{\partial^2 \phi}{\partial t^2} - c^2 \frac{\partial^2 \phi}{\partial x^2} = 0, \tag{4.60}$$

subject to the initial conditions

$$\phi(x, 0) = h_1(x) \quad \text{and} \quad \phi_t(x, 0) = k_1(x) \quad \text{for } x \geq 0, \tag{4.61}$$

and the homogeneous free boundary condition

$$\phi_x(0, t) = 0. \tag{4.62}$$

It is easily shown that the purely mathematical solution for this problem is obtained by using the D'Alembert solution for the pure initial value problem in which $h_1(x)$ and $k_1(x)$ are extended as *even functions* for $x < 0$, and then restricting the solution to the semi-infinite region $x \geq 0$, $t \geq 0$. In this case the free boundary condition at $x = 0$ can again be interpreted physically as a

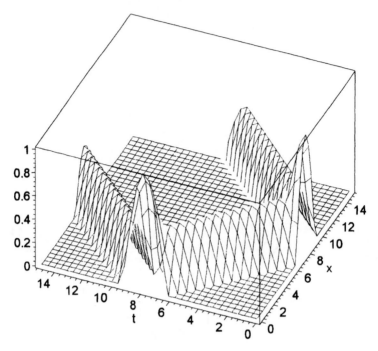

Figure 4.9   Reflection at a free boundary $x = 0$.

reflecting barrier, although this time with the property that a wave moving to the left is reflected back from the barrier *without* a change of sign. The proof of this result is left as an exercise. A computer plot of the solution for a wave problem using the same initial boundary conditions as before with $c = 1$, but using the free boundary condition $\phi_x(0, t) = 0$ is shown in Fig. 4.9. This plot shows the reflection of the leftward-moving wave at the boundary $x = 0$ where the wave experiences *no* change of sign.

In many physical applications a different type of problem arises where a wave passes through a medium comprising two different layers, in one of which the wave speed is c_1 while in the other it is c_2. A typical case occurs when a wave moves through a composite metal rod in the positive x direction, where for $x < a$ the metal has density ρ_1, and for $x > a$ it has density ρ_2, with $\rho_1 \neq \rho_2$. In this case the respective wave speeds c_1 and c_2 to the left and right of $x = a$ are different, because they are proportional to $1/\sqrt{\rho_1}$ and $1/\sqrt{\rho_2}$. In what follows, without loss of generality we will set $a = 0$.

Let us consider the effect on a wave moving with speed c_1 in the positive sense along the x axis, when at $x = 0$ it enters a medium in which the wave speed is c_2. The wave with speed c_1, which moves to the right when $x < 0$, will be called the **incident wave,** and the wave that moves to the right through the region $x > 0$ with speed c_2 will be called the **transmitted wave.** It will also be necessary to allow for the possibility that the change of wave speed across $x = 0$ will cause some part of the incident wave to be reflected back into the region $x < 0$, where it will then move to the left along the x axis with speed c_1. The solution for $x > 0$ will only involve the transmitted wave, but because the wave equation is linear, the solution for $x < 0$ will be the sum of the incident and reflected waves.

If the incident wave is denoted by ϕ_I, the transmitted wave by ϕ_T, and the reflected wave by ϕ_R, we can write

$$\phi_I = f_I(t - x/c_1), \quad \phi_T = f_T(t - x/c_2), \quad \text{and} \quad \phi_R = g_R(t + x/c_1), \quad (4.63)$$

where the incident wave ϕ_I can be considered to be known, while the transmitted wave ϕ_T and the reflected wave ϕ_R must be determined in terms of ϕ_I.

For $x < 0$ we have

$$\phi(x, t) = f_I(t - x/c_1) + g_R(t + x/c_1), \quad (4.64)$$

while for $x > 0$ we have

$$\phi(x, t) = f_T(t - x/c_2). \quad (4.65)$$

We now consider the problem

$$\frac{\partial^2 \phi}{\partial t^2} - c^2 \frac{\partial^2 \phi}{\partial x^2} = 0, \quad \text{where } c = c_1 \quad \text{for } x < 0 \quad \text{and} \quad c = c_2 \quad \text{for } x > 0$$

subject to the initial conditions

$$\phi(x, 0) = 0 \quad \text{and} \quad \phi_t(x, 0) = 0 \quad \text{for } x > 0,$$
$$\phi(x, 0) = f_I(-x/c_1) \quad \text{and} \quad \phi_t(x, 0) = f_I'(-x/c_1) \quad \text{for } x < 0$$
$$f_I(u) = 0 \quad \text{for } u < 0.$$

At the interface $x = 0$ across which the wave speed changes, both ϕ and ϕ_x must be continuous. These conditions can be understood by considering the case of a wave propagating through a composite string with line density ρ_1 for $x < 0$ and ρ_2 for $x > 0$. The string is joined across $x = 0$, so one condition is that the wave displacements to the immediate left and right of $x = 0$ must be equal. The other condition follows from the fact that if ϕ_x were to be discontinuous across $x = 0$, a finite force would act on an arbitrarily small element of the string located at $x = 0$, causing a nonphysical infinite acceleration.

Using (4.64) and (4.65), the continuity of ϕ across $x = 0$ gives

$$f_I(t) + g_R(t) = f_T(t), \quad (4.66)$$

while the continuity of ϕ_x across $x = 0$ gives

$$-\frac{1}{c_1} f_I'(t) + \frac{1}{c_1} g_R'(t) = -\frac{1}{c_2} f_T'(t). \tag{4.67}$$

Differentiating (4.66) with respect to t, and using the result with (4.67) to solve for $f_T'(t)$ and $g_R'(t)$ in terms of $f_I'(t)$, gives

$$f_T'(t) = \left(\frac{2c_2}{c_1 + c_2}\right) f_I'(t) \quad \text{and} \quad g_R'(t) = \left(\frac{c_2 - c_1}{c_1 + c_2}\right) f_I'(t). \tag{4.68}$$

At $x = 0$, $t = 0$ there is no transmitted wave, so $f_T(0) = 0$, and as ϕ is continuous across $x = 0$ and at $t = 0$ the reflected wave has not yet been formed so $g_R(0) = 0$. Replacing the variable t by τ in results (4.68) and integrating then from $\tau = 0$ to $\tau = t$ shows that

$$f_T(t) = \left(\frac{2c_2}{c_1 + c_2}\right) f_I(t) \quad \text{and} \quad g_R(t) = \left(\frac{c_2 - c_1}{c_1 + c_2}\right) f_I(t). \tag{4.69}$$

It can be seen from (4.69) that when c_2/c_1 is large, the amplitude of the transmitted wave is almost twice that of the incident wave, while the amplitude of the reflected wave is almost equal to that of the incident wave. If, however, c_2/c_1 is small, then almost all of the incident wave is reflected, though with a change of sign. Finally, if $c_2/c_1 = 1$, so no change of wave speed occurs across $x = 0$ then, as would be expected, the incident wave passes through $x = 0$ unchanged, and no reflection takes place.

The numbers

$$T = \frac{2c_2}{c_1 + c_2} \quad \text{and} \quad R = \frac{c_2 - c_1}{c_1 + c_2}, \tag{4.70}$$

which are defined at $x = 0$, describe, respectively, the amplitude of the transmitted wave and the reflected wave relative to the amplitude of the incident wave.

In summary:

(i) When a wave with speed c_1 advances into a region with wave speed c_2, where $c_2 > c_1$, the amplitude of the transmitted wave is increased and neither the transmitted wave nor the reflected wave experiences a change of sign.

(ii) When a wave with speed c_1 advances into a region with wave speed c_2, where $c_2 < c_1$, the transmitted wave preserves its sign while the reflected wave experiences a change of sign.

These results show that if $c_2 = 0$ no wave will be transmitted beyond $x = 0$, and the incident wave will undergo a perfect reflection, though with a change of sign. This corresponds to the case where $x = 0$ is a fixed boundary on which $\phi = 0$, and apart from the direction of wave propagation, it describes the situation in Fig. 4.8.

EXERCISES 4.3

1. Let A, B, C, and D be the corners of a parallelogram in the (x, t) plane, with its sides formed by straight line segments from the two families of characteristics of the wave equation $\phi_{tt} - c^2\phi_{xx} = 0$. By using the general solution of the wave equation $\phi(x, t) = f(x - ct) + g(x + ct)$, prove result (4.46) that $\phi(A) + \phi(C) = \phi(B) + \phi(D)$, where $\phi(\alpha)$ is the value of ϕ at the point α in the (x, t) plane.

2. Use the D'Alembert solution together with the result of Exercise 1 to show how the solution of the wave equation $\phi_{tt} - c^2\phi_{xx} = 0$ in the semi-infinite strip $0 \leq x \leq a, t > 0$ can be constructed, given that $\phi(x, 0) = h(x), \phi_t(x, 0) = k(x)$ for $0 \leq x \leq a$, with $h(0) = h(a) = k(0) = k(a) = 0$ and $\phi(0, t) = \phi(a, t) = 0$.

3. Let the wave equation $\phi_{tt} - c^2\phi_{xx} = 0$ be defined for $x > 0, t > 0$ subject to the initial conditions $\phi(x, 0) = h_1(x), \phi_t(x, 0) = k_1(x)$ for $0 \leq x < \infty$, and the free boundary condition $\phi_x(0, t) = 0$ at $x = 0$. Show the solution can be obtained by extending $h_1(x)$ and $k_1(x)$ as even functions to $x < 0$, and then restricting the solution to $x > 0, t > 0$.

4. Derive results (4.70) by considering a wave $y_{\text{incident}} = A_1 \exp 2\pi i(nt - k_1 x)$ to be incident from the left of a discontinuity in line density located at $x = 0$, a wave $y_{\text{reflected}} = A_R \exp 2\pi i(nt + k_1 x)$ reflected back from the discontinuity, and a wave $y_{\text{transmitted}} = A_T \exp 2\pi i(nt - k_2 x)$ transmitted beyond it to the right.

4.4 The Poisson Formula for the Wave Equation, the Method of Descent, and the Difference between Waves in Two and Three Space Dimensions

In previous sections the solution of the homogeneous wave equation $\phi_{tt} = c^2\phi_{xx}$ in one space dimension and time was examined. It is now necessary to consider corresponding solutions of the homogeneous wave equation when two and three space dimensions are involved, from which it will become apparent that there is a fundamental difference between solutions in even and odd numbers of space dimensions.

We first consider the wave equation in *three* space dimensions and time,

$$\frac{\partial^2 \phi}{\partial t^2} = c^2 \left(\frac{\partial^2 \phi}{\partial x^2} + \frac{\partial^2 \phi}{\partial y^2} + \frac{\partial^2 \phi}{\partial z^2} \right), \quad \text{with } c = \text{constant}, \qquad (4.71)$$

subject to the Cauchy conditions

$$\phi(P, 0) = \phi_0(P) \quad \text{and} \quad \phi_t(P, 0) = \phi_1(P), \qquad (4.72)$$

where P is a general point in space. The functions ϕ_0 and ϕ_1 will be assumed either to vanish outside a finite region (to have compact support) or to be differentiable as many times as necessary. If the point P has coordinates (x, y, z) it is usual to say the solution $\phi(x, y, z, t)$, or more simply $\phi(P, t)$, is in the four-dimensional space $R^3 \times T$, where R denotes a space dimension and T is the time.

Our approach will involve using what are called *spherical means,* and instead of working with the solution $\phi(P, t)$ we will use its average over a sphere S_P^ρ of radius ρ with its center at P. The **spherical mean** of $\phi(P, t)$ at P, written $\bar{\phi}(\rho, t)$, is defined as

$$\bar{\phi}(\rho, t) = \frac{1}{4\pi\rho^2} \int_{S_P^\rho} \phi(Q, t) dS_Q, \qquad (4.73)$$

where Q is the dummy variable of integration and dS_Q is an element of area of the spherical surface S_Q^ρ.

This result can be expressed more conveniently by using the concept of a *solid angle,* which is defined as follows. The solid angle ω subtended at a point O by a surface σ is defined as the area on the surface of a unit sphere centered on O enclosed by the curve traced out on its surface by the intersection of radial lines drawn from O to all points on the boundary of σ, as shown in Fig. 4.10.

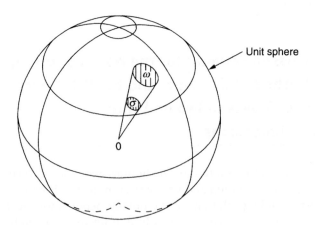

Figure 4.10   The geometrical interpretation of the solid angle of area σ at O.

Solid angles are measured in **steradians,** so the solid angle of any closed surface surrounding P must be 4π, the area of a unit spherical surface.

Letting $\rho \to 0$, it follows from the mean value theorem that $\bar{\phi}(0, t) = \phi(P, t)$, showing that the solution $\phi(P, t)$ at a general point P at time t will be known if an expression for the spherical mean $\bar{\phi}(\rho, t)$ at P can be found. To accomplish this we start by showing that $\bar{\Delta\phi} = \Delta_\rho\bar{\phi}$. In this result the Laplacian $\bar{\Delta\phi}$ is understood to be evaluated with respect to the coordinates (x, y, z) of ϕ at the point P, while in the expression on the right the use of spherical means causes the space dependence of $\bar{\phi}$ to be only through the single radial coordinate ρ, so that

$$\Delta_\rho\bar{\phi} = \frac{1}{\rho^2}\frac{\partial}{\partial\rho}\left(\rho^2\frac{\partial\bar{\phi}}{\partial\rho}\right).$$

We now use the Gauss divergence theorem

$$\iiint_V \operatorname{div}\mathbf{q}\,dV = \iint_S \mathbf{q}\cdot\mathbf{n}\,dS,$$

where $\mathbf{q}$ is a continuous vector function with continuous first-order partial derivatives inside a volume V bounded by a surface S, and $\mathbf{n}$ is the unit normal to S pointing *out* of V. Setting $\mathbf{q} = \operatorname{grad}\phi$ and identifying the volume V with the volume V_P^ρ contained within the surface S_P^ρ, and recognizing that the derivative $\partial\phi/\partial n$ normal to S_P^ρ is then $\partial\phi/\partial\rho$, it follows that

$$\iiint_{V_P^\rho}\Delta\phi\,dV = \iint_{S_P^\rho}\frac{\partial\phi}{\partial n}\,dS = \iint_{S_P^\rho}\frac{\partial\phi}{\partial\rho}\,dS = \rho^2\iint_{S_P^\rho}\frac{\partial\phi}{\partial\rho}\,d\omega$$

$$= \rho^2\frac{\partial}{\partial\rho}\iint_{S_P^\rho}\phi\,d\omega.$$

Using (4.73) to rewrite the integral on the right converts this last result to

$$\iiint_{V_P^\rho}\Delta\phi\,dV = 4\pi\rho^2(\partial\bar{\phi}/\partial\rho).$$

However,

$$\iiint_{V_P^\rho}\Delta\phi\,dV = \int_0^\rho\left(\iint_{S_r^\zeta}\Delta\phi\,dS\right)d\zeta = \int_0^\rho 4\pi\zeta^2(\bar{\Delta\phi})d\zeta,$$

where ζ is a dummy variable of integration, so

$$\int_0^\rho \zeta^2(\bar{\Delta\phi})d\zeta = \rho^2(\partial\bar{\phi}/\partial\rho).$$

Differentiation of this result with respect to ρ establishes the required property, because it yields

$$\bar{\Delta}\phi = \frac{1}{\rho^2}\frac{\partial}{\partial\rho}\left(\rho^2\frac{\partial\bar{\phi}}{\partial\rho}\right) = \Delta_\rho\bar{\phi}.$$

Provided the solution of the Cauchy problem (4.71) and (4.72) exists, forming the spherical mean of (4.71) over the sphere S_P^ρ and using the last result gives $c^2\bar{\Delta}\phi = \bar{\phi}_{tt}$. In terms of the single radial spherical coordinate ρ this becomes

$$\frac{c^2}{\rho^2}\frac{\partial}{\partial\rho}\left(\rho^2\frac{\partial\bar{\phi}}{\partial\rho}\right) = \frac{\partial^2\bar{\phi}}{\partial t^2},$$

from which it follows that $\bar{\phi}$ satisfies the equation

$$c^2\left(\rho\frac{\partial^2\bar{\phi}}{\partial\rho^2} + 2\frac{\partial\bar{\phi}}{\partial\rho}\right) = \rho\frac{\partial^2\bar{\phi}}{\partial t^2}.$$

Finally, by setting $\Phi = \rho\bar{\phi}$, the function Φ becomes the solution of the one-dimensional wave equation $c^2\Phi_{\rho\rho} = \Phi_{tt}$.

Taking the spherical means of the Cauchy data in (4.72) gives

$$\bar{\phi}(\rho,0) = \bar{\phi}_0(\rho) = \frac{1}{4\pi\rho^2}\iint_{S_r^\rho}\phi_0(P)dS$$

and

$$\bar{\phi}_t(\rho,0) = \bar{\phi}_1(\rho) = \frac{1}{4\pi\rho^2}\iint_{S_P^\rho}\phi_1(P)dS. \qquad (4.74)$$

Writing $\varphi_0(\rho) = \rho\bar{\phi}_0(\rho)$ and $\varphi_1(\rho) = \rho\bar{\phi}_1(\rho)$ shows that $\Phi(\rho,t)$ must be such that

$$\Phi(\rho,0) = \varphi_0(\rho), \Phi_t(\rho,0) = \varphi_1(\rho,0) \quad \text{and} \quad \Phi(0,t) = 0, \quad \text{with } 0 \le \rho < \infty.$$

Hence to find $\Phi(\rho,t)$ it is necessary to solve the following initial value problem on the real half-line $0 \le \rho < \infty$

$$c^2\Phi_{\rho\rho} = \Phi_{tt} \quad \text{with } \Phi(\rho,0) = \varphi_0(\rho), \Phi_t(\rho,0) = \varphi_1(\rho,0) \quad \text{and} \quad \Phi(0,t) = 0.$$

This problem was solved in Section 4.3, where it was shown that the D'Alembert formula can be used to find the solution. To do this it is necessary to extend the functions $\varphi_0(\rho)$ and $\varphi_1(\rho)$ as *odd* functions onto the entire initial line $-\infty < \rho < \infty$, and then to confine attention to the solution in the interval $0 \le \rho < \infty$. Denoting the odd extensions of $\varphi_0(\rho)$ and $\varphi_1(\rho)$ by $\psi_0(\rho)$ and $\psi_1(\rho)$, and recalling that $\Phi = \rho\bar{\phi}$, it follows from the D'Alembert formula that

$$\bar{\phi}(\rho, t) = \Phi/\rho = \frac{\psi_0(\rho - ct) + \psi_0(\rho + ct)}{2\rho} + \frac{1}{2c\rho} \int_{\rho-ct}^{\rho+ct} \psi_1(s) ds.$$

The function $\bar{\phi}(0, t)$ needed to determine $\phi(P, t)$ is an indeterminate form, so applying L'Hôpital's rule and using the definitions of ψ_0 and ψ_1 we obtain

$$\bar{\phi}(0, t) = \frac{1}{2}[\bar{\phi}_0(-ct) + \bar{\phi}_0(ct) + ct\bar{\phi}'_0(ct) - ct\bar{\phi}'_0(-ct)]$$
$$+ \frac{1}{2c}[ct\bar{\phi}_1(-ct) + ct\bar{\phi}'_1(ct)].$$

As $\bar{\phi}_0(x)$ and $\bar{\phi}_1(x)$ are even functions of x, and $\bar{\phi}'_0(x)$ is an odd function, it follows directly that

$$\bar{\phi}(0, t) = \bar{\phi}_0(ct) + ct\bar{\phi}'_0 t + t\bar{\phi}'_1(ct) = \frac{d}{dt}\{t\bar{\phi}_0(ct)\} + t\bar{\phi}_1(ct). \quad (4.75)$$

Using results (4.74) in (4.75) we arrive at the **Poisson's formula** for the solution of the Cauchy problem (4.71) and (4.72)

$$\phi(P, t) = \frac{1}{4\pi c^2} \frac{\partial}{\partial t} \left[\frac{1}{t} \int\int_{S_P^{ct}} \phi_0(P) dS \right] + \frac{1}{4\pi c^2 t} \int\int_{S_P^{ct}} \phi_1(P) dS. \quad (4.76)$$

This formula has important implications for wave propagation in three space dimensions and time, because it shows the solution at a point P at time t is completely determined by t and the values of $\phi_0(P)$ and $\phi_1(P)$ on a sphere of radius ct with its center at P. The physical interpretation of this result can be understood by considering an initial disturbance like a sound wave that originates from a volume V bounded by a surface Σ.

In Fig. 4.11 let Q be a fixed point in space outside the surface Σ, with d_1 being the distance of the point on Σ nearest to Q and d_2 the distance of the point on Σ furthest from Q.

The Poisson formula (4.76) shows that for a spherical integration surface centered on Q with radius D less than d_1, both integrals in (4.76) are zero, because in this case the surface over which integration is to be performed does not intersect Σ and so $\phi(Q, t) \equiv 0$.

When the radius D of the spherical integration surface is such that $d_1 < D < d_2$, the surfaces over which the integrals in (4.76) are to be evaluated intersect Σ so then $\phi(Q, t) \neq 0$. If, however, $D > d_2$, the spherical surfaces over which the integrals in (4.76) are to be evaluated no longer intersect Σ, so that once again $\phi(Q, t) \equiv 0$.

Another way of explaining the consequences of the Poisson formula is by saying that waves in three space dimensions and time, which over a finite

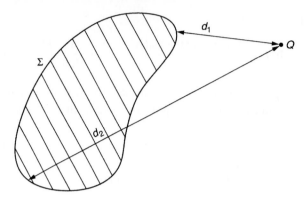

Figure 4.11   Waves originating from inside surface Σ received at point Q.

interval of time originate from a localized source in space, have sharp leading and trailing edges. This explains why an observer distant from a localized sound source of finite duration first hears nothing and then, when the leading edge of the wave arrives suddenly hears the sound and, finally, after the trailing edge of the wave has passed, there is again silence. This sudden "switching on" of a disturbance governed by the wave equation in $R^3 \times T$, followed by its equally sudden "switching off," is known as **Huygens' principle**, after the Dutch physicist who in the 17th century was the first to postulate this property in connection with his study of optics.

Sometimes it is necessary to solve the nonhomogeneous wave equation in $R^3 \times T$

$$\frac{\partial^2 \phi}{\partial t^2} + f(P, t) = c^2 \left(\frac{\partial^2 \phi}{\partial x^2} + \frac{\partial^2 \phi}{\partial y^2} + \frac{\partial^2 \phi}{\partial z^2} \right), \tag{4.77}$$

subject to the Cauchy conditions

$$\phi(P, 0) = \phi_0(P) \quad \text{and} \quad \phi_t(P, 0) = \phi_1(P), \tag{4.78}$$

where P is an arbitrary point in space. This can be accomplished by using the linearity of the wave equation to separate the solution into two distinct parts by setting $\phi = u + v$, where u is the solution of the homogeneous wave equation

$$\frac{\partial^2 u}{\partial t^2} = c^2 \left(\frac{\partial^2 u}{\partial x^2} + \frac{\partial^2 u}{\partial y^2} + \frac{\partial^2 u}{\partial z^2} \right) \tag{4.79}$$

subject to the Cauchy conditions

$$u(P, 0) = \phi_0(P) \quad \text{and} \quad u_t(P, 0) = \phi_1(P), \tag{4.80}$$

and v is the solution of the nonhomogeneous wave equation

$$\frac{\partial^2 v}{\partial t^2} + f(P, t) = c^2 \left(\frac{\partial^2 v}{\partial x^2} + \frac{\partial^2 v}{\partial y^2} + \frac{\partial^2 v}{\partial z^2} \right) \tag{4.81}$$

subject to the homogeneous Cauchy conditions

$$v(P, 0) = 0 \quad \text{and} \quad v_t(P, 0) = 0. \tag{4.82}$$

The solution of problem (4.79) and (4.80) is given by the Poisson formula (4.76), so it only remains to solve problem (4.81) and (4.82). Let $V(P, t, \tau)$ be the solution of (4.79), which satisfies the initial conditions $V|_{t=\tau} = 0$ and $V_t|_{t=\tau} = f(P, \tau)$ at the time $t = \tau$. The solution of problem (4.81) and (4.82) must then be of the form

$$v(P, t) = \int_0^t V(P, t, \tau) d\tau,$$

where from the Poisson formula

$$V(P, t, \tau) = \frac{1}{4\pi c} \iint_{S_P^{c(t-\tau)}} \frac{f(Q, \tau)}{c(t - \tau)} dS,$$

with Q being the dummy variable of integration, and so

$$v(P, t) = \frac{1}{4\pi c} \int_0^t \left[\iint_{S_P^{c(t-\tau)}} \frac{f(Q, \tau)}{c(t - \tau)} dS \right] d\tau.$$

Setting $x = c(t - \tau)$, with $\xi = |P - Q|$ being the distance between point P and the dummy variable of integration Q, this becomes

$$v(P, t) = \frac{1}{4\pi c^2} \int_0^{ct} \left[\iint_{S_P^x} \frac{f(Q, t - \xi/c)}{\xi} dS \right] d\xi.$$

However, this integral can be written as an integral over the volume V_P^{ct} interior to S_P^{ct}, when it becomes

$$v(P, t) = \frac{1}{4\pi c^2} \iiint_{V_P^{ct}} \frac{f(Q, t - \xi/c)}{\xi} dV. \tag{4.83}$$

Thus the solution $\phi(P, t)$ of the nonhomogeneous wave equation (4.77) subject to the Cauchy conditions (4.78) is thus $\phi(P, t) = u(P, t) + v(P, t)$, where $u(P, t)$ is given by the Poisson formula (4.76), and $v(P, t)$ by (4.83).

Note that in the nonhomogeneous case, while $u(P, t)$ depends only on the values of $\phi_0(\mathbf{r})$ and $\phi_1(\mathbf{r})$ on the spherical surface S_P^{ct}, the solution $v(P, t)$ depends on the values of $v_0(\mathbf{r})$ and $v_1(\mathbf{r})$ throughout the volume V_P^{ct}. Furthermore, in the integrand of the integral determining $v(P, t)$, the function

f involves a *delayed* time $t - \xi/a$ called the **retarded time**. Physically, the retarded time represents the time at which a disturbance moving with speed c must start from its point of origin if it is to arrive at the point P at time t.

We now use the Poisson formula (4.76) to examine the solution of the two-dimensional wave equation

$$c^2\left(\frac{\partial^2 \phi}{\partial x^2} + \frac{\partial^2 \phi}{\partial y^2}\right) = \frac{\partial^2 \phi}{\partial t^2}, \quad \text{with } c = \text{constant}, \qquad (4.84)$$

subject to the Cauchy conditions

$$\phi(x, y, 0) = \phi_0(x, y) \quad \text{and} \quad \phi_t(x, y, 0) = \phi_1(x, y). \qquad (4.85)$$

The method we will use is called the **method of descent**, and it is due to the French mathematician Jacque Hadamard (1865–1963). In essence this method involves using the Poisson formula (4.76) in which the initial data are functions of three independent space variables, but with the third variable z missing. Hence the term *descent* refers to descending from three to two space variables in order to solve the wave propagation problem in $R^2 \times T$.

As the initial conditions in (4.85) are independent of z, and P now lies in the plane $z = 0$, the surface integrals over S_P^{ct} in the Poisson formula (4.76) reduce to integrals over the area $\tilde{S}_P^{ct}$ inside the circle C_P^{ct} formed by the intersection of the surface S_P^{ct} and the plane $z = 0$, as shown in Fig. 4.12.

The spherical surface S_P^{ct} is divided into two parts by the plane $z = 0$, comprising an upper hemisphere U_P^{ct} and a lower hemisphere L_P^{ct}. Evaluating the first integral in (4.76) over the upper hemisphere gives

$$\iint_{U_P^{ct}} \frac{\phi_0(Q)}{ct}\, dS = \iint_{S_P^{ct}} \frac{\phi_0(Q)}{ct \cos \gamma}\, dS,$$

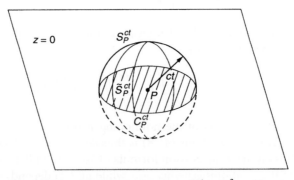

Figure 4.12   The spherical surface S_P^{ct} in $R^3 \times T$ and the area $\tilde{S}_P^{ct}$ in $R^2 \times T$.

where γ is the angle between the z axis and the surface of the sphere S_P^{ct} at the point T shown in Fig. 4.12. Consequently, the element of area $d\tilde{S}$ in the plane $z = 0$ in Fig. 4.12 is the projection onto the $z = 0$ plane of the area dS. Inspection of Fig. 4.12 shows that if T_1 is the projection of T onto the plane $z = 0$, and it has the coordinates (ξ, η), then

$$\cos \gamma = \frac{|TT_1|}{|PT_1|} = \frac{\sqrt{c^2t^2 - (x - \xi)^2 - (y - \eta)^2}}{ct}.$$

Substituting for $\cos \gamma$ in the integral over $U_{S_P^{ct}}$ gives

$$\iint_{U_P^{ct}} \frac{\phi_0(Q)}{ct} \, d\tilde{S} = \iint_{S_P^{ct}} \frac{\phi_0(\xi, \eta)}{\sqrt{c^2t^2 - (x - \xi)^2 - (y - \eta)^2}} \, d\xi \, d\eta,$$

while a similar argument involving the integral over the lower hemisphere gives

$$\iint_{L_P^{ct}} \frac{\phi_0(Q)}{ct} \, d\tilde{S} = \iint_{S_P^{ct}} \frac{\phi_0(\xi, \eta)}{\sqrt{c^2t^2 - (x - \xi)^2 - (y - \eta)^2}} \, d\xi \, d\eta.$$

Using these results in the first integral in (4.76), with corresponding results in the second integral, shows the solution of the Cauchy problem (4.84) and (4.85) in $R^2 \times T$ is given by

$$\phi(x, y, t) = \frac{1}{2\pi c} \frac{\partial}{\partial t} \iint_{S_P^{ct}} \frac{\phi_0(\xi, \eta)}{\sqrt{c^2t^2 - (x - \xi)^2 - (y - \eta)^2}} \, d\xi \, d\eta$$
$$+ \frac{1}{2\pi c} \iint_{S_P^{ct}} \frac{\phi_1(\xi, \eta)}{\sqrt{c^2t^2 - (x - \xi)^2 + (y - \eta)^2}} \, d\xi \, d\eta. \quad (4.86)$$

The solution described by (4.86) in two space dimensions and time exhibits a striking difference from that in three space dimensions and time described by (4.76). In this case result (4.86) shows that when a disturbance of finite duration is produced in a finite area A, an observer outside A will experience the sudden "switching on" of a disturbance when the leading edge of the wave arrives, but because of the presence of the time t in the denominator of the integrands there will be no sudden "switching off" of the disturbance, and instead it will decay slowly as time increases. Thus *Huygens' principle*, which applies to wave propagation in three space dimensions and time, is *not* valid in two space dimensions and time. In general it can be shown that if the number of space dimensions is n, then Huygens' principle applies to solutions of the wave equation in an odd number of dimensions with $n = 3, 5, \ldots$, but not to solutions in any even number of dimensions $n = 2, 4, \ldots$. The solution of the nonhomogeneous wave equation in $R^2 \times T$ can also be obtained by using the method of descent, although the details are left as an exercise.

When the initial conditions are sufficiently simple for the integrals involved to be evaluated analytically, the results of this section can be used to find solutions of the wave equation in either $R^2 \times T$ or $R^3 \times T$. However, the main significance of these results is the information they provide about the nature of solutions of the wave equation in two and three space dimensions.

EXERCISES 4.4

1. Consider two solutions $\phi(x, y, z, t)$ and $\psi(x, y, z, t)$ of the three-dimensional wave equation $c^2 \Delta_3 w = w_{tt}$ corresponding, respectively, to the two pairs of initial conditions $\phi(x, y, z, 0) = \phi_0(x, y, z), \phi_t(x, y, z, 0) = \phi_1(x, y, z)$, and $\psi(x, y, z, 0) = \psi_0(x, y, z), \psi_t(x, y, z, 0) = \psi_1(x, y, z)$. Use solution (4.76) to prove the solution of this wave equation **depends continuously on the initial conditions**. Explain why the result is also true for the two-dimensional wave equation. (Hint: Show that when $|\phi_0 - \psi_0| < \varepsilon$ and $|\phi_1 - \psi_1| < \varepsilon$ for $\varepsilon > 0$, an arbitrarily small quantity, then for any fixed time $t = T$ it follows that $|\phi - \psi|$ is correspondingly small.)

4.5 Kirchhoff's Solution of the Wave Equation in Three Space Variables and Another Representation of Huygens' Principle

Let us consider the wave equation in three space dimensions and time,

$$\frac{\partial^2 \phi}{\partial t^2} = c^2 \left(\frac{\partial^2 \phi}{\partial x^2} + \frac{\partial^2 \phi}{\partial y^2} + \frac{\partial^2 \phi}{\partial z^2} \right), \quad \text{with } c = \text{constant}, \quad (4.87)$$

and derive an important result due to the German mathematical physicist G. R. Kirchhoff at the end of the 19th century.

Let P be an arbitrary point in space at time t, and consider a closed surface S that may, or may not, enclose P. We will assume that the functions $\phi, \partial\phi/\partial t$, and $\partial\phi/\partial n$, the derivative of ϕ in the direction of the unit normal $\mathbf{n}$ directed *out* of S, are known on S at times differing from the time t at which the solution $\phi(P, t)$ is required. As the wave equation (4.87) is linear, we can consider the behavior of a specific component of ϕ with frequency f, and then use linear superposition to construct the solution for the general case when other frequencies are involved. Accordingly, we start by considering a solution of the form

$$\phi(x, y, z, t) = \psi(x, y, z) \exp(ikct), \quad \text{where } k = 2\pi f/c, \quad (4.88)$$

where the space variation $\psi(x, y, z)$ has been separated from the periodic time variation $\exp(ikct)$. Substitution of (4.88) into wave equation (4.87) shows that $\psi(x, y, z)$ must be a solution of the **Helmholtz equation**

$$(\Delta_3 + k^2)\psi = 0, \tag{4.89}$$

where the suffix 3 has been added to the Laplacian to emphasize that three space dimensions are involved.

We now apply Green's theorem, which asserts that if ψ_1 and ψ_2 are any two continuous functions with continuous first- and second-order partial derivatives inside a closed surface S, which may be composed of several smooth parts, then

$$\iiint_V \{\psi_2 \Delta_3 \psi_1 - \psi_1 \Delta_3 \psi_2\} dV = \iint_S \left\{ \psi_2 \frac{\partial \psi_1}{\partial n} - \psi_1 \frac{\partial \psi_2}{\partial n} \right\} dS, \tag{4.90}$$

where V is the volume enclosed by S and $\partial/\partial n$ denotes the derivative in the direction of a unit vector $\mathbf{n}$ directed *out* of S.

The functions ψ_1 and ψ_2 in (4.90) are arbitrary, so we may replace ψ_1 by the solution ψ of the Helmholtz equation (4.89), and set $\psi_2 = (1/r) \exp(-ikr)$, where r is the radial distance from the origin to point P. Let us now take P to be the center of a small sphere Σ strictly inside S, as shown in Fig. 4.13, and exclude the origin from Σ to ensure that ψ_2 remains finite at P. Next we take for S the area $\tilde{S}$ bounding the volume $\tilde{V}$ between the surfaces S and Σ. A simple calculation shows that ψ_2 is a solution of the Helmholtz equation, so $\Delta_3 \psi_2 = -k^2 \psi_2$, while from (4.89) we have $\Delta_3 \psi = -k^2 \psi$. Using these results in (4.90) reduces its left side to 0, leading to the result

$$\iint_{\tilde{S}} \left\{ (1/r) \exp(-ikr) \frac{\partial \psi}{\partial \tilde{n}} - \psi \frac{\partial}{\partial \tilde{n}} [(1/r) \exp(-ikr)] \right\} d\tilde{S} = 0, \tag{4.91}$$

where $\partial/\partial \tilde{n}$ indicates differentiation along the unit normal directed *out* of $\tilde{S}$, which now involves the outward drawn unit normals to both S and Σ.

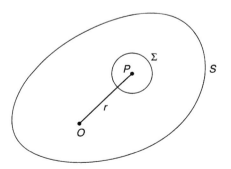

Figure 4.13 Point P at the center of the sphere Σ enclosed in surface S.

On Σ the *outward* drawn unit normal is directed along the radius r from Σ to P, so in this case the differentiation operation $\partial/\partial\tilde{n}$ becomes $-\partial/\partial r$. Thus the contribution to the integral in (4.91) made by integrating over the surface Σ becomes

$$\iint_{\Sigma} \left\{ (1/r)\exp(-ikr)\left(-\frac{\partial\psi}{\partial r}\right) - \psi\left(-\frac{\partial}{\partial r}[(1/r)\exp(-ikr)]\right) \right\} d\Sigma.$$

Allowing the radius of Σ to tend to 0 causes the integral of the first term in this expression to vanish, because the integrand is finite and it is integrated over a vanishingly small volume. Thus, all that remains is the term

$$-\iint_{\Sigma} \psi(1/r^2)\exp(-ikr)d\Sigma. \tag{4.92}$$

To simplify this expression we must now make use of the solid angle $d\omega$ subtended at P by the surface element $d\Sigma$ (see Fig. 4.10). Then, in terms of solid angles, this becomes

$$-\iint_{\Sigma} \psi(1/r^2)\exp(-ikr)d\Sigma = -\iint_{\Sigma} \psi(1/r^2)\exp(-ikr)r^2 d\omega,$$

and proceeding to the limit as $r \to 0$ and using the mean value theorem this reduces to $-4\pi\psi(P,t)$. Substituting this result into (4.91), where now only the integral over S remains, we find that

$$\psi(P,t) = \frac{1}{4\pi}\iint_{S} \left\{ (1/r)\exp(-ikr)\frac{\partial\psi}{\partial n} - \psi\frac{\partial}{\partial n}[(1/r)\exp(-ikr)] \right\} dS$$

$$= \frac{1}{4\pi}\iint_{S} \left\{ (1/r)\exp(-ikr)\frac{\partial\psi}{\partial n} - \psi\exp(-ikr)\frac{\partial}{\partial r}\left(\frac{1}{r}\right) \right.$$

$$\left. + ik\psi(1/r)\exp(-ikr)\frac{\partial r}{\partial n} \right\} dS, \tag{4.93}$$

with $\partial/\partial n$ being the directional derivative in the direction of the *outward* drawn unit normal $\mathbf{n}$ to S.

As $\phi(x, y, z, t) = \psi(x, y, z)\exp(ikct)$, in terms of ϕ solution (4.93) becomes

$$\phi(P,t) = \frac{1}{4\pi}\iint_{S} (1/r)\exp\{ik(ct - r)\}\frac{\partial\psi}{\partial n} dS$$

$$- \frac{1}{4\pi}\iint_{S} \psi\exp\{ik(ct - r)\}\frac{\partial}{\partial n}(1/r) dS$$

$$+ \frac{1}{4\pi}\iint_{S} ik\psi(1/r)\exp\{ik(ct - r)\}\frac{\partial r}{\partial n} dS. \tag{4.94}$$

Note that the time variation of ϕ determined by the factor $\exp\{ik(ct - r)\}$ corresponds to the **retarded time** $t - r/c$ when the disturbance left S in order to reach P at time t.

If we define $[\phi]_{t-r/c}$, $[\partial\phi/\partial n]_{t-r/c}$, and $[\partial\phi/\partial t]_{t-r/c}$ to mean the quantity in square brackets is to be evaluated at the time $t - r/c$, and not at time t, the solution can be written

$$\phi(P, t) = \frac{1}{4\pi} \iint_S F \, dS, \tag{4.95}$$

with

$$F = \frac{1}{r}\left[\frac{\partial\phi}{\partial n}\right]_{t-r/c} - \frac{\partial}{\partial n}\left(\frac{1}{r}\right)[\phi]_{t-r/c} + \frac{1}{cr}\frac{\partial r}{\partial n}\left[\frac{\partial\phi}{\partial t}\right]_{t-r/c}. \tag{4.96}$$

Result (4.95) is, in effect, another statement of **Huygens' principle**, because it shows how the solution at P and time t can be regarded as the sum of small elements of spherical waves of magnitude $F/(4\pi)$ coming from each part of surface S.

Although the topic will not be developed here, we mention that (4.96) forms the basis for the explanation of the **diffraction** of waves. The term *diffraction* describes the observed physical phenomenon that waves (like light), when encountering a slit in an obstacle, not only pass through the slit, but also penetrate around the sides of the slit into what, if only straight line propagation is allowed, would be the "*shadow region*" on the other side. A simple explanation of this effect is that each point of the slit acts as a wave source from which spherical waves emanate, so waves on the other side of the slit are able to enter with diminished intensity into what would otherwise be a shadow region.

4.6 Uniqueness of Solutions of the Wave Equation

So far no consideration has been given to the question of the uniqueness of the solution of the wave equation. The arguments used when formulating the equation in terms of a vibrating string were reasonable, and certainly appeared to indicate that the specification of boundary and initial conditions will lead to only one solution. In addition, the correctness of the formulation is suggested by the practical observation that when different violinists are asked to play a particular note on the *same* violin, the result is the same. However, despite the plausibility of these arguments, it is necessary to prove mathematically that the solution of the wave equation subject to boundary and initial conditions is unique. In what follows we offer one of several possible forms of proof of uniqueness for the nonhomogeneous wave equation in one space dimension and time, subject to homogeneous boundary conditions on the interval $0 \le x \le L$.

We will consider the one-dimensional *nonhomogeneous* wave equation

$$\phi_{tt} = c^2\phi_{xx} + f(x, t) \tag{4.97}$$

on the finite interval $0 \le x \le L$ subject to the homogeneous boundary conditions

$$\phi(0, t) = 0 \quad \text{and} \quad \phi(L, t) = 0, \tag{4.98}$$

and the initial conditions

$$\phi(x, 0) = \phi_0(x) \quad \text{and} \quad \phi_t(x, 0) = \phi_1(x). \tag{4.99}$$

If two different solutions $\varphi(x, t)$ and $\psi(x, t)$ are possible, each satisfying the same boundary and initial conditions, the linearity of the wave equation shows that $u(x, t) = \varphi(x, t) - \psi(x, t)$ must be a solution of the *homogeneous* wave equation

$$u_{tt} = c^2 u_{xx}, \tag{4.100}$$

subject to the homogeneous boundary conditions

$$u(0, t) = 0 \quad \text{and} \quad u(L, t) = 0, \tag{4.101}$$

and the *homogeneous* initial conditions

$$u(x, 0) = 0 \quad \text{and} \quad u_t(x, 0) = 0. \tag{4.102}$$

Multiplying (4.100) by u_t and integrating with respect to x over the interval $0 \le x \le L$ gives

$$\int_0^L u_t u_{tt}\, dx = c^2 \int_0^L u_t u_{xx}\, dx. \tag{4.103}$$

However,

$$\int_0^L u_t u_{tt}\, dx = \frac{1}{2}\frac{d}{dt}\int_0^L (u_t)^2\, dx,$$

and integrating the expression on the right of (4.103) by parts gives

$$c^2 \int_0^L u_t u_{xx}\, dx = c^2 \left[(u_t u_x)|_{x=0}^{x=L} - \int_0^L u_x u_{tx}\, dx \right].$$

The second condition in (4.102) causes the first term on the right to vanish, while the second term can be written $-\frac{1}{2}\frac{d}{dt}\int_0^L (u_x)^2\, dx$, so after combining results Eq. (4.103) becomes

$$\frac{1}{2}\frac{d}{dt}\left[\int_0^L \{(u_t)^2 + c^2(u_x)^2\}dx \right] = 0. \tag{4.104}$$

Setting $E(t) = \frac{1}{2}\int_0^L\{(u_t)^2 + c^2(u_x)^2\}\,dx$, it follows from (4.104) that $E(t) =$ constant for all $t > 0$. However, from the initial conditions $E(0) = 0$, since $u(x,0) = 0$ implies $u_x(x,0) = 0$, from which we conclude that $E(t) \equiv 0$ for all $t > 0$. As the integrand of $E(t)$ is a sum of squares it is nonnegative for all $t > 0$, so the result $E(t) \equiv 0$ is only possible if $u_t(x,t) = u_x(x,t) = 0$. Consequently $u(x,t) =$ constant, thus from (4.102) $u(x,t) \equiv 0$, from which it follows that $\varphi(x,t) \equiv \psi(x,t)$, and the uniqueness is proved.

This form of proof makes use of what is called the **energy method**, and to see one reason for this name we return to the vibrating string problem in Chapter 1 and consider the equation in the absence of any distributed force $f(x,t)$ acting on the string. So, in this case, the governing equation is (4.100) subject to the homogeneous boundary conditions $u(0,t) = u(L,t) = 0$ and initial conditions of the form $u(x,0) = h(x)$ and $u_t(x,0) = k(x)$, where now h and k are arbitrary functions of x apart from the requirement that $h(0) = h(L) = 0$ and $k(0) = k(L) = 0$, because the string is clamped at its ends. From Chapter 1 we know that if the string has a uniform tension T and a uniform line density ρ the square of the wave speed $c^2 = T/\rho$, so

$$E(t) = \frac{1}{2}\int_0^L\{(u_t)^2 + c^2(u_x)^2\}dx = \frac{1}{2}\int_0^L\{\rho(u_t)^2 + T(u_x)^2\}\,dx/\rho.$$

To interpret this last result we need to recognize that as $u(x,t)$ is the transverse displacement of the string, the kinetic energy of the string is KE $= \int_0^L \frac{1}{2}\rho(u_t)^2\,dx$. The potential energy of the string is the integral of the product of the tension T of the string and its extension, so from elementary calculus PE $= \int_0^L T(\sqrt{1 + (u_x)^2} - 1)\,dx$, but when u_x is small this last result can be approximated by PE $= \frac{1}{2}\int_0^L T(u_x)^2\,dx$.

This has established that $E(t) = (\text{KE} + \text{PE})/\rho$, showing that $E(t)$ is proportional to the **total energy** contained in the string, and hence the name *energy method*. The initial conditions are not now homogeneous so $E(0) > 0$, and as $E(t) =$ constant it follows that the total energy in the vibrating string is *conserved*, as would be expected since energy is neither dissipated nor added to the system.

This argument provides a different justification for the assumption made in Chapter 1 that to this order of magnitude the string tension remains constant. The result follows from the fact that as the change in tension is proportional to the change in the length of the string, and we have seen that an element of length δx becomes one of length $\frac{1}{2}(u_x)^2\delta x$, when u_x is small the change of tension is of a smaller magnitude, and so in this approximation can be neglected.

In practical terms, the smallness of u_x is equivalent to the requirement that the radius of curvature of the string is always large. From elementary calculus the radius of curvature of the string $r = \{1 + (u_x)^2\}^{2/3}/(u_{xx})$, so when u_x is small this can be approximated by $r = 1/(u_{xx})$, and this in turn leads to the wave equation in the form $u_{tt} = c^2 u_{xx}$.

The uniqueness proof also applies to strings of infinite length provided the initial disturbance to the string is localized with the rest of the string at rest, because then the energy in the string is finite. Uniqueness proofs for the wave equation in more than one space dimension can be established along similar lines, although out of necessity the details will be somewhat different. Accounts of such methods involve arguments inappropriate for this introduction to PDEs, so they will be omitted.

EXERCISES 4.6

1. Construct a uniqueness proof using the D'Alembert solution (4.31) for the homogeneous wave equation $u_{tt} = c^2 u_{xx}$ on the unbounded line $-\infty < x < \infty$, subject to the Cauchy conditions (4.25).

2. Use the energy method to establish the uniqueness of the solution of the wave equation $u_{tt} = c^2 u_{xx}$ on the bounded interval $0 \le x \le L$ subject to the mixed homogeneous boundary conditions

$$\alpha u(0, t) + \beta u_x(0, t) = 0 \quad \text{and} \quad \gamma u(L, t) + \delta u_x(L, t) = 0 \quad \text{for } t \ge 0,$$

with $\alpha, \beta, \gamma, \delta$ being constants, where not both α and β or γ and δ vanish, and the initial conditions

$$u(x, 0) = h(x) \quad \text{and} \quad u_t(x, 0) = k(x).$$

Fourier Series, Legendre and Bessel Functions

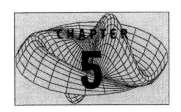

5.1 An Introduction to Fourier Series

Before discussing Fourier series and some of its generalizations we first outline the basic ideas involved. Linear constant coefficient PDEs are solved in Chapter 6 by the *method of separation of variables*, and when using this approach it becomes necessary to represent an arbitrary function $f(x)$ of a single real variable x, defined over some interval $\alpha \leq x \leq \beta$, in the form of an infinite sum $f(x) = \sum_{n=0}^{\infty} c_n \varphi_n(x)$, where the sequence of functions $\varphi_n(x)$ are of a special kind.

In the case of Fourier series, the set of functions $\varphi_n(x)$ is the infinite set of trigonometric functions

$$\sin \pi x/L, \sin 2\pi x/L, \ldots, \sin n\pi x/L, \ldots, 1, \cos \pi x/L,$$
$$\cos 2\pi x/L, \ldots, \cos n\pi x/L, \ldots. \tag{5.1}$$

Let us now consider an arbitrary function $f(x)$ defined over the interval $-L \leq x \leq L$, where $f(x)$ may have finite jump discontinuities at a finite number of points, and *define* the **Fourier series** of $f(x)$ in terms of the set of functions in (5.1) by the relationship

$$f(x) = a_0 + \sum_{n=1}^{\infty} \left(a_n \cos \frac{n\pi x}{L} + b_n \sin \frac{n\pi x}{L} \right), \tag{5.2}$$

where the coefficients a_n and b_n are given by

$$a_0 = \frac{1}{2L} \int_{-L}^{L} f(x)\, dx, \quad a_n = \frac{1}{L} \int_{-L}^{L} f(x) \cos \frac{n\pi x}{L} dx, \quad \text{for } n = 1, 2, \ldots,$$

$$\tag{5.3}$$

and

$$b_n = \frac{1}{L} \int_{-L}^{L} f(x) \sin \frac{n\pi x}{L} dx, \quad \text{for } n = 1, 2, \ldots . \tag{5.4}$$

The reason for defining a_n and b_n in this way will become clear later. The interval $-L \leq x \leq L$ is called the **fundamental interval** of the Fourier series, and the numbers a_n and b_n are called the **Fourier coefficients** of $f(x)$.

Some simple properties of Fourier series

A number of fundamental properties of Fourier series can be deduced directly from the definition of Fourier series given in (5.2) to (5.4), together with an important question that must be answered.

1. The functions in (5.1) are periodic with period $2L$, and sine and cosine functions are defined for all x, so Fourier series (5.2) must be periodic with period $2L$. Here the term **periodicity** is used in the sense that a function $g(x)$ defined for all x will be periodic with **period** X if X is the smallest number such that $g(x + X) = g(x)$. The implication of periodicity is illustrated in Fig. 5.1.

2. As the Fourier series of $f(x)$ depends only on the behavior of $f(x)$ in the fundamental interval, the periodicity of Fourier series means that its behavior in each of the intervals $(2n + 1)L < x < (2n + 3)L$ for $n = 0, 1, \pm 2, \ldots$ outside the fundamental interval will be the same as in the fundamental interval.

    It follows from this that the behavior of the Fourier series of a function $f(x)$ in any one of these intervals outside the fundamental interval will not necessarily be the same as the behavior of the function $f(x)$ itself in that same interval. The Fourier series of a function $f(x)$ in any one of the intervals outside the fundamental interval is called a **periodic extension** of the Fourier series in the fundamental interval (see Fig. 5.2).

3. If $f(x)$ is an *even function* in the fundamental interval, so that $f(-x) = f(x)$ in the interval, then the definite integrals in (5.3) and (5.4) reduce to

$$a_0 = \frac{1}{L} \int_{0}^{L} f(x) \, dx, \quad \text{and} \quad a_n = \frac{2}{L} \int_{0}^{L} f(x) \cos \frac{n\pi x}{L} \, dx,$$

$$\text{for } n = 1, 2, \ldots \tag{5.5}$$

and

$$b_n = 0, \quad \text{for } n = 1, 2, \ldots . \tag{5.6}$$

4. If $f(x)$ is an *odd function* in the fundamental interval, so that $f(-x) = -f(x)$ in the interval, the definite integrals (5.3) and (5.4) reduce to

$$a_n = 0 \quad \text{for } n = 0, 1, \ldots \tag{5.7}$$

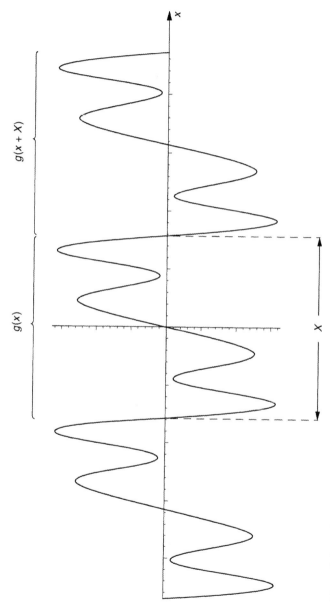

Figure 5.1   A function $g(x)$ with period X.

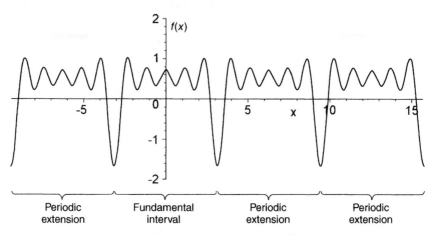

Figure 5.2   Periodic extensions of a Fourier series for $f(x)$.

and

$$b_n = \frac{2}{L} \int_0^L f(x) \sin \frac{n\pi x}{L} \, dx, \quad \text{for } n = 1, 2, \dots. \tag{5.8}$$

5. The coefficients a_n and b_n involve integrals depending on $f(x)$, and these are defined even when $f(x)$ has finite jump discontinuities at one or more points in the fundamental interval. Consequently, as the sine and cosine functions on the right of (5.2) are defined for all x, it follows that the Fourier series of a function $f(x)$ with a finite jump discontinuity at a point $x = \xi$ in the fundamental interval will be defined at that point, even though the function $f(x)$ itself is not defined at $x = \xi$. This fact calls into question the meaning of the equality sign in definition (5.2) at points where $f(x)$ is discontinuous. This matter will be resolved in the next section when the convergence property of Fourier series is examined. It is because of this need for interpretation of the precise nature of the relationship between a function and its Fourier series that more advanced accounts of Fourier series replace the equality in (5.2) by the tilde symbol $\sim$. The understanding is then that $\sim$ means "the expression on the left is related to the expression on the right" where the nature of the relationship must still be resolved.

Orthogonality and the Euler formulas

To understand how for a given function $f(x)$ the Fourier coefficients a_n and b_n in (5.3) and (5.4) are derived, we need to make use of the following results, which follow from elementary integration:

$$\int_{-L}^{L} \sin\frac{m\pi x}{L}\sin\frac{n\pi x}{L}dx = \begin{cases} 0, & m \neq n \\ L, & m = n, \end{cases} \tag{5.9}$$

$$\int_{-L}^{L} \cos\frac{m\pi x}{L}\cos\frac{n\pi x}{L}dx = \begin{cases} 0, & m \neq n \\ L, & m = n \neq 0 \\ 2L, & m = n = 0, \end{cases} \tag{5.10}$$

$$\int_{-L}^{L} \sin\frac{m\pi x}{L}\cos\frac{n\pi x}{L}dx = 0, \quad \text{for all } m, n. \tag{5.11}$$

The integrals (5.9) to (5.11) express what in mathematical terms is called the **orthogonality** of the set of functions

$$\left\{1, \cos\frac{\pi x}{L}, \sin\frac{\pi x}{L}, \cos\frac{2\pi x}{L}, \sin\frac{2\pi x}{L}, \cdots\right\} \tag{5.12}$$

over the interval $-L \leq x \leq L$ on which Fourier series (5.2) is based. This form of orthogonality can be thought of as an extension to an infinite-dimensional space involving the trigonometric functions of the concept of the orthogonality of mutually orthogonal (perpendicular) geometrical vectors in three-dimensional space. However now, instead of orthogonality in terms of the dot (scalar) product in vector analysis, in Fourier series the orthogonality property takes the form of a definite integral over the interval $-L \leq x \leq L$. This integral is zero when the product of two *different* functions from (5.12) is integrated, and a positive number when the *square* of any of the functions is integrated. The numbers $\|\cos\frac{m\pi x}{L}\|$ and $\|\sin\frac{n\pi x}{L}\|$, called the **norms** of the functions $\cos\frac{m\pi x}{L}$ and $\sin\frac{n\pi x}{L}$, respectively, are defined by the integrals

$$\left\|\cos\frac{m\pi x}{L}\right\|^2 = \int_{-L}^{L}\cos^2\frac{m\pi x}{L}dx = \begin{cases} 2L, & m = 0 \\ L, & m \neq 0 \end{cases}$$

$$\text{and} \quad \left\|\sin\frac{m\pi x}{L}\right\|^2 = \int_{-L}^{L}\sin^2\frac{m\pi x}{L}dx = L. \tag{5.13}$$

The reason why the Fourier coefficient a_0 in (5.2) is derived differently from the coefficients a_n for $n = 1, 2, \ldots$ is seen to be because it compensates for the factor $2L$, which appears in result (5.10) when $m = n = 0$.

The norms can be used to extend to the infinite set of sine and cosine functions in (5.12) the property of unit geometrical vectors $\mathbf{i}, \mathbf{j}$, and $\mathbf{k}$ parallel to the x, y, and z rectangular Cartesian coordinate axes for which $\mathbf{i}\cdot\mathbf{i} = \mathbf{j}\cdot\mathbf{j} = \mathbf{k}\cdot\mathbf{k} = 1$. This extension can be achieved by dividing each trigonometric function in (5.12) by its norm to obtain the equivalent set of **normalized trigonometric** functions

$$\left\{ \frac{1}{\sqrt{2L}}, \frac{1}{\sqrt{L}} \cos\frac{\pi x}{L}, \frac{1}{\sqrt{L}} \sin\frac{\pi x}{L}, \frac{1}{\sqrt{L}} \cos\frac{2\pi x}{L}, \frac{1}{\sqrt{L}} \sin\frac{2\pi x}{L}, \cdots \right\}, \quad (5.14)$$

each of which now has a unit norm. This normalized, but otherwise equivalent set of functions, is said to form an **orthonormal** set of functions over the interval $-L \leq x \leq L$.

We now return to the task of showing how the coefficients a_n and b_n in (5.2) are related to a given function $f(x)$. If both sides of (5.2) are multiplied by $\cos\frac{n\pi x}{L}$ and the result is integrated over the interval $-L \leq x \leq L$, results (5.10) and (5.11) show that

$$\int_{-L}^{L} f(x)dx = 2La_0 \quad \text{and} \quad \int_{-L}^{L} f(x) \cos\frac{n\pi x}{L} dx = La_n \quad \text{for } n = 1, 2, \ldots .$$

$$(5.15)$$

Similarly, if both sides of (5.2) are multiplied by $\sin\frac{n\pi x}{L}$ and integrated over the interval $-L \leq x \leq L$, results (5.9) and (5.11) show that

$$\int_{-L}^{L} f(x) \sin\frac{n\pi x}{L} dx = b_n L, \quad n = 1, 2, \ldots . \quad (5.16)$$

Consequently, from (5.15) and (5.16), we find that

$$a_0 = \frac{1}{2L} \int_{-L}^{L} f(x)dx, \quad a_n = \frac{1}{L} \int_{-L}^{L} f(x) \cos\frac{n\pi x}{L} dx \quad \text{for } n = 1, 2, \ldots,$$

$$(5.17)$$

and

$$b_n = \frac{1}{L} \int_{-L}^{L} f(x) \sin\frac{n\pi x}{L} dx, \quad n = 1, 2, \ldots . \quad (5.18)$$

Results (5.17) and (5.18) are called the **Euler formulas** for the Fourier coefficients in the Fourier series representation of $f(x)$ on the right of (5.2). It is obvious that any multiple of the sine and cosine functions in (5.12), equivalently in (5.14), is its own Fourier series comprising a single term.

Attention has already been drawn to the fact that $f(x)$ may have a finite discontinuity in its fundamental interval, and Fig. 5.3 shows a typical example of such a discontinuous function, together with the limiting notation that will be used later. In Fig. 5.3 the conventions $f(\xi - 0)$ and $f(\xi + 0)$ are used to denote the values of $f(x)$ to the immediate left and right of $x = \xi$, so that $f(\xi - 0) = \lim_{h \to 0} f(\xi - h) = c_1$, say, and $f(\xi + 0) = \lim_{k \to 0} f(\xi + k) = c_2$, say.

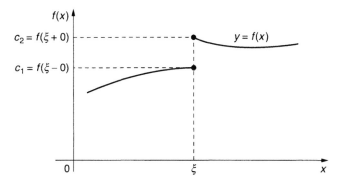

Figure 5.3 A function $f(x)$ with a finite jump discontinuity at $x = \xi$.

The presence of a discontinuity at $x = \xi$ in the interval $-L \leq x \leq L$ causes $f(x)$ to be nondifferentiable at that point, so for such functions no Taylor series representation of $f(x)$ is possible about the point. However, as the coefficients a_n and b_n can still be computed, it is seen that a Fourier series representation of a function $f(x)$ over the interval $-L \leq x \leq L$ is more general than a Taylor series expansion.

Some examples will help to demonstrate the nature and properties of Fourier series before their convergence properties are investigated in the next section.

Example 5.1. Find the Fourier series representation of

$$f(x) = (x + \pi) \sin x \quad \text{for } -\pi \leq x \leq \pi.$$

Plot the Fourier series representation in the fundamental interval by approximating the infinite sum by finite sums involving different numbers of terms, and plot the periodic extension in the interval $\pi \leq x \leq 3\pi$.

Solution: In this example the interval is $-\pi \leq x \leq \pi$, so $L = \pi$, and substitution into the Euler formula for a_n gives

$$a_n = \frac{1}{\pi} \int_{-\pi}^{\pi} (x + \pi) \sin(x) \cos(nx) dx = -2 \cos(n\pi)/(n^2 - 1)$$
$$= 2(-1)^{n+1}/(n^2 - 1),$$

where we have used the results that $\cos(n\pi) = (-1)^n$ and $\sin(n\pi) = 0$.

This expression for a_n is true for $n = 0, 2, 3, \ldots$, but not for $n = 1$ because of the term $(n^2 - 1)$ in the denominator. To find a_1 it is necessary to set $n = 1$ in the integral for a_n *before* integration, as a result of which we find that

$$a_1 = \frac{1}{\pi} \int_{-\pi}^{\pi} (x + \pi) \sin(x) \cos(x) dx = -\frac{1}{2}.$$

The general expression for a_n is valid for $n = 0$ so we see that $a_0 = 2$.

The Euler formula for b_n is

$$b_n = \frac{1}{\pi} \int_{-\pi}^{\pi} (x + \pi) \sin(x) \sin(nx)dx = \begin{cases} 0, & n = 2, 3, \ldots \\ \pi, & n = 1. \end{cases}$$

Consequently, from (5.2), the Fourier series representation of $f(x)$ becomes

$$f(x) = 1 - \frac{1}{2} \cos x + \pi \sin x + 2 \sum_{n=2}^{\infty} \frac{(-1)^{n+1} \cos(nx)}{n^2 - 1}.$$

We now approximate the infinite sum on the right by the finite sum

$$s_N(x) = 1 - \frac{1}{2} \cos x + \pi \sin x + 2 \sum_{n=2}^{N} \frac{(-1)^{n+1} \cos(nx)}{n^2 - 1}.$$

This is called the **Nth partial sum** of the Fourier series.

Figure 5.4a shows a plot of $f(x)$ in the fundamental interval $-\pi \le x \le \pi$ and also in the interval $\pi \le x \le 3\pi$. Figure 5.4b shows a plot of the partial sum $s_5(x)$ with terms up to and including $\cos(5x)$ in both the fundamental interval, and its periodic extension to $\pi \le x \le 3\pi$, while Fig. 5.4c shows a plot of the partial sum $s_{10}(x)$ in the fundamental interval with terms up to and including $\cos(10x)$.

This is an example where $f(x)$ is everywhere continuous with $f(-\pi) = f(\pi)$, and it shows how as the number of terms in partial sum approximations increase, so the approximations converge rapidly to the function $f(x)$ throughout the entire fundamental interval. A comparison of Figs. 5.4a and 5.4b show how in this example the function and its periodic extension to the interval $\pi \le x \le 3\pi$ differ.   ■

Example 5.2. Find the Fourier series representation of

$$f(x) = \begin{cases} 0, & -2 \le x < 0 \\ 2 - x, & 0 < x \le 2. \end{cases}$$

Plot the approximation to $f(x)$ obtained by using different numbers of terms in partial sums of the Fourier series.

Solution: The function is defined over the interval $-2 \le x \le 2$ where it is piecewise smooth with a finite discontinuous jump at the origin. Setting $L = 2$ in the Euler formulas for the Fourier coefficients we find that

$$a_0 = \frac{1}{4} \int_0^2 (2 - x)dx, \qquad a_n = \frac{1}{2} \int_0^2 (2 - x) \cos \frac{n\pi x}{2} dx, \quad n = 1, 2, \ldots,$$

and

$$b_n = \frac{1}{2} \int_{-2}^2 f(x) \sin \frac{n\pi x}{2} dx = \frac{1}{2} \int_0^2 (2 - x) \sin \frac{n\pi x}{2} dx, \quad n = 1, 2, 3, \ldots.$$

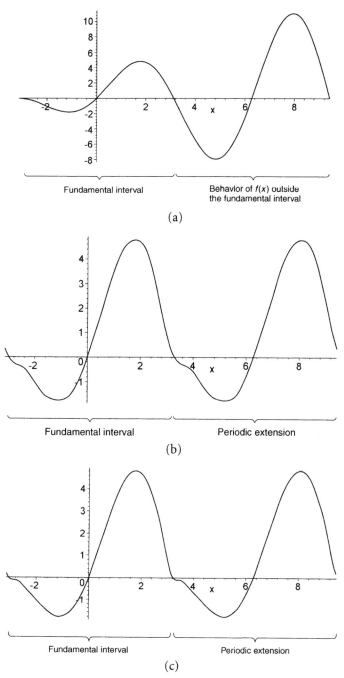

(a)

(b)

(c)

Figure 5.4 (a) $f(x)$ in the fundamental interval $-\pi \leq x \leq \pi$ and its behavior in the interval $\pi \leq x \leq 3\pi$. (b) $s_5(x)$ in the fundamental interval and its periodic extension to $\pi \leq x \leq 3\pi$. (c) $s_{10}(x)$ in the fundamental interval.

Evaluating these integrals gives

$$a_0 = \frac{1}{2}, \quad a_n = \frac{2}{n^2 \pi^2}[1 - (-1)^n] \quad \text{and} \quad b_n = \frac{2}{n\pi} \quad \text{for } n = 1, 2, \ldots,$$

where use has been made of the fact that $\cos(n\pi) = (-1)^n$ and $\sin(n\pi) = 0$. Thus the Fourier series becomes

$$f(x) = \frac{1}{2} + \frac{2}{\pi} \sum_{n=1}^{\infty} \left(\frac{[1 - (-1)^n]}{n^2 \pi} \cos \frac{n\pi x}{2} + \frac{1}{n} \sin \frac{n\pi x}{2} \right).$$

Let us now approximate the infinite series on the right by the partial sum

$$s_N(x) = \frac{1}{2} + \frac{2}{\pi} \sum_{n=1}^{N} \left(\frac{[1 - (-1)^n]}{n^2 \pi} \cos \frac{n\pi x}{2} + \frac{1}{n} \sin \frac{n\pi x}{2} \right).$$

The consequences of summing different numbers of terms in $s_N(x)$ can be seen from the plots of $s_N(x)$ in Fig. 5.5 over the interval $-2 \le x \le 2$ for $N = 20, 30, 40$. These plots indicate that the partial sums for Fourier series are well defined in the neighborhood of the finite jump discontinuity at the origin, and also at the discontinuity itself. They also show how elsewhere, as N increases, the partial sums converge to $f(x)$. In particular, at the jump discontinuity, the Nth partial sums are all seen to pass through the *midpoint* of the discontinuity. In the next section we prove this is always the case.

Note also that overshoot and undershoot oscillations occur on either side of the discontinuity, and that they persist as N increases. This is a feature of Fourier series in the neighborhood of a discontinuity, and the effect is called the **Gibbs phenomenon**. This effect is due to the failure of a Fourier series to represent a function at a point where the function is *not* uniformly continuous. ∎

Example 5.3. Find the Fourier series representation of

$$f(x) = x^2 \quad \text{for } -\pi \le x \le \pi.$$

Plot $f(x)$ in the fundamental interval $-\pi \le x \le \pi$, and also in the interval $-3\pi \le x \le -\pi$. Plot a partial sum approximation in the fundamental interval and also its periodic extension to $-3\pi \le x \le -\pi$. Hence show how in this case a periodic extension of the Fourier series of $f(x)$ and $f(x)$ itself differ outside the fundamental interval.

Solution: As $f(x)$ is an even function all the coefficients b_n vanish, while $a_0 = \frac{1}{\pi} \int_0^{\pi} x^2 dx$ and $a_n = \frac{2}{\pi} \int_0^{\pi} x^2 \cos nx \, dx$. This gives $a_0 = \pi^2/3$, and routine integration shows that

$$a_n = (-1)^n (4/n^2).$$

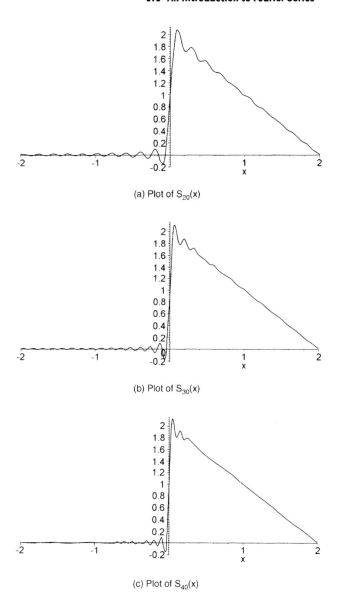

(a) Plot of $S_{20}(x)$

(b) Plot of $S_{30}(x)$

(c) Plot of $S_{40}(x)$

Figure 5.5 Plots of $s_N(x)$ for (a) $N = 20$, (b) $N = 30$, and (c) $N = 40$.

Thus the Fourier series representation of $f(x)$ over the interval $-\pi \le x \le \pi$ is given by

$$f(x) = \frac{\pi^2}{3} + 4 \sum_{n=1}^{\infty} (-1)^n \frac{\cos nx}{n^2}.$$

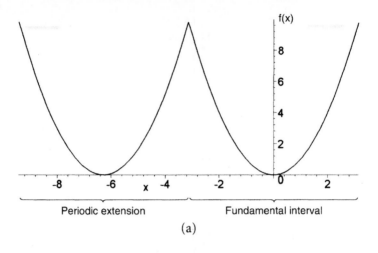

(a)

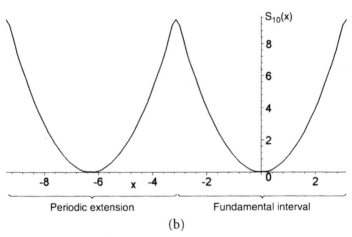

(b)

Figure 5.6   (a) $f(x)$ in the fundamental interval $-\pi \leq x \leq \pi$ and in the interval $-3\pi \leq x \leq -\pi$. (b) $s_{10}(x)$ in the fundamental interval $-\pi \leq x \leq \pi$ and its periodic extension to $-3\pi \leq x \leq -\pi$.

Now consider the partial sum approximation

$$s_{10}(x) = \frac{\pi^2}{3} + 4\sum_{n=1}^{10}(-1)^n\frac{\cos nx}{n^2}.$$

The result of this approximation is shown in Fig. 5.6, where Fig. 5.6a shows the function $f(x)$ itself in the fundamental interval $-\pi \leq x \leq \pi$ and in the interval $-3\pi \leq x \leq -\pi$, and Fig. 5.6b shows $s_{10}(x)$ in the fundamental interval and its periodic extension to the interval $-3\pi \leq x \leq -\pi$.   ∎

It will be seen later that different sets of orthogonal functions occur when solving PDEs by the method of separation of variables, although for them the definition of orthogonality may need to be generalized as follows. Let $\varphi_m(x)$ and $\varphi_n(x)$ be any two members of an infinite set of functions $\{\varphi_1(x), \varphi_2(x), \varphi_3(x), \cdots\}$ that are defined over an interval $a \leq x \leq b$. Then the functions will be said to be mutually **orthogonal with respect to a weight function** $w(x) > 0$ if

$$\int_a^b w(x)\varphi_m(x)\varphi_n(x)dx = \begin{cases} 0, & m \neq n \\ \|\varphi_n(x)\|^2, & m = n, \end{cases} \tag{5.19}$$

where $\|\varphi_n(x)\|$ is the norm of the function $\varphi_n(x)$, and

$$\|\varphi_n(x)\|^2 = \int_a^b w(x)\varphi_n^2(x)dx. \tag{5.20}$$

As with Fourier series, for suitable sets of functions $\{\varphi_1(x), \varphi_2(x), \varphi_3(x), \cdots\}$, a **generalized Fourier series** representation of a function $f(x)$ defined over the interval $a \leq x \leq b$ can be obtained by setting

$$f(x) = \sum_{n=1}^{\infty} c_n\varphi_n(x), \tag{5.21}$$

where the coefficients c_n are obtained from the definite integrals

$$c_n = \int_a^b w(x) f(x)\varphi_n(x)dx, \quad n = 1, 2, \ldots. \tag{5.22}$$

If required, a set of functions $\{\varphi_1(x), \varphi_2(x), \varphi_3(x), \cdots\}$, which are orthogonal over an interval $a \leq x \leq b$ with respect to a weight function $w(x)$, can be converted to an **orthonormal** set with respect to the weight function $w(x)$ by dividing each function by its norm to arrive at the new set $\{\varphi_1(x)/\|\varphi_1(x)\|, \varphi_2(x)/\|\varphi_2(x)\|, \cdots\}$. Each of these normalized functions now has a unit norm.

EXERCISES 5.1

1. Prove the results given in (5.9), (5.10), and (5.11).

2. Prove that when $f(x)$ is an even function over $-L \leq x \leq L$, $a_0 = \frac{1}{L}\int_0^L f(x)dx$, $a_n = \frac{2}{L}\int_0^L f(x)\cos\frac{n\pi x}{L}dx$ for $n = 1, 2, \ldots$ and $b_n = 0$ for $n = 1, 2, \ldots$.

3. Prove that when $f(x)$ is an odd function over the interval $-L \leq x \leq L$, $b_n = \frac{2}{L}\int_0^L f(x)\sin\frac{n\pi x}{L}dx$ for $n = 1, 2, \ldots$ and $a_n = 0$ for $n = 0, 1, \ldots$.

4. Find an easy way of writing down the Fourier representation for $f(x) = \sin^2(x) + \sin^3(x)$ for $-\pi \leq x \leq \pi$, and explain why the symbol $\sim$ can be replaced by an equality.

In Exercises 5 through 9 sketch the function $f(x)$, and derive the stated Fourier series representation. By making computer plots of some partial sum approximations $s_N(x)$ with increasing values of N, examine the convergence of the partial sums to $f(x)$ and, where appropriate, the nature of the Gibbs phenomenon at any point of discontinuity.

5. If $f(x) = |x|, -\pi \leq x \leq \pi$, then $f(x) = \frac{1}{2}\pi - \frac{4}{\pi}\sum_{n=1}^{\infty}\frac{\cos(2n-1)x}{(2n-1)^2}$.

6. If $f(x) = \begin{cases} \sin x, & -\pi \leq x < 0 \\ 0, & 0 \leq x \leq \pi \end{cases}$, then $f(x) = \frac{1}{2}\sin x + \frac{2}{\pi}\sum_{n=1}^{\infty}\frac{\cos 2nx}{(4n^2-1)} - \frac{1}{\pi}$.

7. If $f(x) = \cos ax, -\pi < x < \pi, a \neq 0, \pm 1, \pm 2, \ldots$, then

$$f(x) = \frac{2\alpha}{\pi}\sin(\alpha\pi)\left\{\frac{1}{2\alpha^2} + \sum_{n=1}^{\infty}\frac{(-1)^{n+1}\cos(nx)}{n^2 - \alpha^2}\right\}.$$

8. If $f(x) = \begin{cases} 0, & -\pi \leq x < -\pi/2 \\ 1, & -\pi/2 < x < \pi/2, \\ 0, & \pi/2 < x \leq \pi \end{cases}$ then $f(x) = \frac{1}{2} + \frac{2}{\pi}\sum_{n=1}^{\infty}\frac{(-1)^{n+1}\cos(2n-1)x}{2n-1}$.

9. If $f(x) = \begin{cases} \sin x - 1, & -\pi < x < 0 \\ \sin x + 1, & 0 < x < \pi, \end{cases}$ then

$$f(x) = \left(\frac{4+\pi}{\pi}\right)\sin x + \frac{2}{\pi}\sum_{n=2}^{\infty}\frac{[1-(-1)^n]\sin nx}{n}.$$

10. Prove that when $f(x)$ is periodic with period $2L$, the Fourier coefficients a_n and b_n can be determined from the Euler formulas (5.3) and (5.4) by integrating over any interval of the form $a \leq x \leq a + 2L$.

11. It follows from Exercise 2 that any function $f(x)$ defined over the interval $0 \leq x \leq L$ can be represented in terms of cosine functions by a series of the form

$$f(x) = a_0 + \sum_{n=1}^{\infty}a_n\cos\frac{n\pi x}{L}, \quad \text{where } a_0 = \frac{1}{L}\int_0^L f(x)dx$$

and $a_n = \frac{2}{L}\int_0^L f(x)\cos\frac{n\pi x}{L}dx, n = 1, 2, \ldots$.
This is called the **half-range Fourier cosine representation** of $f(x)$ or, more simply, the **cosine series representation** of $f(x)$. Use this to show that if $f(x) = x(2 - x)$ for $0 \leq x \leq 2$, then

$$f(x) = \frac{2}{3} - \frac{8}{\pi^2}\sum_{n=1}^{\infty}\frac{[1+(-1)^n]}{n^2}\cos\frac{n\pi x}{2}.$$

12. It follows from Exercise 3 that any function $f(x)$ defined over the interval $0 \leq x \leq L$ can be represented in terms of sine functions by a series of the form

$$f(x) = \sum_{n=1}^{\infty} b_n \sin \frac{n\pi x}{L}, \quad \text{where } b_n = \frac{2}{L} \int_0^L f(x) \sin \frac{n\pi x}{L} dx,$$
$$n = 1, 2, \ldots.$$

This is called the **half-range Fourier sine representation** of $f(x)$ or, more simply, the **sine series representation** of $f(x)$. Use this to show that if $f(x) = \cos(\pi x)$ for $0 \leq x \leq 1$, then

$$f(x) = \frac{8}{\pi} \sum_{n=1}^{\infty} \frac{n}{4n^2 - 1} \sin 2n\pi x.$$

13. If $f(x)$ is defined for $-L < x < L$, show by setting $\cos nx = \frac{1}{2}(e^{inx} + e^{-inx})$ and $\sin nx = \frac{1}{2i}(e^{inx} - e^{-inx})$ in the Fourier representation (5.2) that it can be represented in the **complex Fourier series form**

$$f(x) = \lim_{m \to \infty} \sum_{n=-m}^{m} c_n \exp(in\pi x/L) \quad \text{where}$$

$$c_n = \frac{1}{2L} \int_{-L}^{L} f(x) \exp(-in\pi x/L) dx.$$

(Hint: group terms derived from $\exp(in\pi x/L)$ and $\exp(-in\pi x/L)$.)

14. Use the results of Exercise 13 to show that if $f(x) = \sin \frac{1}{2}x$ for $-\pi \leq x \leq \pi$, its complex Fourier series representation is

$$f(x) = \lim_{m \to \infty} \sum_{n=-m}^{m} c_n \exp(inx), \quad \text{where}$$

$$c_n = \bar{c}_{-n} = \frac{i}{\pi}\left((-1)^{n+1} \frac{4n}{1 - 4n^2}\right),$$

where the overbar in $\bar{a}$ denotes the complex conjugate of a.

15. The functions $P_0(x) = 1$, $P_1(x) = x$, $P_2(x) = \frac{1}{2}(3x^2 - 1)$, $P_3(x) = \frac{1}{2}(5x^3 - 3x)$, $P_4(x) = \frac{1}{8}(35x^4 - 30x^2 + 3)$ defined over the interval $-1 \leq x \leq 1$ are the first five members of an infinite set of functions called **Legendre polynomials**. Confirm by direct integration that these polynomials are mutually orthogonal over the interval $-1 \leq x \leq 1$ with weight function $w(x) = 1$, and that $\| P_n(x) \|^2 = 2/(2n + 1)$ for $n = 0, 1, \ldots, 4$. Use this orthogonality property to find an approximation to the function $f(x) = 1/(1 + x^2)$ in the interval $-1 \leq x \leq 1$ by writing $f(x) \approx \sum_{n=0}^{4} c_n P_n(x)$.

Make computer plots of $f(x)$ and its Legendre polynomial approximation and compare the results.

16. The functions $T_0(x) = 1$, $T_1(x) = x$, $T_2(x) = 2x^2 - 1$, $T_3(x) = 4x^3 - 3x$, $T_4(x) = 8x^4 - 8x^2 + 1$ defined over the interval $-1 \leq x \leq 1$ are the first five members of an infinite set of functions called **Chebyshev polynomials**. Show by direct integration that these polynomials are mutually orthogonal with weight function $w(x) = 1/(1 - x^2)^{1/2}$, and that $\| T_n(x)\|^2 = \begin{cases} \pi, & n = 0 \\ \pi/2, & n = 1, 2, 3, 4. \end{cases}$

Use this orthogonality property to find an approximation to the function $f(x) = (1 - x^2)^{1/2}$ in the interval $-1 \leq x \leq 1$ by writing $f(x) \approx \sum_{n=0}^{4} c_n T_n(x)$. Make computer plots of $f(x)$ and this Chebyshev polynomial approximation and compare the results.

5.2 Major Results Involving Fourier Series

The purpose of this section is to establish a number of important general results concerning the Fourier series representation of a function $f(x)$ over a finite interval. In Section 5.1 the representations were purely formal, as no proof was given that the series obtained were convergent or indeed, if they were convergent, the sense in which they converged to the function $f(x)$. The success of a Fourier series, and of its generalizations to be used later, rests on the fact that the infinite set of functions used in any representation is **complete**. This means there is *no* function $\phi(x)$, say, that is orthogonal to *every* member of the set of functions being used to represent the Fourier or generalized Fourier representation of $f(x)$. The set of sine and cosine functions $\{1, \cos x, \sin x, \cos 2x, \sin 2x, \ldots, \cos nx, \sin nx, \ldots\}$ used in Fourier series can be shown to be complete, as can the other expansions to be used later with the method of separation of variables when solving linear partial differential equations. We will not, however, attempt to prove these statements, and instead we will always assume the completeness of any system used.

Examples 5.1 and 5.3 illustrated the behavior of Fourier series representations of *continuous* functions where the *same* functional value occurred at each end of the fundamental interval, while Example 5.2 illustrated the behavior of a Fourier series at an internal point of the fundamental interval where a finite jump discontinuity occurs. Whereas in Examples 5.1 and 5.3, as the number of terms in the partial sum $s_N(x)$ increased, so the approximation to $f(x)$ by $s_N(x)$ was seen to become ever closer to $f(x)$, the situation in Example 5.2 was quite different. In that case the approximation improved with increasing N apart for the neighborhood of the jump discontinuity, where overshoot and undershoot oscillations persisted. These situations are examples of two different types of convergences called, respectively, *uniform convergence* and *pointwise convergence*.

A sequence of functions $s_N(x)$ defined for x in an interval I is said to converge **pointwise** to a function $F(x)$, if

$$\lim_{N\to\infty} |s_N(x) - F(x)| \to 0 \quad \text{for all } x \text{ in interval } I.$$

A sequence of functions $s_N(x)$ defined for x in an interval I is said to converge **uniformly** to a function $F(x)$, if

$$\lim_{N\to\infty} \max_{x\in I} |s_N(x) - F(x)| \to 0.$$

Thus a sequence of functions $s_N(x)$ converges pointwise to a function $F(x)$ at each point of I, but the rate at which it does so can depend on x. This means, for example, that if for some arbitrarily small number $\varepsilon > 0$ it is required that $|s_N(x) - F(x)| \le \varepsilon$, the choice of N will depend on x. This is because to achieve this accuracy of approximation by $s_N(x)$ for a particular value $x = x_1$, a greater value of N may be required than for a different value $x = x_2$. An illustration of the pointwise convergence of a partial sum $s_N(x)$ of a Fourier series to its defining function $f(x)$ can be seen in Fig. 5.5. In this case the accuracy of the approximation provided by $s_N(x)$ depends on whether x is close to or well removed from the jump discontinuity, because to achieve a given accuracy at two different points it is clear that two different values of N are necessary.

The uniform convergence of a sequence of functions $s_N(x)$ to $F(x)$ has a very different property, because if it is required that for some arbitrarily small number $\varepsilon > 0, |s_N(x) - F(x)| \le \varepsilon$, then a *fixed* value of N can be found such that this accuracy of approximation can be attained for *all* points x in the interval I. Hence the term *uniform convergence*. Examples of uniform convergence are illustrated in Figs. 5.4 and 5.6, where the approximation to $f(x)$ is seen to be uniformly good over the entire fundamental interval, since no Gibbs phenomenon occurs. It is clear that while uniform convergence implies pointwise convergence, the converse is not necessarily true.

When working with Fourier series approximations, and its generalizations, another type of convergence called **mean-square convergence** needs to be introduced. This is a weaker form of convergence, but still one of great practical importance. It is defined by the requirement that a sequence of functions $s_N(x)$ defined over an interval $a \le x \le b$ **converges in a mean-square sense** to a function $F(x)$ if

$$\lim_{N\to\infty} \int_a^b [s_N(x) - F(x)]^2 \, dx \to 0.$$

Thus mean-square convergence is a form of average convergence over the interval $a \le x \le b$, because it only requires the limit of the integral of the squared quantity $[s_N(x) - F(x)]^2$ over the interval to vanish as $N \to \infty$. Hence, in mean square convergence, it is *not* necessary that in the limit as $N \to \infty$

the approximation $s_N(x)$ should tend to every value of $F(x)$. This form of convergence is particularly useful when working with Fourier series and its generalizations involving functions $f(x)$ with finite jump discontinuities. This is because as $N \to \infty$, so the spikes due to the Gibbs phenomenon in the neighborhood of a jump discontinuity become arbitrarily thin, so in the limit this does not influence this type of convergence. It must be recognized that pointwise, uniform, and mean-square convergences are all quite different types of property proved by the **Riemann–Lebesgue** lemma, which describes the behavior of the Fourier coefficients of a piecewise smooth function $f(x)$ as $n \to \infty$.

Lemma 5.1. (The Riemann–Lebesgue Lemma)
Let $f(x)$ be a piecewise smooth function over the finite interval $a \le x \le b$, then for any real number μ

$$\lim_{\lambda \to \infty} \int_a^b f(x) \cos(\lambda x + \mu)dx = 0,$$

or equivalently

$$\lim_{\lambda \to \infty} \int_a^b f(x) \cos(\lambda x)dx = 0 \quad and \quad \lim_{\lambda \to \infty} \int_a^b f(x) \sin(\lambda x)dx = 0.$$

Proof: Suppose $f(x)$ and its derivative $f'(x)$ have finite jump discontinuities at the $n - 1$ points $x_1 < x_2 < \cdots < x_{n-1}$, where $x_0 = a < x_1 < x_2 < \cdots < x_{n-1} < x_n = b$. Then, before proceeding to the limit in the lemma, using integration by parts we have

$$\int_a^b f(x) \cos(\lambda x + \mu)dx$$

$$= \sum_{r=0}^{n-1} \int_{x_r}^{x_{r+1}} f(x) \cos(\lambda x + \mu)dx$$

$$= \sum_{r=0}^{n-1} \left[\frac{f(x) \sin(\lambda x + \mu)}{\lambda} \Big|_{x=x_r}^{x_{r+1}} - \frac{1}{\lambda} \int_{x_r}^{x_{r+1}} f'(x) \sin(\lambda x + \mu)dx \right]$$

$$= \frac{1}{\lambda} \sum_{r=0}^{n-1} [f(x_{r+1} - 0) \sin(\lambda x_{r+1} + \mu) - f(x_r + 0) \sin(\lambda x_r + \mu)]$$

$$- \frac{1}{\lambda} \sum_{r=0}^{n-1} \int_{x_r}^{x_{r+1}} f'(x) \sin(\lambda x + \mu)dx.$$

The piecewise smoothness of $f(x)$, and the finiteness of its discontinuous jumps, ensures that some finite number $M > 0$ that allows us to write $|f(a + 0)| \le M, |f(b - 0)| \le M, |f(x_r \pm 0)| \le M$ for $r = 1, 2, \ldots, n-1$, and, in addition, $|f'(x)| \le M$ at each of the $n - 1$ points of discontinuity exists. Using these upper bounds, and forming the absolute value of the integral, gives

$$\left| \int_a^b f(x) \cos(\lambda x + \mu) dx \right| \leq \frac{2Mn}{\lambda} + \frac{M}{\lambda} \sum_{r=0}^{n-1} (x_{r+1} - x_r)$$

$$= \frac{M}{\lambda}(2n + b - a).$$

As M and n are finite the first result in the lemma now follows, because the expression on the right vanishes as $\lambda \to \infty$. The last two results in the lemma follow from main result by first setting $\mu = 0$ and then $\mu = \frac{\pi}{2}$. ∎

This lemma has many uses, one of the simplest of which is showing that if the coefficients a_n and b_n of the terms $\cos nx$ and $\sin nx$ in an infinite trigonometric series do not tend to 0 as $n \to \infty$, then the series *cannot* be the Fourier series of a function. This lemma provides a *necessary* condition that must be satisfied by the coefficients of a Fourier series, but *not* a sufficient one for, say, the coefficients of the trigonometric series $\sum_{n=1}^{\infty} \sin nx/\sqrt{n}$, which is convergent for all x, but is not the Fourier series of any Riemann integrable function. The justification of this assertion will be given after the derivation of the Bessel inequality, which is satisfied by the coefficients of a Fourier series. Another more important use will arise when we establish the convergence property of a Fourier series.

In order to establish the convergence properties of Fourier series the following theorem will be needed, and without loss of generality we will set $L = \pi$ in (5.2) to (5.4), causing the fundamental interval to become $-\pi \leq x \leq \pi$.

Theorem 5.1. (The Partial Sum $s_N(x)$ of a Fourier Series and the Dirichlet Kernel)
Let $f(x)$, defined over the interval $-\pi \leq x \leq \pi$, have the Fourier series repre sentation

$$f(x) = a_0 + \sum_{n=1}^{\infty} (a_n \cos nx + b_n \sin nx),$$

with

$$a_0 = \frac{1}{2\pi} \int_{-\pi}^{\pi} f(x) dx, \qquad a_n = \frac{1}{\pi} \int_{-\pi}^{\pi} f(x) \cos nx \, dx$$

$$b_n = \frac{1}{\pi} \int_{-\pi}^{\pi} f(x) \sin nx \, dx, \qquad n = 1, 2, \ldots.$$

Then the partial sum

$$s_N(x) = a_0 + \sum_{r=1}^{N} (a_r \cos rx + b_r \sin rx)$$

is given by the integral

$$s_N(x) = \frac{1}{2\pi} \int_{-\pi}^{\pi} f(x-u) \frac{\sin\left[(N+\frac{1}{2})u\right]}{\sin\frac{1}{2}u} \, du,$$

where $f(x)$ is determined outside the fundamental interval $-\pi \le x \le \pi$ by periodic extension.

Proof: The proof starts from the easily proved trigonometric identity (see Exercise 1 in Exercise Set 5.2)

$$\frac{1}{2} + \sum_{r=1}^{N} \cos rx = \frac{\sin\left(N+\frac{1}{2}\right)x}{2\sin\frac{1}{2}x}.$$

Integrating this identity over the intervals $-\pi \le x \le 0$ and $0 \le x \le \pi$ gives

$$\frac{1}{\pi}\int_{-\pi}^{0} \frac{\sin\left(N+\frac{1}{2}\right)x}{2\sin\frac{1}{2}x}\,dx = \frac{1}{\pi}\int_{0}^{\pi}\frac{\sin\left(N+\frac{1}{2}\right)x}{2\sin\frac{1}{2}x}\,dx = \frac{1}{2}. \quad (5.23)$$

Using the definitions of a_n and b_n in $s_N(x)$ gives

$$s_N(x) = \frac{1}{2\pi}\int_{-\pi}^{\pi} f(u)\,du + \frac{1}{\pi}\sum_{r=1}^{N}\left(\int_{-\pi}^{\pi} f(u)\cos rx \cos ru \, du\right.$$
$$\left. + \int_{-\pi}^{\pi} f(u)\sin rx \sin ru \, du\right),$$

where u is a dummy variable. Taking the summation under the integral sign, which is permissible since only a finite number of terms is involved, gives

$$s_N(x) = \frac{1}{\pi}\int_{-\pi}^{\pi} f(u)\left[\frac{1}{2} + \sum_{r=1}^{N}\cos r(x-u)\right] du.$$

Using the previous results obtained by integration of the trigonometric identity this now becomes

$$s_N(x) = \frac{1}{\pi}\int_{-\pi}^{\pi} f(u)\frac{\sin\left[(N+\frac{1}{2})(x-u)\right]}{2\sin\frac{1}{2}(x-u)}\,du.$$

This result is called the **Dirichlet integral representation** of $s_N(x)$.

To convert the expression for $s_N(x)$ to the form given in the theorem we make the substitution $t = x - u$, and define $f(u)$ outside its fundamental interval by periodic extension. Then, after a slight change of notation, $s_N(x)$

becomes

$$s_N(x) = \frac{1}{2\pi} \int_{-\pi}^{\pi} f(x - u) \frac{\sin\left[\left(N + \frac{1}{2}\right)u\right]}{\sin\frac{1}{2}u} du.$$

The factor

$$D_N(x - u) = \frac{1}{2} \frac{\sin\left[\left(N + \frac{1}{2}\right)(x - u)\right]}{\sin\frac{1}{2}(x - u)}$$

is called the **Dirichlet kernel**. The function $D_N(x)$ serves as a weight function for $f(u)$ when integrating over the interval $-\pi \le u \le \pi$, where as $N \to \infty$ it has the effect of concentrating the integration of $f(u)$ over an increasingly small interval about the point $u = x$ (see Exercise 12 of Exercise Set 5.2). ∎

To determine the convergence properties of a Fourier series let us suppose, for convenience, that $f(x)$ is defined over the interval $-\pi \le x \le \pi$ where it is assumed to have properties (i) to (iv) listed below. These four properties are one form of the **Dirichlet conditions for Fourier series** which are sufficient to ensure the existence of a Fourier series:

(i) $\int_{-\pi}^{\pi} [f(x)]^2 dx$ is finite. This property of $f(x)$ is identified by saying it is **square integrable** over the interval $-\pi \le x \le \pi$. The square integrability of $f(x)$ is *sufficient* to ensure $f(x)$ has a Fourier series representation, although it is *not* a necessary condition. This is because it is possible to find Fourier series representations of functions that are not square integrable. However, for all practical purposes, the assumption that $f(x)$ is square integrable is not restrictive.

(ii) $f(x)$ is piecewise continuous over the interval $-\pi \le x \le \pi$ with only a finite number of bounded jump discontinuities.

(iii) The derivatives $f'(x)$ are bounded to the immediate left and right of each jump discontinuity of $f(x)$.

(iv) $f(x)$ is assumed to be defined outside the fundamental interval $-\pi \le x \le \pi$ by its periodic extension.

We have the following theorem.

Theorem 5.2. (The Fourier Theorem)
Subject to the stated Dirichlet conditions for the existence of a Fourier series representation of $f(x)$ over the interval $-\pi \le x \le \pi$, at each point ξ of this interval

$$f(x) = \frac{1}{2}[f(x+0) + f(x-0)] = a_0 + \sum_{n=1}^{\infty} (a_n \cos nx + b_n \sin nx),$$

where

$$a_0 = \frac{1}{2\pi} \int_{-\pi}^{\pi} f(x)\,dx, \qquad a_n = \frac{1}{\pi} \int_{-\pi}^{\pi} f(x) \cos nx\,dx,$$

$$b_n = \frac{1}{\pi} \int_{-\pi}^{\pi} f(x) \sin nx\,dx, \qquad n = 1, 2, \ldots.$$

Proof: The proof makes use of the Dirichlet integral representation for $s_N(x)$ and the Riemann–Lebesgue lemma.

Let a finite jump discontinuity occur at $x = \xi$; then from Theorem 5.1

$$s_N(\xi) = \frac{1}{2\pi} \int_{-\pi}^{\pi} f(\xi - u) \frac{\sin\left[(N + \frac{1}{2})u\right]}{\sin \frac{1}{2}u}\,du.$$

Let the values of $f(x)$ to the immediate left and right of $x = \xi$ be $f(\xi - 0)$ and $f(\xi + 0)$, respectively. Then multiplication of the integrals in (5.23) by the respective constants $f(\xi - 0)$ and $f(\xi + 0)$ gives

$$\frac{1}{2} f(\xi - 0) = \frac{1}{\pi} \int_{-\pi}^{0} f(\xi - 0) \frac{\sin\left[(N + \frac{1}{2})x\right]}{2\sin \frac{1}{2}x}\,dx,$$

and

$$\frac{1}{2} f(\xi + 0) = \frac{1}{\pi} \int_{0}^{\pi} f(\xi + 0) \frac{\sin\left[(N + \frac{1}{2})x\right]}{2\sin \frac{1}{2}x}\,dx.$$

Combining these results we have

$$s_N(\xi) - \frac{1}{2}[f(\xi - 0) + f(\xi + 0)]$$

$$= \frac{1}{\pi} \int_{-\pi}^{0} [f(\xi - x) - f(\xi - 0)] \frac{\sin\left[(N + \frac{1}{2})x\right]}{\sin \frac{1}{2}x}\,dx$$

$$+ \frac{1}{\pi} \int_{0}^{\pi} [f(\xi - x) - f(\xi + 0)] \frac{\sin\left[(N + \frac{1}{2})x\right]}{\sin \frac{1}{2}x}\,dx.$$

We now write this as

$$s_N(\xi) - \frac{1}{2}[f(\xi - 0) + f(\xi + 0)]$$

$$= \frac{1}{2\pi} \int_{-\pi}^{0} \left(\frac{f(x - \xi) - f(\xi - 0)}{x}\right) \frac{x}{\sin \frac{1}{2}x} \sin\left[\left(N + \frac{1}{2}\right)x\right]dx$$

$$+ \frac{1}{2\pi} \int_{0}^{\pi} \left(\frac{f(x + \xi) - f(\xi + 0)}{x}\right) \frac{x}{\sin \frac{1}{2}x} \sin\left[\left(N + \frac{1}{2}\right)x\right]dx.$$

In the limit as $\xi \to 0$ the bracketed term in the first integral becomes the left-hand derivative of $f(x)$ at $x = \xi$, while the corresponding term in the second integral becomes the right-hand derivative of $f(x)$ at $x = \xi$, both of which are bounded by supposition, as is the term $x / \sin \frac{1}{2} x$ for all x. Thus, in each integral, the factor multiplying $\sin[(N + \frac{1}{2})x]$ is finite. An application of the Riemann–Lebesgue lemma now shows that in the limit as $N \to \infty$ the expression on the right vanishes, leaving only the result

$$\lim_{N \to \infty} s_N(\xi) - \frac{1}{2}[f(\xi - 0) + f(\xi + 0)] = 0.$$

This shows that when $f(x)$ is continuous at $x = \xi$ the Fourier series converges to $f(\xi)$, but when it is discontinuous it converges to the average of the values to either side of the jump, so the theorem is proved. ∎

From now, although an equality sign will always be used with Fourier series, its interpretation will be that provided by this theorem. We remark that although the general result of Theorem 5.2 was derived for the interval $-\pi \leq x \leq \pi$, a change of variable shows it applies to Fourier series over any interval and, of course, also to half-range Fourier series.

The consequence of this theorem has already been seen in the computer plots for Example 5.2 given in Fig. 5.5. There, for all three values $N = 20, 30, 40$, the partial sum $s_N(x)$ was seen to pass through the *midpoint* of the jump discontinuity located at the origin.

On occasion this theorem can enable infinite series to be summed in closed form. This happens, for example, when the substitution of a value of x, say $x = \eta$, into the Fourier series representation of $f(x)$ leads to a simple numerical series, because the sum of this series equals $f(\eta)$ when the function is continuous at $x = \eta$, and $\frac{1}{2}[f(\eta - 0) + f(\eta + 0)]$ when the function is discontinuous at $x = \eta$.

Example 5.4. Use the results of Example 5.2 to derive series for $\pi/4$ and $\pi^2/8$.

Solution: The function $f(x)$ in Example 5.2 is continuous at $x = 1$, where $f(1) = 1$. Setting $x = 1$ in the Fourier series and equating the result to 1 gives

$$1 = \frac{1}{2} + \frac{2}{\pi} \sum_{n=1}^{\infty} \left(\frac{[1 - (-1)^n]}{n^2 \pi} \cos \frac{n\pi}{2} + \frac{1}{n} \sin \frac{n\pi}{2} \right).$$

A simple calculation shows the first set of terms in the summation all vanishes, because when n is even the factor $[1 - (-1)]^n$ is zero, and when it is odd the factor $\cos n\pi/2$ vanishes. Furthermore, as n increases through the values 1, 2, ..., so the term $\sin(n\pi/2)$ takes on the successive values $1, 0, -1, 0, 1, \ldots$.

Hence, after simplification, the series is seen to reduce to

$$\frac{\pi}{4} = \sum_{n=1}^{\infty} \frac{(-1)^n}{2n+1}.$$

The rate of convergence of this series originally due to Leibniz (1646–1716) is very slow, because when the series is approximated by its first N terms, the consequence of neglecting the infinite tail of the series is to cause an error of magnitude equal to that of the first term of the series to have been omitted (it is an *alternating series*).

Now suppose we set $x = 0$. Then as $f(x)$ is discontinuous at this point the Fourier series must converge to $\frac{1}{2}[0+2] = 1$, while setting $x = 0$ in the Fourier series yields

$$\frac{1}{2} + \frac{2}{\pi} \sum_{n=1}^{\infty} \frac{[1 - (-1)^n]}{n^2 \pi}.$$

Equating these two results, and after simplification, we find that

$$\frac{\pi^2}{8} = \sum_{n=1}^{\infty} \frac{1}{(2n-1)^2}.$$

The rate of convergence of this series is still rather slow, although it is considerably faster than that of the previous one. ∎

Next we derive an inequality satisfied by the coefficients of a Fourier series. Consider the expression $\int_{-\pi}^{\pi} [s_N(x) - f(x)]^2 dx$, where $f(x)$ is assumed to be square integrable and $s_N(x)$ is a Fourier partial sum. Expanding this expression gives

$$\int_{-\pi}^{\pi} [s_N(x) - f(x)]^2 dx = \int_{-\pi}^{\pi} [f(x)]^2 dx - 2 \int_{-\pi}^{\pi} f(x) s_N(x) dx$$
$$+ \int_{-\pi}^{\pi} [s_N(x)]^2 dx.$$

The integral on the left is nonnegative, so we can write

$$0 \le \int_{-\pi}^{\pi} [f(x)]^2 dx - 2 \int_{-\pi}^{\pi} f(x) s_N(x) dx + \int_{-\pi}^{\pi} [s_N(x)]^2 dx.$$

Replacing $f(x)$ in the second integral by its Fourier series representation and using the orthogonality of the sine and cosine functions causes it to reduce to $-2 \int_{-\pi}^{\pi} [s_N(x)]^2 dx$. Combining the last two integrals on the right and remembering the definition of $s_N(x)$ reduces the inequality to

$$\frac{1}{\pi} \int_{-\pi}^{\pi} [f(x)]^2 dx \ge 2a_0^2 + \sum_{n=1}^{N} \left(a_n^2 + b_n^2 \right).$$

As N was arbitrary, proceeding to the limit as $N \to \infty$ we obtain **Bessel's inequality**

$$\frac{1}{\pi} \int_{-\pi}^{\pi} [f(x)]^2 dx \geq 2a_0^2 + \sum_{n=1}^{\infty} (a_n^2 + b_n^2). \tag{5.24}$$

When deriving this inequality we used the obvious fact that $\int_{-\pi}^{\pi} [s_N(x) - f(x)]^2 dx$ is nonnegative, and did not require the pointwise convergence of $s_N(x)$ to $f(x)$. If, however, we had required $s_N(x)$ to converge pointwise to $f(x)$, the result would have been that $\lim_{N \to \infty} \int_{-\pi}^{\pi} [s_N(x) - f(x)]^2 dx = 0$. The inequality in (5.24) would then have been replaced by an equality sign, strengthening the result to

$$\frac{1}{\pi} \int_{-\pi}^{\pi} [f(x)]^2 dx = 2a_0^2 + \sum_{n=1}^{\infty} (a_n^2 + b_n^2), \tag{5.25}$$

known as the **Parseval relation**, or sometimes the **Parseval formula**.

We recall we stated that the trigonometric series $\sum_{n=1}^{\infty} \sin nx/\sqrt{n}$ converges for all x, but is *not* the Fourier series of any square integrable function. We are now in a position to justify this statement. We see from Bessel's inequality that if this trigonometric series were to be the Fourier series of a function $\tilde{f}(x)$, then by Bessel's inequality

$$\frac{1}{\pi} \int_{-\pi}^{\pi} [\tilde{f}(x)]^2 dx \geq \sum_{n=1}^{\infty} \frac{1}{n}.$$

However, the series on the right is a harmonic series, which is divergent, so the Riemann integrals leading to such Fourier coefficients do not exist.

An important property of the Parseval relation is that it provides a *necessary and sufficient condition* for the existence of a Fourier series of a square integrable $f(x)$ and, conversely, when a square integrable function $f(x)$ has a Fourier series its coefficients must satisfy the Parseval relation.

It should be noted that the Parseval relation implies the Riemann–Lebesgue lemma. This follows from the fact that the Parseval relation shows the infinite series $2a_0^2 + \sum_{n=1}^{\infty} (a_n^2 + b_n^2)$ to be convergent, because its sum equals $\frac{1}{\pi} \int_{-\pi}^{\pi} [f(x)]^2 dx$. Consequently the nth term $a_n^2 + b_n^2$ of a convergent series must tend to 0 as $n \to \infty$; this is only possible if $\lim_{n \to \infty} a_n = \lim_{n \to \infty} b_n = 0$, which is simply the statement of the Riemann–Lebesgue lemma. However, the form of the proof given for Lemma 5.1 is stronger than the result obtained from the Parseval relation, because it did not require the assumption that $f(x)$ is square integrable.

A simple use of the Parseval relation is in the summation of certain types of series, as illustrated by the next example.

Example 5.5. Set $f(x) = x^2$ for $-\pi \leq x \leq \pi$, and use the result of Example 5.3 to find a series representation for π^4.

Solution: From Example 5.3 we have $a_0 = \pi^2/3$, $a_n = (-1)^n 4/n^2$, and $b_n = 0$, for $n = 1, 2, \ldots$. Substitution into the Parseval relation gives

$$\frac{1}{\pi} \int_{-\pi}^{\pi} x^2 dx = 2 \left(\frac{\pi^2}{3} \right)^2 + \sum_{n=1}^{\infty} \left(\frac{(-1)^n 4}{n^2} \right)^2, \quad \text{or} \quad \frac{2\pi^4}{5} = \frac{2\pi^4}{9} + 16 \sum_{n=1}^{\infty} \frac{1}{n^4},$$

and after simplification this becomes $\sum_{n=1}^{\infty} \frac{1}{n^4} = \frac{\pi^4}{90}$.

This series converges fairly rapidly, and summing only 25 terms and rounding the result to six decimal places yields the approximation $\pi \approx 3.141578$. ∎

By making a suitable change of variable it is a simple matter to show the Parseval relation is valid for a Fourier series representation over any interval, provided the function $f(x)$ is square integrable.

In conclusion we introduce the notion of a **mean-square approximation** $\phi_N(x)$ to a function $f(x)$ defined over an interval $a \le x \le b$, and relate it to the Fourier partial sum $s_N(x)$ of $f(x)$. Suppose $\phi_N(x) = \sum_{n=1}^{N} c_n \varphi_n(x)$, where $\varphi_n(x)$ belongs to some complete orthogonal set of functions defined over the interval $a \le x \le b$, and the c_n are constants. Then $\phi_N(x)$ is said to provide the best **mean-square approximation** to $f(x)$ in terms of the set of functions $\{\varphi_n(x)\}$ when the coefficients c_n are chosen so the mean-square error

$$E = \int_{-\pi}^{\pi} [f(x) - \phi_N(x)]^2 dx$$

is minimized.

Theorem 5.3. (The Fourier Partial Sum $s_N(x)$ Is the Best Mean-Square Trigonometric Approximation of Order N to $f(x)$)
Let $f(x)$ be defined over the interval $-\pi \le x \le \pi$ where it has the Nth-order Fourier partial sum

$$s_N(x) = a_0 + \sum_{n=1}^{N} (a_n \cos nx + b_n \sin nx).$$

Then $s_N(x)$ provides the best mean-square trigonometric approximation of order N to $f(x)$ over the interval $-\pi \le x \le \pi$, in the sense that for all trigonometric approximations of order N, $s_N(x)$ has the property that the error $E_N = \int_{-\pi}^{\pi} [f(x) - s_N(x)]^2 dx$ is minimized.

Proof: Let $f(x)$ be approximated by its Fourier partial sum $s_N(x)$, and let $\phi_N(x) = A_0 + \sum_{n=1}^{N} (A_n \cos nx + B_n \sin nx)$ be a different trigonometric approximation of order N to $f(x)$. The mean-square error E_N between

$f(x)$ and $\phi_N(x)$ is

$$E_N = \int_{-\pi}^{\pi} [f(x) - \phi_N(x)]^2 dx,$$

so

$$E_N = \int_{-\pi}^{\pi} [f(x)]^2 dx - 2\int_{-\pi}^{\pi} f(x)\phi_N(x)dx + \int_{-\pi}^{\pi} [\phi_N(x)]^2 dx.$$

After using the orthogonality of the sine and cosine functions this becomes

$$E_N = \int_{-\pi}^{\pi} [f(x)]^2 dx - 2\pi \left[2a_0 A_0 + \sum_{n=1}^{N} (a_n A_n + b_n B_n) \right]$$

$$+ \pi \left[2A_0^2 + \sum_{n=1}^{N} \left(A_n^2 + B_n^2 \right) \right],$$

which can be rewritten as

$$E_N = \int_{-\pi}^{\pi} [f(x)]^2 dx + \pi \left\{ 2(A_0 - a_0)^2 + \sum_{n=1}^{N} [(A_n - a_n)^2 + (B_n - b_n)^2] \right\}$$

$$- \pi \left[2a_0^2 + \sum_{n=1}^{N} \left(a_n^2 + b_n^2 \right) \right].$$

This shows E_N will be minimized when $a_0 = A_0, a_n = A_n, b_n = B_n$ for $n = 1, 2, \ldots, N$. Hence minimization of E_N implies that $\phi_N(x) \equiv s_N(x)$, as was to be shown. ∎

EXERCISES 5.2

1. The geometric series $\sum_{r=1}^{n} e^{irx}$ has the sum $\sum_{r=1}^{n} e^{irx} = \frac{e^{inx}-1}{1-e^{-ix}}$. Equate the real parts of this result and deduce the identity

$$\frac{1}{2} + \sum_{r=1}^{n} \cos rx = \frac{\sin\left[\left(n + \frac{1}{2}\right)x\right]}{2 \sin \frac{1}{2}x}.$$

2. Integrate the result of Exercise 1 to show that

$$\frac{1}{\pi} \int_{-\pi}^{0} \frac{\sin\left(N + \frac{1}{2}\right)x}{2 \sin \frac{1}{2}x} dx = \frac{1}{\pi} \int_{0}^{\pi} \frac{\sin\left(N + \frac{1}{2}\right)x}{2 \sin \frac{1}{2}x} dx = \frac{1}{2}.$$

3. Find a Fourier series representation for $f(x) = |x|$ in the interval $-\pi \leq x \leq \pi$ and use it to express π^2 in terms of a series.

4. Find a Fourier sine series representation of $f(x) = x(\pi - x)$ in the interval $0 \le x \le \pi$, and use it to find a series for π^3.

5. Find the Fourier series representation for $f(x) = e^{ax}$ in the interval $-\pi < x < \pi$ and use it to find a series for $\operatorname{cosech} a\pi$ and $\coth a\pi$.

6. Find a Fourier series representation for $f(x) = \begin{cases} \sin x, & 0 \le x \le \pi \\ 0, & \pi < x \le 2\pi \end{cases}$ over the interval $0 \le x \le 2\pi$, and use it with the Parseval relation to find a series for $\pi^2/8$.

7. Find a Fourier cosine series representation for $f(x) = (x - \pi)^2/4$ for $0 \le x \le \pi$, and hence show that $\sum_{n=1}^{\infty} \frac{1}{n^2} = \frac{\pi^2}{6}$.

8. A Fourier series can always be integrated term by term, and if $f(t) = a_0 + \sum_{n=1}^{\infty}(a_n \cos nt + b_n \sin nt)$ for $-\pi \le t \le \pi$, then $\int_{\alpha}^{x} f(t)dt = a_0(x - \alpha) + \sum_{n=1}^{\infty} \int_{\alpha}^{x}(a_n \cos nt + b_n \sin nt)dt$ for $-\pi \le \alpha \le x \le \pi$. However, the expression on the right is not necessarily the Fourier series of the function $\int_{\alpha}^{x} f(t)dt$, because of the presence of the term $a_0(x-\alpha)$ when $a_0 \ne 0$. Find a half-range Fourier sine series for the function $f(t) = 1$ over the interval $0 < t < \pi$. Integrate the Fourier series and use the result $\pi^2/8 = \sum_{n=1}^{\infty} 1/(2n - 1)^2$ to find a series for $\frac{1}{4}\pi(\frac{1}{2}\pi - x)$.

9. Let $f(x)$ be continuous over the interval $-\pi \le x \le \pi$ and such that $f(-\pi) = f(\pi)$. Then if $f'(x)$ is piecewise continuous over $-\pi \le x \le \pi$, at every point where $f''(x)$ exists the Fourier series of $f(x)$ may be differentiated term by term to give the Fourier series for $f'(x)$. Confirm that the function $f(x) = |\sin x|$ for $-\pi \le x \le \pi$ satisfies the above conditions. Find the Fourier series for $f(x)$ and differentiate it to find the Fourier series for $f'(x) = d/dx|\sin x|$. State the values of x for which the result does not converge to $f'(x)$.

10. Prove that the **Parseval relation** for the half-range Fourier series representations for $f(x)$ over the interval $0 \le x \le L$ are

$$\frac{2}{L}\int_0^L f(x)\cos\frac{n\pi x}{L}dx = 2a_0^2 + \sum_{n=1}^{\infty} a_n^2,$$

for the *half-range Fourier cosine series*, and

$$\frac{2}{L}\int_0^L f(x)\sin\frac{n\pi x}{L}dx = \sum_{n=1}^{\infty} b_n,$$

for the *half-range Fourier sine series*, where a_n and b_n have their usual meanings with respect to the interval $0 \le x \le L$.

11. Use the Fourier series representation $x(2 - x) = \frac{2}{3} - \frac{4}{\pi^2}\sum_{n=1}^{\infty}\frac{1}{n^2}\cos n\pi x$ with $0 \le x \le 2$ with the first result in Exercise 10 to find a series representation for π^4.

12. This problem establishes that when the product $(1/\pi)D_N(x)\phi(x)$ is integrated over the interval $-\pi \le x \le \pi$, where $D_N(x)$ is the Dirichlet kernel and $\phi(x)$ is a continuously differentiable function, then in the limit as $N \to \infty$ the function $(1/\pi)D_N(x)$ acts like the Dirac delta function $\delta(x-\xi)$, which has the property that $\int_{-\pi}^{\pi} \delta(x-\xi)\phi(x)dx = \phi(\xi)$ for any ξ in the interval $-\pi \le x \le \pi$. Let $\phi(x)$ be a continuously differentiable function over the interval $-\pi \le x \le \pi$. Prove that

$$\lim_{N\to\infty} \frac{1}{\pi} \int_{-\pi}^{\pi} D_N(x)\phi(x)dx = \phi(0),$$

by showing that

$$\frac{1}{\pi}\int_{-\pi}^{\pi} D_N(x)\phi(x)dx = \frac{1}{\pi}\int_{-\pi}^{\pi} D_N(x)[\phi(x) - \pi\phi(0)]dx,$$

substituting for $D_N(x)$, and then using the Riemann–Lebesgue lemma and L'Hospital's rule.

5.3 A Summary of the Properties of the Legendre and Bessel Differential Equations

When solving linear partial differential equations by the method of separation of variables, the geometry of physical problems may dictate that cylindrical or spherical polar coordinates should be used. These different coordinate systems will be seen to lead to the need to solve two special types of linear second-order differential equations (ODEs). One of these is **Legendre's equation**

$$(1 - x^2)y'' - 2xy' + \alpha(1 + \alpha)y = 0, \tag{5.26}$$

with α a real constant, and the other is **Bessel's equation**

$$x^2 y'' + xy' + (x^2 - \nu^2)y = 0, \tag{5.27}$$

with ν being a real constant. In the applications to PDEs to be considered later, the parameters α and ν are either zero or positive integers, and by convention when this happens it is indicated by replacing α and ν by n, with the understanding that $n = 0, 1, \ldots$.

These linear ODEs have variable coefficients, and in each case the coefficient of y'' vanishes for some value of x, thereby reducing the order of the differential equation from two to one at that point, as a result of which the solutions defined in terms of series are more complicated than solutions of constant coefficient equations. In a linear second-order equation, a point x where the coefficient of the second derivative term vanishes is called a **singular point** of the equation. Thus the Legendre equation has singular points at $x = \pm 1$, and Bessel's equation has a singular point at the origin $x = 0$.

The details of the derivations of the series solutions for the Legendre and Bessel equations are not directly relevant to the study of PDEs, so they will not be given here. However, instead, a summary of the most important and useful properties of these solutions is given below, and for more detailed information the reader is referred to one of the standard ODE reference texts mentioned in the bibliography.

General solutions of Legendre's equation occur in the form series, but it turns out that when the parameter $\alpha = n$ is zero or a positive integer a series terminates after a finite number of terms, leading to special polynomial solutions called *Legendre polynomials*. A polynomial solution of degree n exists for each integral value of n, and the importance of these polynomials is that they form a complete set of mutually orthogonal functions over the interval $-1 \leq x \leq 1$, and it is this property that makes them useful. It enables arbitrary functions defined over the interval $-1 \leq x \leq 1$ to be represented as an infinite series of Legendre polynomials, and this is crucial when seeking analytic solutions of certain types of linear PDE. Such a representation of an arbitrary function, which is similar to the Fourier series representation considered in previous sections, is called either a *Fourier–Legendre representation* or a *Fourier–Legendre expansion* of the function.

Solutions of Bessel's equation are more complicated and can only be obtained in the form of series for $v \neq \pm(2n + 1)/2$. As the equation is linear and second order, it must have two linearly independent solutions. These are called *Bessel functions*, and their form and properties depend on the parameter n. Bessel functions are oscillatory in nature, almost periodic, and each has an infinite number of zeros. Although extensive tabulations of Bessel functions are available, when needed in calculations the value of a Bessel function is usually found by using a standard subroutine found in most computer numerical packages. The functions have orthogonality properties, though of a very different kind from that of trigonometric sine and cosine functions, or Legendre polynomials. It will be seen that, unlike Legendre polynomials where orthogonality is between polynomials of *different* orders, in the case of Bessel functions the orthogonality is between Bessel functions of the *same* order, but involving different *zeros* of the Bessel function. An expansion of an arbitrary function in terms of Bessel functions is called a *Fourier–Bessel representation* or a *Fourier–Bessel expansion* of the function.

Legendre polynomials

When the parameter α in the Legendre equation (5.26) is such that $\alpha = n$, with $n = 0, 1, \ldots$, the special polynomial solutions $P_n(x)$ of degree n are called **Legendre polynomials**, and they are all defined in the interval $-1 \leq x \leq 1$. The parameter n in the Legendre equation with the polynomial solution $P_n(x)$

is called the **order** of the Legendre polynomial. The important property of the infinite set of Legendre polynomials $\{P_n(x)\}$ is that they form a complete set of orthogonal functions over the interval $-1 \le x \le 1$, with weight function $w(x) \equiv 1$.

The first few Legendre polynomials are

$$P_0(x) = 1, \qquad P_1(x) = x, \qquad P_2(x) = \frac{1}{2}(3x^2 - 1),$$

$$P_3(x) = \frac{1}{2}(5x^3 - 3x), \qquad P_4(x) = \frac{1}{8}(35x^4 - 30x^2 + 3),$$

$$P_5(x) = \frac{1}{8}(63x^5 - 70x^3 + 15x), \tag{5.28}$$

$$P_6(x) = \frac{1}{16}(231x^6 - 315x^4 + 105x^2 - 5),$$

$$P_7(x) = \frac{1}{16}(429x^7 - 693x^5 + 315x^3 - 35x).$$

Graphs of some low-order Legendre polynomials are shown in Fig. 5.7.

One of the ways Legendre polynomials of order n (an integer) can be defined is in terms of the **Rodrigues formula**

$$P_n(x) = \frac{1}{2^n n!} \frac{d^n}{dx^n}[(x^2 - 1)^n]. \tag{5.29}$$

As the Legendre weight function is $w(x) = 1$, the orthogonality of Legendre polynomials over the interval $-1 \le x \le 1$ takes the form

$$\int_{-1}^{1} P_m(x) P_n(x) dx = \begin{cases} 0, & m \ne n \\ \frac{2}{2n+1}, & m = n. \end{cases} \tag{5.30}$$

The previous statement, to the effect that the set of Legendre polynomials is *complete*, means that any arbitrary function $f(x)$ defined over the interval $-1 \le x \le 1$ can be expanded in terms of Legendre polynomials by writing

$$f(x) = \sum_{n=0}^{\infty} a_n P_n(x) \quad \text{for } -1 \le x \le 1. \tag{5.31}$$

In this **Fourier–Legendre** expansion of $f(x)$ the coefficient a_n in the expansion is found by multiplying (5.31) by $P_m(x)$, integrating the result over the interval $-1 \le x \le 1$, and then using the orthogonality result in (5.30) to determine the number a_n.

This expansion of a function $f(x)$ at points where the function is continuous can be shown to converge to the value of the function itself, but if $f(x)$ has a finite jump discontinuity at a point $x = c$ in the interval $-1 \le x \le 1$ then, like a trigonometric Fourier series, the Fourier–Legendre expansion of $f(x)$ can be shown to converge to the midpoint of the jump at $x = c$. Thus, as with

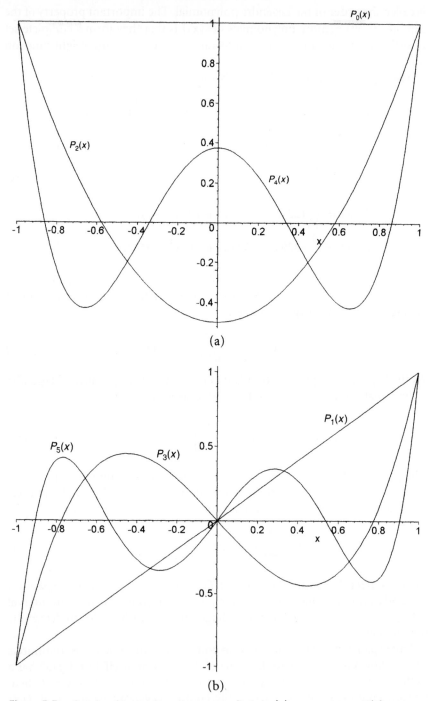

Figure 5.7   Graphs of Legendre polynomials $P_n(x)$: (a) even order and (b) odd order.

Fourier series, the meaning of the equality sign in (5.31) depends on whether the function $f(x)$ is continuous or discontinuous at a point in $-1 \le x \le 1$. The graph of any partial sum of (5.31) in a neighborhood of a jump discontinuity at $x = c$ will exhibit the same Gibbs-type phenomenon as a trigonometric Fourier series, and like Fourier series this oscillatory behavior is caused by the failure of the Fourier–Legendre expansion to converge uniformly at $x = c$.

Example 5.6. Find and plot the partial sum $s_{13}(x)$, up to and including the term in $P_{13}(x)$, in the Fourier–Legendre expansion of the discontinuous function

$$f(x) = \begin{cases} 0, & -1 \le x < 0 \\ 1, & 0 < x \le 1. \end{cases}$$

Solution: Multiplying the Fourier–Legendre expansion $f(x) = \sum_{n=0}^{\infty} a_n P_m(x)$ by $P_n(x)$, integrating over the interval $-1 \le x \le 1$, and using (5.30) gives

$$\int_{-1}^{1} f(x) P_n(x) dx = a_n \frac{2}{2n+1} \quad \text{for } n = 0, 1, \ldots,$$

so substituting for $f(x)$

$$a_n = \frac{2n+1}{2} \int_0^1 P_n(x) dx, \quad \text{for } n = 0, 1, \ldots.$$

Determining the coefficients a_n for $n = 0, 1, \ldots, 13$ gives the Fourier–Legendre approximation

$$f(x) \approx \frac{1}{2} P_0(x) + \frac{3}{4} P_1(x) - \frac{7}{16} P_3(x) + \frac{11}{32} P_5(x) - \frac{75}{256} P_7(x)$$
$$+ \frac{133}{512} P_9(x) - \frac{483}{2048} P_{11}(x) + \frac{891}{4096} P_{13}(x).$$

When simplified by computer this becomes the following polynomial approximation of degree 13

$$f(x) \approx 0.5 + 4.60727x - 52.21576x^3 + 297.62986x^5 - 850.37102x^7$$
$$+ 1267.68272x^9 - 942.90451x^{11} + 276.17617x^{13}.$$

However, to make a computer plot of this approximation, as in Fig. 5.8, this expansion in terms of x is unnecessary, because a computer numerical package makes direct use of the expansion in terms of Legendre polynomials. Figure 5.8 exhibits the typical behavior of a Fourier–Legendre expansion of a function with a jump discontinuity and, like Fourier series approximations, the plot is seen to pass through the midpoint of the discontinuity near which it exhibits localized oscillations. ■

The simplicity of the function $f(x)$ in Example 5.6 allowed the Fourier–Legendre approximation to be found in analytical form. In general this is not

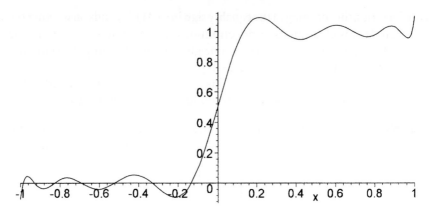

Figure 5.8   The Fourier–Legendre partial sum approximation $s_{13}(x)$.

possible, and for more complicated functions $f(x)$ the coefficients a_n must be determined by numerical integration. This is a simple matter when a computer numerical package is used.

If required, any polynomial in x can always be represented exactly in terms of Legendre polynomials by substituting for powers of x using the results

$$1 = P_0(x), \quad x = P_1(x), \quad x^2 = \frac{1}{3}\,(2\,P_2(x) + P_0(x)),$$

$$x^3 = \frac{1}{5}(2\,P_3(x) + 3\,P_1(x)),$$

$$x^4 = \frac{1}{35}(8\,P_4(x) + 20\,P_2(x) + 7\,P_0(x)), \tag{5.32}$$

$$x^5 = \frac{1}{63}(8\,P_5(x) + 28\,P_3(x) + 27\,P_1(x)), \dots .$$

Legendre polynomials and their derivatives are related by numerous **recurrence (recursion) relations**, the most important and useful of which are as follows:

1. $(n+1)\,P_{n+1}(x) = (2n+1)x\,P_n(x) - n\,P_{n-1}(x)$

2. $(x^2-1)\dfrac{d}{dx}[\,P_n(x)] = nx\,P_n(x) - n\,P_{n-1}(x)$

$$= \frac{n(n+1)}{2n+1}[\,P_{n+1}(x) - P_{n-1}(x)] \tag{5.33}$$

3. $\dfrac{d}{dx}[\,P_{n+1}(x)] - x\dfrac{d}{dx}[\,P_n(x)] = (n+1)\,P_n(x)$

4. $\dfrac{d}{dx}[\,P_{n+1}(x) - P_{n-1}(x)] = (2n+1)\,P_n(x).$

In the event that more Legendre polynomials are required than are listed in (5.28), they are easily generated by using result 1.

For the sake of completeness, we mention that for each n, associated with the polynomial solution $P_n(x)$, there is another linearly independent solution $Q_n(x)$ of the Legendre equation of order n with logarithmic singularities at $x = \pm 1$. These functions $Q_n(x)$ are called **Legendre functions of the second kind of order** n, and because of their singularities they are only defined in the open interval $-1 < x < 1$. The general solution of the Legendre differential equation of order n on the interval $-1 < x < 1$ is thus

$$y(x) = A P_n(x) + B Q_n(x), \tag{5.34}$$

with A and B arbitrary constants.

In what follows the functions $Q_n(x)$ will not be needed, so no further mention will be made of them. However, to illustrate the nature of the logarithmic singularities possessed by all Legendre functions of the second kind, we mention that

$$Q_2(x) = \frac{1}{4}(3x^2 - 1) \ln\left(\frac{1+x}{1-x}\right) - \frac{3}{2}x.$$

Bessel functions

When the order ν of Bessel's equation is not an integer the equation has two linearly independent solutions denoted by $J_\nu(x)$ and $J_{-\nu}(x)$, where

$$J_\nu(x) = x^\nu \sum_{k=0}^{\infty} \frac{(-1)^k x^{2k}}{2^{2k+\nu} k! \Gamma(\nu + k + 1)}, \tag{5.35}$$

from which $J_{-\nu}(x)$ follows by reversing the sign of ν.

In (5.35), $\Gamma(\nu + k + 1)$ is the **gamma function** $\Gamma(x)$ with $x = \nu + k + 1$. The gamma function is defined by the improper integral

$$\Gamma(x) = \int_0^\infty t^{x-1} e^{-t} dt, \tag{5.36}$$

and it satisfies the recurrence relation

$$\Gamma(x + 1) = x \Gamma(x). \tag{5.37}$$

When x is an integer n, the gamma function has the property that

$$\Gamma(n + 1) = n!, \tag{5.38}$$

so it provides an extension of the factorial function $n! = 1 \cdot 2 \cdot 3 \cdots n$ to the case when n is replaced by an arbitrary real number x.

More details of the gamma function can be found in the references quoted in the bibliography in connection with series solutions of ODEs. For reference

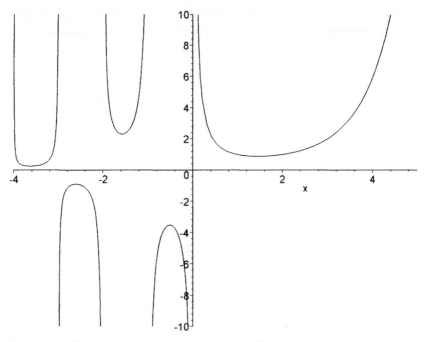

Figure 5.9   The Gamma function for $-4 < x \leq 5$.

purposes a plot of the gamma function is shown in Fig. 5.9, from which it can be seen it is defined for both positive and negative values of x, although it becomes infinite when x is zero or a negative integer.

Returning to the Bessel function $J_\nu(x)$ when ν is an integer n, the two solutions $J_n(x)$ and $J_{-n}(x)$ are linearly dependent, because

$$J_{-n}(x) = (-1)^n J_n(x). \tag{5.39}$$

To obtain a second solution that is *always* linearly independent of $J_n(x)$, irrespective of the value of ν, a new function $Y_\nu(x)$ is defined as

$$Y_\nu(x) = \frac{J_\nu(x)\cos(\nu\pi) - J_{-\nu}(x)}{\sin(\nu\pi)}. \tag{5.40}$$

Thus the general solution of Bessel's equation

$$x^2 y'' + x y' + (x^2 - \nu^2) y = 0$$

for any ν can *always* be written

$$y(x) = A J_\nu(x) + B Y_\nu(x), \tag{5.41}$$

with A and B arbitrary constants. In some books the Bessel function $Y_\nu(x)$ is denoted by $N_\nu(x)$ in recognition of the fact that this definition, which is valid for all ν, was introduced by Carl Neumann (1832–1925).

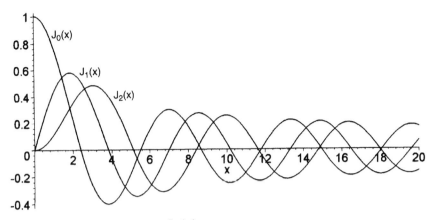

Figure 5.10   Bessel functions $J_n(x)$.

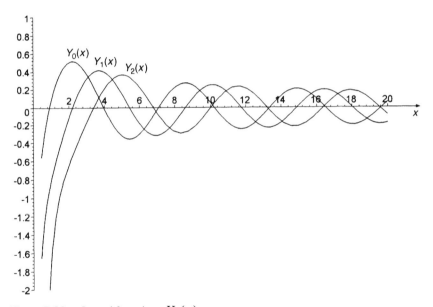

Figure 5.11   Bessel functions $Y_n(x)$.

The functions $J_\nu(x)$ are called **Bessel functions of the first kind of order** ν, and the functions $Y_\nu(x)$ are called **Bessel functions of the second kind of order** ν.

Plots of $J_0(x)$, $J_1(x)$, and $J_2(x)$ are shown in Fig. 5.10, and plots of $Y_0(x)$, $Y_1(x)$, and $Y_2(x)$ are shown in Fig. 5.11. The functions $J_n(x)$ and $Y_n(x)$ are oscillatory, with $J_n(x)$ remaining finite for all x with

$$\lim_{x \to \infty} J_n(x) = 0, \qquad (5.42)$$

and

$$\lim_{x\to 0} Y_n(x) = -\infty \quad \text{and} \quad \lim_{x\to\infty} Y_n(x) = 0 \quad \text{for } n = 0, 1, 2, \ldots . \quad (5.43)$$

Approximations for $J_\nu(x)$ and $Y_\nu(x)$ when x is large are

$$J_\nu(x) \sim \sqrt{\frac{2}{\pi x}} \left[\cos\left(x - \frac{1}{2}\nu\pi - \frac{1}{4}\pi \right) \right] \quad \text{and}$$

$$Y_\nu(x) \sim \sqrt{\frac{2}{\pi x}} \left[\sin\left(x - \frac{1}{2}\nu\pi - \frac{1}{4}\pi \right) \right] \quad (x \gg 0). \qquad (5.44)$$

The tilde symbol $\sim$ is used to indicate that these are **asymptotic representations** of $J_\nu(x)$ and $Y_\nu(x)$. This means the ratio of each Bessel function and its asymptotic representation tends to 1 as x increases, so the larger x becomes, the closer the approximations in (5.44) approach the true functional values of $J_\nu(x)$ and $Y_\nu(x)$. These asymptotic approximations show $J_\nu(x)$ and $Y_\nu(x)$ each possess an infinite number of real zeros closely related to those of the sine and cosine functions, with each zero occurring with multiplicity 1, so they are all *simple* zeros.

Some special values of the Bessel function $J_n(x)$ at the origin $x = 0$, which are useful when working with PDEs, are

$$J_0(0) = 1, \qquad J_n(0) = 0 \quad [n = 1, 2, \ldots] \qquad\qquad (5.45)$$

$$J_0'(0) = 0, \qquad J_1'(0) = \frac{1}{2}, \qquad J_n'(0) = 0 \quad [n = 2, 3, \ldots]. \quad (5.46)$$

In applications, Bessel's equation of order ν frequently occurs in the slightly different form

$$x^2 y'' + xy' + (\lambda^2 x^2 - \nu^2)y = 0, \qquad (5.47)$$

when a simple change of variable shows its general solution to be given by

$$y(x) = AJ_\nu(\lambda x) + BY_\nu(\lambda x). \qquad (5.48)$$

In certain boundary value problems involving Bessel functions it is necessary to know the zeros of $J_n(x)$ for $x > 0$. Table 5.1 lists the first few of these zeros, using the notation $j_{n,m}$ to denote the mth zero of $J_n(x)$.

The zeros $j_{n,m}$ have the property that $j_{n,m} < j_{n,m+1}$ for $m = 1, 2, \ldots$, and, in addition, the zeros *interlace* in the sense that

$$j_{n-1,m} < j_{n,m} < j_{n-1,m+1}. \qquad (5.49)$$

m	$j_{0,m}$	$j_{1,m}$	$j_{2,m}$	$j_{3,m}$	$j_{4,m}$
1	2.4048	3.8317	5.1356	6.3802	7.5883
2	5.5201	7.0156	8.4172	9.7610	11.0647
3	8.6537	10.1735	11.6198	13.0152	14.3725
4	11.7915	13.3237	14.7960	16.2235	17.6160
5	14.9309	16.4706	17.9598	19.4094	20.8269
6	18.0711	19.6159	21.1170	22.5827	24.0190
7	21.2116	22.7601	24.2701	25.7482	27.1991
8	24.3525	25.9037	27.4206	28.9084	30.3710
9	27.4935	29.0468	30.5692	32.0649	33.5371

Table 5.1 The Zeros $j_{n,m}$

Bessel functions and their derivatives are related by many *recurrence (recursion) relations*, the most important and useful of which are as follows:

Relationships involving $J_n(x)$	Relationships involving $Y_n(x)$
1. $\frac{d}{dx}[J_0(x)] = -J_1(x)$	$\frac{d}{dx}[Y_0(x)] = -Y_1(x)$
2. $\frac{d}{dx}[x^n J_n(x)] = x^n J_{n-1}(x)$	$\frac{d}{dx}[x^n Y_n(x)] = x^n Y_{n-1}(x)$
3. $2\frac{d}{dx}[J_n(x)] = J_{n-1}(x) - J_{n+1}(x)$	$2\frac{d}{dx}[Y_n(x)] = Y_{n-1}(x) - Y_{n+1}(x)$
4. $x\frac{d}{dx}[J_n(x)] + nJ_n(x) = xJ_{n-1}(x)$	$x\frac{d}{dx}[Y_n(x)] + nY_n(x) = xY_{n-1}(x)$
5. $x\frac{d}{dx}[J_n(x)] - nJ_n(x) = -xJ_{n+1}(x)$	$x\frac{d}{dx}[Y_n(x)] - nY_n(x) = -xY_{n+1}(x)$
6. $xJ_{n-1}(x) + xJ_{n+1}(x) = 2nJ_n(x)$	$xY_{n-1}(x) + xY_{n+1}(x) = 2nY_n(x)$

$$(5.50)$$

The following are some useful integrals, although for conciseness the arbitrary additive constants have been omitted from the indefinite integrals:

1. $\displaystyle\int xJ_0(ax)dx = \frac{x}{a}J_1(ax)$

2. $\displaystyle\int J_1(ax)dx = -\frac{1}{a}J_0(ax)$

3. $\displaystyle\int x^n J_{n-1}(ax)dx = \frac{x^n}{a}J_n(ax)$

4. $\displaystyle\int x^{-n} J_{n+1}(ax)dx = -\frac{1}{ax^n}J_n(ax)$

5. $\displaystyle\int x^3 J_0(ax)dx = (1/a^3)[a^2x^3 J_1(ax) - 4xJ_1(ax) + 2ax^2 J_0(ax)]$

   $(a > 0)$

6. $\displaystyle\int x[J_0(ax)]^2 dx = \frac{1}{2}x^2([J_0(ax)]^2 + [J_1(ax)]^2)$ $(a > 0)$ (5.51)

7. $\displaystyle\int_0^\infty J_n(ax)dx = \frac{1}{a}$ $(n > -1, a > 0)$

8. $\displaystyle\int_0^\infty \frac{J_n(ax)}{x} dx = \frac{1}{n}$ $(n = 1, 2, \ldots)$.

The orthogonality of Bessel functions $J_n(x)$ over the interval $0 \le x \le R$ is between Bessel functions $J_n(x)$ of the *same* order (that is, with n fixed) and with respect to a weight function $w(x) = x$, although the arguments involve *different* zeros of $J_n(x)$. The orthogonality condition for Bessel functions of the first kind takes the form

$$\int_0^R x J_n\left(\frac{j_{n,r}x}{R}\right) J_n\left(\frac{j_{n,s}x}{R}\right)dx = \begin{cases} 0, & r \ne s \\ \frac{R^2}{2}[J_{n+1}(j_{n,r})]^2, & r = s. \end{cases}$$ (5.52)

This result is of fundamental importance when expanding an arbitrary function $f(x)$ over the interval $0 \le x \le R$ using a *Fourier–Bessel* expansion in terms of $J_n(x)$. The sets of functions $J_n(j_{n,1}x/R), J_n(j_{n,2}x/R), J_n(j_{n,3}x/R), \ldots$, for $n = 0, 1, 2, \ldots$ form a complete set of orthogonal functions over the interval $0 \le x \le R$ with weight function $w(x) = x$, so for any fixed n a **Fourier–Bessel expansion** of a function $f(x)$ takes the form

$$f(x) = \sum_{m=1}^\infty a_m J_n\left(\frac{j_{n,m}x}{R}\right) = a_1 J_n\left(\frac{j_{n,1}x}{R}\right) + a_2 J_n\left(\frac{j_{n,2}x}{R}\right)$$

$$+ a_3 J_n\left(\frac{j_{n,3}x}{R}\right) + \cdots.$$ (5.53)

The Fourier–Bessel coefficients a_m are obtained by multiplying (5.53) by $x J_n$ $(j_{n,m}x)$, integrating over the interval $0 \le x \le R$, and using (5.52) to show that

$$a_m = \frac{2}{R^2[J_{n+1}(j_{n,m})]^2} \int_0^R x f(x) J_n\left(\frac{j_{n,m}x}{R}\right)dx \quad \text{for } m = 1, 2, \ldots.$$ (5.54)

Example 5.7. Find the Fourier–Bessel expansion of the function $f(x) = \sin \pi x$ over the interval $0 \le x \le 3$ in terms of an appropriate Bessel function $J_n(x)$, and plot the result.

Solution: It is possible to work with any fixed positive integral value of n, but an inspection of Fig. 5.10 suggests the result can be expected to converge rapidly if we choose $n = 1$, because after scaling, the graph of J_1 resembles that of $\sin \pi x$. The coefficients a_m follow from (5.54) by setting $f(x) = \sin \pi x$ and $R = 3$. As the integrals determining the coefficients a_n cannot be found

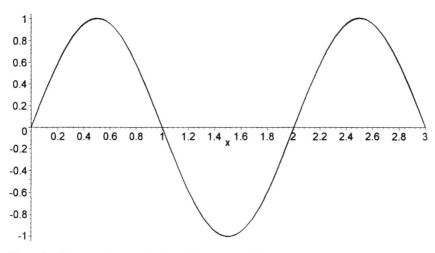

Figure 5.12   Superimposed plots of $\sin \pi x$ and the Fourier–Bessel approximation.

analytically, the calculations must be performed numerically. The results obtained from a computer numerical package are $a_1 = -0.0416$, $a_2 = -0.8917$, $a_3 = 2.6171$, $a_4 = -0.1390$, $a_5 = 0.0464$. Thus the Fourier–Bessel approximation over the interval $0 \le x \le 3$ using five terms becomes

$$\sin \pi x \approx -0.0416 J_1(3.8317x/3) - 0.8917 J_1(7.0156x/3)$$
$$+ 2.6171 J_1(10.1735x/3) - 0.1390 J_1(13.3237x/3)$$
$$+ 0.0464 J_1(16.4706x/3).$$

A plot of this approximation superimposed on one of $f(x) = \sin \pi x$ is shown in Fig. 5.12, where the plots are virtually indistinguishable. To illustrate the rapidity with which this particular approximation converges to $f(x)$, Fig. 5.13 shows superimposed plots of $f(x)$ and the result of using only the first three terms of the Fourier–Bessel approximation. When $x = 0.5$ the three-term approximation gives 1.0616, when $x = 1.5$ it gives -1.0232, and when $x = 2.5$ it gives 0.9741, which with only three terms are close to the true values $1, -1$, and 1. This rate of convergence of a Fourier–Bessel approximation to $f(x) = \sin \pi x$ is unusually rapid, and in general many terms are needed if the error is to be small, particularly when $f(x)$ has a discontinuity.   ∎

Two other Bessel functions of importance are the functions $I_n(x)$ and $K_n(x)$ called, respectively, **modified Bessel functions of the first and second kind of order** ν. These are linearly independent solutions of the **modified Bessel equation** of order ν,

$$x^2 y'' + x y' + (x^2 + \nu^2) y = 0, \tag{5.55}$$

with the general solution

$$y(x) = A I_\nu(x) + B K_\nu(x).$$

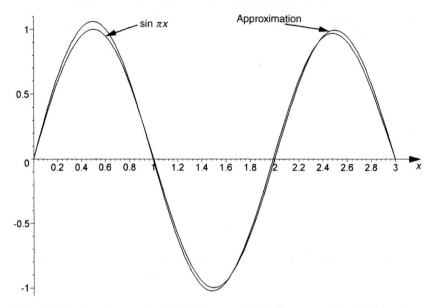

Figure 5.13   Superimposed plots of $\sin \pi x$ and the three-term Fourier–Bessel approximation.

When the modified Bessel equation arises in the slightly more general form

$$x^2 y'' + xy' + (\lambda^2 x^2 + v^2)y = 0, \tag{5.56}$$

its general solution becomes

$$y(x) = A I_v(\lambda x) + B K_v(\lambda x). \tag{5.57}$$

The functions $I_v(x)$ and $K_v(x)$ are defined as

$$I_v(x) = i^{-v} J_v(ix) = \sum_{k=0}^{\infty} \frac{x^{v+2k}}{2^{v+2k} k! \Gamma(v+k+1)} \tag{5.58}$$

and

$$K_v(x) = \frac{\frac{1}{2}\pi [I_{-v}(x) - I_v(x)]}{\sin v\pi}, \tag{5.59}$$

where $K_n(x)$ follows from (5.59) in the limit as $v \to n$.

It can be seen from (5.58), and the relationship between $I_n(x)$ and $J_n(x)$, that $I_n(x)$ has no real zeros, but an infinite number of purely imaginary ones. The function $K_v(x)$ has a singularity at the origin of the form x^{-v} when $v \neq 0$, and a logarithmic singularity at the origin when $v = 0$.

Plots of $I_0(x)$ and $I_1(x)$ are shown in Fig. 5.14, while plots of $K_0(x)$ and $K_1(x)$ are shown in Fig. 5.15.

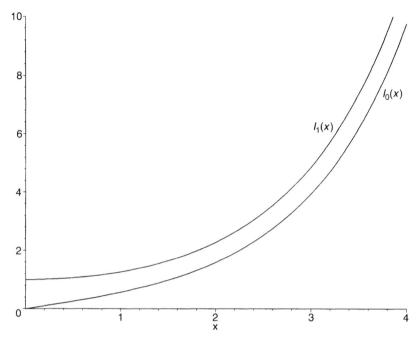

Figure 5.14   Modified Bessel functions $I_0(x)$ and $I_1(x)$.

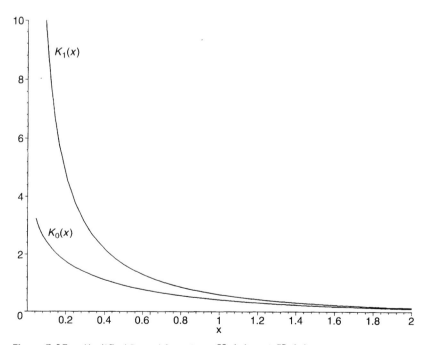

Figure 5.15   Modified Bessel functions $K_0(x)$ and $K_1(x)$.

Some useful limits involving modified Bessel functions are

$$\lim_{\substack{x \to 0 \\ \nu \neq 0}} I_\nu(x) = 0, \quad \lim_{x \to \infty} I_\nu(x) = \infty, \quad \lim_{x \to 0} K_\nu(x) = \infty, \quad \lim_{x \to \infty} K_\nu(x) = 0.$$

$$\tag{5.60}$$

Asymptotic representations of modified Bessel functions when x is large are

$$I_\nu(x) \sim \frac{e^x}{\sqrt{2\pi x}} \left[1 - \frac{4\nu^2 - 1}{8x} \right] \quad \text{and}$$

$$K_\nu(x) \sim \sqrt{\frac{\pi}{2x}} e^{-x} \left[1 + \frac{4\nu^2 - 1}{8x} \right] \quad (x \gg 0). \tag{5.61}$$

Modified Bessel functions satisfy recurrence relations similar to those in (5.50), although these are not listed here as they are not required later.

EXERCISES 5.3

Using a numerical software package find the following approximations and plot the results.

1. Find the Fourier–Legendre approximation up to and including the term in $P_9(x)$ of $f(x) = \begin{cases} 0, & -1 < x \leq 0 \\ \sin \frac{1}{2}\pi x, & 0 < x < 1 \end{cases}$ in the interval $-1 \leq x \leq 1$. Plot the approximation and compare it with the plot of $f(x)$.

2. Find the Fourier–Legendre approximation up to and including the term in $P_{10}(x)$ of $f(x) = 1 - |x|$ in the interval $-1 \leq x \leq 1$. Plot the approximation and compare it with the plot of $f(x)$.

3. Find the Fourier–Legendre approximation up to and including the term in $P_8(x)$ of $f(x) = \sin \pi x + \cos 2\pi x$ in the interval $-1 \leq x \leq 1$. Plot the approximation and compare it with the plot of $f(x)$.

4. Find the Fourier–Bessel approximation of $f(x) = 1 - x$ in the interval $0 \leq x \leq 1$ in terms of J_0, up to and including the term involving $j_{0,9}$. Plot the approximation and compare it with the plot of $f(x)$.

5. Find the Fourier–Bessel approximation of $f(x) = x$ in the interval $0 \leq x \leq 1$ in terms of J_1, up to and including the term involving $j_{1,8}$. Plot the approximation and compare it with the plot of $f(x)$.

6. Find the Fourier–Bessel approximation of $f(x) = \sin \pi x + 1.5 \cos 2\pi x$ in the interval $0 \leq x \leq 1$ in terms of J_0, up to and including the term in $j_{0,8}$. As each term in the approximation vanishes at $x = 1$, but $f(1) = 1.5$, comment on the way you would expect an approximation up to the term in $j_{0,n}$ to represent $f(x)$ in the neighborhood of $x = 1$ as n increases.

Background to Separation of Variables with Applications

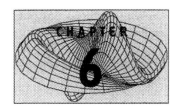

6.1 A General Approach to Separation of Variables

The method known as *separation of variables* is used to find an analytical solution of a linear PDE in the form of a series solution for Cauchy problems, boundary value problems in space, and problems satisfying both initial conditions (usually with $t = 0$) and space boundary conditions. A problem with initial and boundary conditions is called an initial boundary value problem, although for conciseness the name is often abbreviated to an IBVP.

The method of separation of variables can be applied to PDEs with independent variables belonging to one of the most frequently used systems of orthogonal coordinates, when the physical boundaries in space all coincide with constant coordinate lines or surfaces. In time-dependent problems the solution is found in terms of an infinite series, each term of which is a product of a function of time and a function of space variables. The arguments of each of these functions depend on constants called the *eigenvalues* of the problem, and the function of space is called an *eigenfunction* of the PDE. A corresponding form of solution arises, although without functions of time entering into the terms when, like the Laplace equation, the PDE is independent of the time.

Typically, in one space dimension, a region D may be the finite interval $a \leq x \leq b$; in two space dimensions it may be a rectangle, a sector of a circle, or a sector of an annulus, while in three space dimensions it may, for example, be a rectangular parallelepiped, a cylinder, a spherical sector, or the region between two concentric spheres.

To motivate the arguments to follow, we first solve a simple boundary value problem by applying the separation of variables method to the two-dimensional

Laplace equation, and only afterward will the individual steps involved be examined in detail.

Example 6.1. Solve the following boundary value problem for the Laplace equation

$$\frac{\partial^2 u}{\partial x^2} + \frac{\partial^2 u}{\partial y^2} = 0 \quad \text{for } 0 < x < a,\ 0 < y < b,$$

with $u(x, 0) = 0$, $u(x, b) = \sin^2 \frac{\pi x}{a}$, $\frac{\partial u}{\partial x}\big|_{x=0} = 0$, and $\frac{\partial u}{\partial x}\big|_{x=a} = 0$.

Solution: Before attempting to solve a PDE subject to auxiliary conditions it is necessary to check that the problem is well posed. Here the PDE is the Laplace equation, and this is an elliptic equation. It will be recalled from Section 3.3 that sufficient conditions for an elliptic equation and its auxiliary conditions to be well posed are that the region involved is closed, and the boundary conditions are of Dirichlet or Neumann type or, possibly, a linear combination of these conditions. In this problem all of these conditions are satisfied, so we may proceed with the solution.

The Laplace equation is expressed in terms of Cartesian coordinates with boundary conditions imposed on lines parallel to the coordinate axes, so a solution by the method of separation of variables can be used. A possible physical interpretation of this problem is that it determines the steady-state temperature distribution in a uniform sheet or block of metal, in which case it can be considered to represent either of the following situations.

The first interpretation involves the determination of the steady-state temperature distribution in a uniform and very long bar of metal of constant rectangular cross section, subject to the given boundary conditions imposed on its plane sides. In this case the bar is assumed to extend sufficiently far in the z direction for the conditions at the ends of the bar to be neglected, in which case the solution represents the situation in any cross section, normal to the axis of the bar, that is well removed from the ends of the bar. Such a solution could describe the steady-state temperature distribution in any rectangular cross section of the bar that depends only on x and y, and so is determined by a Laplace equation independent of z. The boundary conditions correspond to two faces of the bar being thermally insulated, a third face at zero temperature while the fourth face has a sinusoidal temperature distribution across it.

In the second interpretation, the problem can be considered to describe the steady-state temperature distribution in a uniform thin rectangular sheet of metal with its plane faces perpendicular to the z axis thermally insulated, while the stated boundary conditions are imposed along the edges parallel to the axes in the (x, y) plane.

The connection between the two interpretations follows from the fact that since there is no heat flow in the z direction in either case, the thin sheet of metal can be considered to be a thin slice of the bar contained between any two close parallel rectangular cross sections perpendicular to the axis of the bar.

The method of separation of variables starts by seeking special solutions of the form $u(x, y) = X(x)Y(y)$, where $X(x)$ is a function only of x and $Y(y)$ is a function only of y. When $u(x, y)$ is substituted into the Laplace equation, as x and y are independent variables, the partial derivatives $u_x(x, y) = \partial/\partial x[u(x, y)] = Y(y)dX/dx$, with similar results for $u_y(x, y), u_{xx}(x, y)$ and $u_{yy}(x, y)$. As a result the Laplace equation simplifies to

$$Y(y)\frac{d^2 X(x)}{dx^2} + X(x)\frac{d^2 Y(y)}{dy^2} = 0.$$

For convenience, using a prime to indicate differentiation, this can be written more concisely as

$$YX'' + XY'' = 0,$$

where, of course, $X'' = d^2 X/dx^2$ and $Y'' = d^2 Y/dy^2$. In the next step, leading to the derivation of ordinary differential equations to be satisfied by $X(x)$ and $Y(y)$, we now divide this result by XY and rewrite it as

$$\frac{X''}{X} = -\frac{Y''}{Y}.$$

There now follows a fundamental step in the method of separation of variables, in that as the expression X''/X depends only on x, and the expression Y''/Y depends only on y, the only way these results can be equal in the rectangle $0 < x < a, 0 < y < b$ for all x and y is for each expression to equal the same absolute constant μ, say, whose sign has yet to be determined. This means we can now write

$$\frac{X''}{X} = -\frac{Y''}{Y} = \mu.$$

The introduction of the constant μ enables the variables X and Y to be *separated*, or uncoupled, so the last result must be equivalent to the two separate ordinary differential equations

$$X'' - \mu X = 0 \quad \text{and} \quad Y'' + \mu Y = 0.$$

We must now determine the constant μ, which for obvious reasons is called a **separation constant**. Solving these ordinary differential equations, and for the moment assuming the separation constant $\mu > 0$, gives

$$X(x) = A\exp(\sqrt{\mu}x) + B\exp(-\sqrt{\mu}x) \quad \text{and}$$
$$Y(y) = C\cos(\sqrt{\mu}y) + D\sin(\sqrt{\mu}y).$$

As solutions of the form $u(x, y) = X(x)Y(y)$ are required, the boundary condition $\partial u/\partial x|_{x=0} = 0$ implies that $Y(y)X'(0) = 0$, while the boundary condition $\partial u/\partial x|_{x=a} = 0$ implies that $Y(y)X'(a) = 0$. The function $Y(y)$ cannot be identically zero, because then $u(x, y) \equiv 0$, violating the boundary condition on $y = b$, so we conclude $X(x)$ must satisfy the boundary conditions $X'(0) = 0$ and $X'(a) = 0$.

Of all four boundary conditions, only those at $x = 0$ and $x = a$ are simple enough to be used directly, and they allow us to solve for X. When this is done, we find that imposing the boundary condition $X'(0) = 0$ on the expression for $X(x)$ found previously leads to the result

$$\sqrt{\mu}\,A - \sqrt{\mu}\,B = 0,$$

while imposing the boundary condition $X'(a) = 0$ leads to the result

$$\sqrt{\mu}\,A \exp(\sqrt{\mu}\,a) - \sqrt{\mu}\,B \exp(-\sqrt{\mu}\,a) = 0.$$

Assuming $\sqrt{\mu} \neq 0$, and canceling the factor $\sqrt{\mu}$ from the first equation, shows $B = A$, so from the second equation we find that $\exp(\sqrt{\mu}\,a) = \exp(-\sqrt{\mu}\,a)$. This last condition is only possible if $\sqrt{\mu} = 0$, contradicting our original assumption that $\sqrt{\mu} \neq 0$, so the assumption that $\mu > 0$ was incorrect. Henceforth we will assume that $\mu \leq 0$ and, for convenience, set $\mu = -\lambda^2$, so that now with this change of sign of μ we must set

$$X(x) = A\cos(\lambda x) + B\sin(\lambda x) \quad \text{and} \quad Y(y) = C\exp(\lambda y) + D\exp(-\lambda y).$$

Applying the boundary condition $X'(0) = 0$ to this new expression for $X(x)$ gives

$$(-\lambda A \sin(\lambda x) + \lambda B \cos(\lambda x))\big|_{x=0} = 0, \quad \text{so } B = 0,$$

and so

$$X(x) = A\cos(\lambda x).$$

Imposing the second boundary condition $X'(a) = 0$ on $X(x)$ leads to the result

$$-\lambda A \sin(\lambda a) = 0.$$

Clearly $A \neq 0$, because then the solution would vanish identically, so this last result can only be possible if $\lambda = 0$ or $\lambda a = n\pi$ for $n = 1, 2, \ldots$. This argument shows λ can have any one of the infinite set of values $\lambda_n = n\pi/a$ for $n = 0, 1, 2, \ldots$. Later we will see the numbers λ are called the *eigenvalues* of the differential equation for $X(x)$, and that the differential equation for $X(x)$, together with its boundary conditions, constitutes what will be called a *Sturm–Liouville problem* for $X(x)$.

Setting $\lambda = 0$ in the differential equation for $X(x)$ reduces it to $X'' = 0$, from which it follows that $X(x) = c_0 + c_1 x$. An application of the boundary conditions $X'(0) = X'(a) = 0$ to this result shows that $c_1 = 0$, so corresponding to $\lambda = 0$ we have the solution $X_0(x) = c_0$.

When $\lambda \neq 0$ there is a different solution $X_n(x)$ for each value of n so, as the solutions can be scaled by an arbitrary nonzero constant c_n, they can be written $X_n(x) = c_n \cos(n\pi x/a)$ for $n = 1, 2, \ldots$. We have now established that for arbitrary constants $c_0, c_1, \ldots,$

$$X_0(x) = c_0 \quad \text{and} \quad X_n(x) = c_n \cos \frac{n\pi x}{a}, \quad \text{for } n = 1, 2, \ldots.$$

Subsequently, the solutions $X_n(x)$ with the arbitrary constants set equal to unity will be called the *eigenfunctions* of the Sturm–Liouville problem for $X(x)$, corresponding to the respective *eigenvalues* λ_n.

As the permissible values of the separation constant λ (the eigenvalues) are now known, we can proceed to solve for $Y(y)$. Setting $\lambda = \lambda_0 = 0$ in the differential equation for y reduces it to $Y'' = 0$. The boundary condition $u(x, 0) = 0$ implies that $X(x)Y(0) = 0$, but $X(x)$ cannot vanish identically because then $u(x, y) \equiv 0$, so we conclude that $Y(0) = 0$. Solving the equation $Y'' = 0$, and using the boundary condition $Y(0) = 0$, shows that $Y_0(y) = d_0 y$, with d_0 an arbitrary constant. Correspondingly, when $\lambda = \lambda_n$ for $n = 1, 2, \ldots$, we see that from the general solution for $Y(y)$ that the boundary condition $Y_n(0) = 0$ implies that $C_n = -D_n$, so we may write this result in the more convenient form involving hyperbolic functions

$$Y_n(y) = D_n \sinh \frac{n\pi y}{a}, \quad \text{for } n = 1, 2, \ldots.$$

Recalling now that $u(x, y) = X(x)Y(y)$, we find that elementary partial solutions of the boundary value problem are given by

$$u_n(x, y) = X_n(x)Y_n(y) \quad \text{for } n = 0, 1, \ldots.$$

These are only *partial solutions*, because they satisfy the boundary conditions at $x = 0$ and $x = a$, but not those at $y = b$. Because Laplace's equation is linear and homogeneous, if $u_n(x, y)$ is a partial solution then so also is $ku_n(x, y)$, with k an arbitrary constant. The next step in the method of separation of variables depends on the fact that, due to the linearity of the Laplace equation, any sum of multiples of partial solutions of the equation will itself be a partial solution.

At this stage of the argument the constant coefficients $c_0, c_1, \ldots$ and $d_0, D_1, \ldots$ are arbitrary, so products of these coefficients must also be arbitrary. Forming the products $X_n Y_n$, and setting $a_0 = c_0 D_0$, $a_1 = c_1 D_1$, $a_2 = c_2 D_2, \ldots$ we see that we must seek a solution of the full problem of the form

$$u(x, y) = \sum_{n=0}^{\infty} a_n u_n(x, y) = a_0 y + \sum_{n=1}^{\infty} a_n \sinh \frac{n\pi y}{a} \cos \frac{n\pi x}{a},$$

where the coefficients $a_0, a_1, \ldots$ must still be determined.

To find these coefficients we must appeal to the boundary condition $u(x, b) = \sin^2(\pi x/a)$, which so far has not been used. Setting $y = b$ in the above expression for $u(x, y)$ and replacing $u(x, b)$ by the boundary condition $\sin^2(\pi x/a)$

gives

$$\sin^2 \frac{\pi x}{a} = a_0 b + \sum_{n=1}^{\infty} a_n \sinh \frac{n\pi b}{a} \cos \frac{n\pi x}{a},$$

where the expression on the right must be the Fourier cosine series expansion of $\sin^2(\pi x/a)$ over the interval $0 \le x \le a$, with the Fourier coefficients $a_n \sinh(n\pi b/a)$.

In more general cases the coefficients $a_n \sinh(n\pi b/a)$ would be found by using the orthogonality of the set of cosine functions $\{1, \cos(\pi x/a), \cos(2\pi x/a), \ldots\}$ over the interval $0 \le x \le a$. This would involve multiplying the result by $\cos(m\pi x/a)$ and integrating the result over the interval $0 \le x \le a$. However, the simplicity of the boundary condition $u(x, b) = \sin^2(\pi x/a)$ makes this unnecessary, because

$$\sin^2(\pi x/a) = \frac{1}{2}[1 - \cos(2\pi x/a)],$$

so we can write

$$\frac{1}{2}[1 - \cos(2\pi x/a)] = a_0 b + \sum_{n=1}^{\infty} a_n \sinh \frac{n\pi b}{a} \cos \frac{n\pi x}{a}.$$

As this result must be an identity, the unknown constants a_n can be found by equating the coefficients of corresponding cosine terms on either side of the equality sign, when we find that

$$\frac{1}{2} = a_0 b, \quad a_1 = 0, \quad -\frac{1}{2} = a_2 \sinh \frac{2\pi b}{a}, \quad \text{and} \quad a_n = 0 \quad \text{for } n = 3, 4, \ldots.$$

The only nonzero coefficients are thus $a_0 = 1/(2b)$ and $a_2 = -1/[2\sinh(2\pi b/a)]$, so substituting into the series solution for $u(x, y)$ it reduces to the simple expression

$$u(x, y) = \frac{y}{2b} - \frac{\sinh(2\pi y/a)\cos(2\pi x/a)}{2\sinh(2\pi b/a)}.$$

It is easily verified that this expression satisfies both Laplace's equation and the boundary conditions, so this **closed form solution** (an explicit solution comprising a finite number of terms) is a **classical solution** of the boundary value problem. Figure 6.1 shows a plot of the solution when $a = b = 1$. ∎

We now examine more generally, and in greater detail, the individual steps used in the above application of the method of separation of variables.

Consideration of the examples of second-order linear PDEs derived in Section 1.2 shows they are all special cases of one of the more general PDEs

$$\text{div}(k \text{ grad } u) - pu = wu_t, \tag{6.1}$$

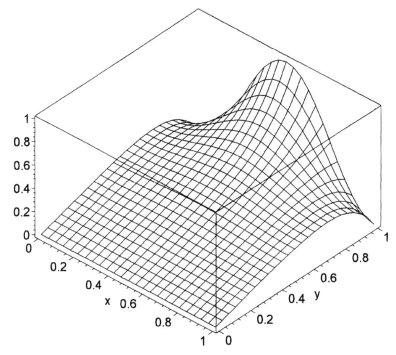

Figure 6.1   A plot of the solution of the Laplace boundary value problem.

or

$$\text{div}(k \text{ grad } u) - pu = wu_{tt}, \tag{6.2}$$

for a scalar function $u(\mathbf{r}, t)$, where each PDE is defined for time $t > 0$ in a region of space D enclosed by a closed surface S (an interval in one space dimension or a curve in two space dimensions), with $\mathbf{r}$ being a general position vector in D (a point on a line, or in space) and the functions $k = k(\mathbf{r}) > 0$, $p = p(\mathbf{r}) \geq 0$, $w = w(\mathbf{r}) \geq 0$. In each PDE the vector $\mathbf{r}$, the unknown scalar $u(\mathbf{r}, t)$, and the vector operators grad and div will involve one, two, or three space variables, accordingly as region D is an interval on a line, an area, or a volume.

When $w \neq 0$ in (6.1) the equation for $u(\mathbf{r}, t)$ is *parabolic*, and so represents a general heat or diffusion equation, but if $w \equiv 0$ it simplifies to a steady-state *elliptic* equation independent of the time for the unknown function $u = u(\mathbf{r})$. Similarly, when $w \neq 0$ in (6.2), the equation for $u(\mathbf{r}, t)$ is hyperbolic, and so represents a general wave-type equation, but when $w \equiv 0$ time is absent and the PDE again reduces to an *elliptic* equation for the unknown function $u = u(\mathbf{r})$.

Equations (6.1) and (6.2) will describe specific physical problems once appropriate initial conditions are given and suitable boundary conditions have been imposed on the surface (interval or curve) S bounding the region D.

As (6.1) is first order in time, and (6.2) is second order in time, the appropriate initial conditions will be:

Initial Dirichlet-type condition for the parabolic equation (6.1):

$$u(\mathbf{r}, 0) = U(\mathbf{r}), \tag{6.3}$$

Initial Cauchy-type conditions for the hyperbolic equation (6.2):

$$u(\mathbf{r}, 0) = U(\mathbf{r}), u_t(\mathbf{r}, 0) = V(\mathbf{r}). \tag{6.4}$$

The space boundary condition for both types of PDE will be taken to be the following general homogeneous condition:

Space boundary condition for (6.1) and (6.2):

$$\left(\alpha u + \beta \frac{\partial u}{\partial n} \right)_S = 0, \tag{6.5}$$

where u_S represents the solution $u(\mathbf{r}, t)$ on S for all $t > 0$, and $(\partial u/\partial n)_S$ represents the directional derivative of $u(\mathbf{r}, t)$ on S, normal to the surface, for all $t > 0$. Condition (6.5) simplifies to a homogeneous Dirichlet condition when $\beta = 0$, and to a homogeneous Neumann condition when $\alpha = 0$.

To simplify the arguments that follow, it will be convenient to use the operator notation introduced in Section 1.3 and to write $L[u] \equiv \text{div}(k \text{ grad } u) - pu$, when (6.1) becomes $L[u] = wu_t$ and (6.2) becomes $L[u] = wu_{tt}$. The linearity of differential equations (6.1) and (6.2) means that if φ_1 and φ_2 are any two linearly independent solutions, then so also is the **linear superposition** of solutions $u = c_1\varphi_1 + c_2\varphi_2$, where c_1 and c_2 are arbitrary constants. This follows because substituting $u = c_1\varphi_1 + c_2\varphi_2$ into the parabolic equation (6.1) gives

$$L[c_1\varphi_1 + c_2\varphi_2] = w\frac{\partial}{\partial t}\{c_1\varphi_1 + c_2\varphi_2\},$$

but c_1 and c_2 are constants, and L is a linear operator, so this becomes

$$c_1 L[\varphi_1] + c_2 L[\varphi_2] = c_1 w\varphi_{1t} + c_2 w\varphi_{2t}.$$

Thus, as $L[\varphi_1] = w\varphi_{1t}$ and $L[\varphi_2] = w\varphi_{2t}$, we see that $u = c_1\varphi_1 + c_2\varphi_2$ satisfies $L[u] = wu_t$ identically, confirming that it is a solution. A similar argument applied to the hyperbolic equation (6.2) shows that there, also, if φ_1 and φ_2 are linearly independent solutions, $u = c_1\varphi_1 + c_2\varphi_2$ is also a solution.

Developing this idea further, it will be necessary to use infinite sets of linearly independent solutions $\varphi_1, \varphi_2, \ldots$, of the PDEs, and then to seek solutions of specific problems by writing $u = \sum_{n=1}^{\infty} c_n\varphi_n$, where the c_n are constants. Solutions of this type are called **formal solutions**, and they only become classical solutions when the series can be shown to converge to a function that satisfies the PDE, the initial conditions (when time is involved), and boundary conditions that are continuous and smooth. The formal solution u becomes a **generalized solution** when the series converges to a solution of the PDE, satisfying the initial conditions (when time is involved) and boundary conditions that are piecewise smooth with at most a finite number of jump discontinuities.

As illustrated in Example 6.1, the basic idea involved in the *method of separation of variables* is to try to construct a solution u by seeking special solutions of the PDE of the form

$$u(\mathbf{r}, t) = X(\mathbf{r}) T(t). \tag{6.6}$$

In (6.6) the function $X(\mathbf{r})$ depends only the space variable $\mathbf{r}$, and $T(t)$ depends only on the time t. We now show how an infinite sequence $u_1, u_2, \ldots,$ of special solutions of this type can be constructed by setting $u_n(\mathbf{r}, t) = X_n(\mathbf{r}) T_n(t)$, where the functions $X_n(\mathbf{r})$ and $T_n(t)$ are determined by the nature of the problem. It will be these special solutions that will be superimposed and matched to the initial and boundary conditions when determining the required solution of the PDE.

The functions $X_n(\mathbf{r})$ involving the space variables will depend on the boundary conditions and the coordinate system used, while each of the functions of time $T_n(t)$ will contain one or two undetermined coefficients (integration constants), depending on whether PDE (6.1) or (6.2) is involved. Thus the special solutions $u_n(\mathbf{r}, t)$ will satisfy the boundary conditions, but *not* the initial condition(s) of the problem. The linearity of the PDE will then be used to represent a formal solution $u(\mathbf{r}, t)$ as a linear superposition of the special solutions

$$u(\mathbf{r}, t) = \sum_{n=1}^{\infty} u_n(\mathbf{r}, t). \tag{6.7}$$

When PDE (6.1) is involved, the arbitrary integration constant in each time-dependent function $T_n(t)$ will be found by setting $t = 0$ in (6.7), requiring $u(\mathbf{r}, 0)$ to satisfy initial condition (6.3), and then using the orthogonality property of the functions $X_n(\mathbf{r})$ to be established later. Similarly, if PDE (6.2) is involved, the arbitrary integration constants in each time-dependent function $T_n(t)$ will be found by setting $t = 0$ in (6.7), and requiring $u(\mathbf{r}, 0)$ and $u_t(\mathbf{r}, 0)$ (found by differentiating (6.7) and setting $t = 0$) to satisfy the initial conditions in (6.4), and again using the orthogonality properties of functions $X_n(\mathbf{r})$.

Substituting (6.7) into (6.1), and abbreviating $X(\mathbf{r})$ and $T(t)$ by X and T, respectively, gives

$$L[XT] = w \frac{\partial}{\partial t}(XT). \tag{6.8}$$

Operator L acts only on the space variables, so $L[XT] = XT[X]$ and $\partial/\partial t$ acts only on T, so $\partial/\partial t(XT) = XT'$, where a prime indicates differentiation with respect to t. Using these results in (6.8) and dividing by wXT gives

$$\frac{L[X]}{wX} = \frac{T'}{T}. \tag{6.9}$$

A similar argument applied to (6.2) leads to the corresponding result

$$\frac{L[X]}{wX} = \frac{T''}{T}. \tag{6.10}$$

Examination of Eqs. (6.9) and (6.10) shows that in each case the expression on the left depends only on the space variables, while that on the right depends only on the time t. As the space and time variables are independent, the only way a function of space variables can always equal a function of the time is when each of these expressions is equal to the same *numerical* constant. Setting this constant equal to $-\lambda$, with $\lambda > 0$ (the reason for the negative sign will become apparent later), (6.9) becomes

$$\frac{L[X]}{wX} = \frac{T'}{T} = -\lambda, \tag{6.11}$$

and (6.10) becomes

$$\frac{L[X]}{wX} = \frac{T''}{T} = -\lambda. \tag{6.12}$$

The space and time variables now uncouple, because the parabolic equation that gave rise to (6.11) is equivalent to the respective PDE and ODE

$$L[X] + \lambda w X = 0 \quad \text{and} \quad T' + \lambda T = 0, \tag{6.13}$$

while the hyperbolic equation that gave rise to (6.12) uncouples in a similar manner to give the respective PDE and ODE

$$L[X] + \lambda w X = 0 \quad \text{and} \quad T'' + \lambda T = 0. \tag{6.14}$$

The first equation in both (6.13) and (6.14) will be a PDE if two or three space variables are involved, and an ODE if only one space variable is involved. As in Example 6.1, the positive constant λ introduced into (6.11) and (6.12) is called a **separation constant**. It is because of this separation of variables that this method of solving linear homogeneous PDEs is called the **separation of variables method**.

The reason for defining the separation constant as $-\lambda$, with $\lambda > 0$, can be seen from an examination of the equation for $T(t)$ in (6.13). In this parabolic case, by taking $\lambda > 0$ the effect of the time variation will be an exponential decay that causes the solution to remain finite as $t \to \infty$, as would be expected of a physical solution. If, instead, we had $\lambda < 0$, the solution would behave nonphysically, because it would increase exponentially with time. A similar argument applies to the equation for $T(t)$ in (6.14), where if $\lambda < 0$ the solution will behave nonphysically and increase exponentially with time, but if $\lambda > 0$ it will satisfy the physical requirement that the solution remains bounded and oscillatory for all time.

The next step involves finding the permissible values of the separation constant λ, and to do this it is necessary to consider the implication of the boundary conditions (6.5) to be satisfied by u on S. The boundary conditions on u apply only to the space variables on S and they must be satisfied for all t so, as $u = XT$ and T cannot vanish, it follows that X must satisfy the boundary condition

$$\left(\alpha X + \beta \frac{\partial X}{\partial n} \right)_S = 0. \tag{6.15}$$

When this boundary condition is applied to the first equation in either (6.13) or (6.14), it will determine the permissible values $\lambda_1, \lambda_2, \ldots$ of the separation constant λ, called the **eigenvalues** of the PDE $L[X] + \lambda w X = 0$. These eigenvalues will, in turn, lead to a corresponding sequence of functions $X_1(\mathbf{r}), X_2(\mathbf{r}), \ldots$. When two or three space dimensions are involved these functions are called the **multidimensional eigenfunctions** of the PDE $L[X] + \lambda w X = 0$, but when only one space variable is involved, as in Example 6.1, they are simply called the **eigenfunctions** of what is then an ODE. Once the eigenvalues are known, the corresponding time variations of the functions $T_1(t), T_2(t), \ldots$ follow from the second ODE in (6.13) or (6.14), with $\lambda = \lambda_m$ in $T_m(t)$. These results lead to *partial solutions* of the PDE of the form $u_m(\mathbf{r}, t) = X_m(\mathbf{r}) T_m(t)$, which through the $X_m(\mathbf{r})$ satisfy the boundary conditions for all time, but not the initial conditions, because so far these have not been used.

When two or three space dimensions are involved, the PDE $L[X(\mathbf{r})] + \lambda w X(\mathbf{r}) = 0$, subject to boundary conditions (6.5), equivalently (6.15), is called a **multidimensional Sturm–Liouville problem**. The situation simplifies when only one space variable x is involved, with $a \leq x \leq b$, because then the PDE $L[X(\mathbf{r})] + \lambda w X(\mathbf{r}) = 0$ reduces to an ODE of the form

$$\frac{d}{dx}\left(k \frac{dX}{dx} \right) + (w\lambda - p) X = 0, \tag{6.16}$$

subject to the homogeneous boundary conditions derived from (6.15)

$$\alpha_1 X(a) + \beta_1 \left(\frac{dX}{dx} \right)_{x=a} = 0 \quad \text{and} \quad \alpha_2 X(b) + \beta_2 \left(\frac{dX}{dx} \right)_{x=b} = 0. \tag{6.17}$$

Equation (6.16), subject to the mixed homogeneous boundary conditions in (6.17), is called a (one-dimensional) *Sturm–Liouville system*, and the ODE in (6.16) is said to be written in the **standard form** for a one-dimensional Sturm–Liouville equation. The importance of this form will become apparent in Section 6.2. Sturm–Liouville problems for which $k(x) > 0$ for $a \leq x \leq b$ are called **regular problems**, while ones for which $k(a) = 0$ and $\alpha(a) = 0$, $\beta(a) = 0$, or $k(b) = 0$ and $\alpha(b) = 0, \beta(b) = 0$ are called **singular problems**.

Sturm–Liouville problems often involve *periodic boundary conditions* at the ends of the interval $a \le x \le b$, in which case (6.17) is replaced by the boundary conditions

$$X(a) = X(b) \quad \text{and} \quad \left(\frac{dX}{dx}\right)_{x=a} = \left(\frac{dX}{dx}\right)_{x=b}. \tag{6.18}$$

A simple one-dimensional example of a regular Sturm–Liouville problem is

$$X'' + \lambda X = 0, \quad \text{with } X'(0) = 0 \quad \text{and} \quad X(\pi/2) = 0,$$

where a prime indicates differentiation with respect to x. The general solution of the ODE is

$$X(x) = A \cos \sqrt{\lambda}x + B \sin \sqrt{\lambda}x,$$

so as $X'(0) = 0$ we must have $B = 0$, and from $X(\pi/2) = 0$ it then follows that $\cos(\sqrt{\lambda}\pi/2) = 0$. This last result is only possible if $\sqrt{\lambda} = 2m - 1$, with $m = 1, 2, \ldots$, so the general term in the infinite set of eigenvalues is $\lambda_m = (2m - 1)^2$, with $m = 1, 2, \ldots$, and the corresponding infinite set of eigenfunctions will have as its general term $X_m(x) = \cos[(2m - 1)x]$, with $m = 1, 2, \ldots$. When giving the eigenfunction $X_m(x)$ we have set $A = 1$. This involves no loss of generality because the ODE is homogeneous, so its solution can be scaled by any constant and still remain a solution. In this example the boundary condition at the left was a homogeneous Neumann condition and that at the right was a homogeneous Dirichlet condition. In the Sturm–Liouville problem which arose in Example 6.1, both boundary conditions for $X(x)$ were homogeneous Neumann conditions.

Our concern will be with orthogonal coordinate systems like rectangular Cartesian coordinates, plane polar coordinates, cylindrical polar coordinates, and spherical polar coordinates, all of which allow the first of the equations in (6.13) and (6.14) to undergo further separation of variables, until only an ODE occurs involving each of the separated variables. This process will introduce additional separation constants, and coordinate systems of this type simple enough to allow such a reduction are said to be **separable**. When the first equations in (6.13) and (6.14) involve two or three space dimensions, a separable coordinate system will reduce them, respectively, to two or three one-dimensional Sturm–Liouville problems.

To illustrate matters, let us consider a one-dimensional time-dependent problem involving the ODE in (6.16), subject to the boundary condition (6.15) and the initial condition $u(x,0) = U(x)$. After separating the variables, the solution of the IBVP will be of the form

$$u(x, t) = \sum_{m=1}^{\infty} X_m(x) T_m(t), \tag{6.19}$$

where $X_m(x)$ is a solution of (6.16) with $\lambda = \lambda_m$, and when a parabolic problem is involved $T_m(t)$ will be a solution of $T_m' + \lambda_m T_m = 0$ (see (6.13)), from which it follows that $T_m(t) = A_m \exp(-\lambda_m t)$, with A_m being an arbitrary integration constant. In Section 6.2 it will be shown that, for both regular and singular Sturm–Liouville problems, the set of functions X_n satisfying the homogeneous boundary conditions in (6.18) is always mutually orthogonal over the interval $a \le x \le b$ with weight function w, so

$$\int_a^b w\, X_m X_n\, dx = \begin{cases} 0, & m \neq m \\ \|X_n\|^2, & m = n, \end{cases} \qquad \text{where } \|X_n\|^2 = \int_a^b w\,[X_n]^2 dx.$$

$$(6.20)$$

Inserting $T_m(t) = A_m \exp(-\lambda_m t)$ in (6.19) gives

$$u(x, t) = \sum_{m=1}^{\infty} A_m X_m(x) \exp(-\lambda_m t), \qquad (6.21)$$

where the initial condition requires that $u(x, 0) = U(x)$. Setting $t = 0$ in (6.21) and using the initial condition leads to the result

$$U(x) = \sum_{n=1}^{\infty} A_m X_m(x). \qquad (6.22)$$

To determine the coefficients A_m in this generalized Fourier series we multiply (6.22) by $w X_n$ and integrate over the interval $a \le x \le b$. Then, from the orthogonality condition (6.20),

$$\int_a^b U(x) w(x) X_n(x)\, dx = A_n \|X_n\|^2, \quad \text{for } n = 1, 2, \ldots,$$

and so

$$A_n = \frac{1}{\|X_n\|^2} \int_a^b U(x) w(x) X_n(x)\, dx, \quad \text{for } n = 1, 2, \ldots. \qquad (6.23)$$

Using this expression for A_n in (6.21), with n replaced by m, gives the formal solution of the IVBP.

In an application, in order to prove rigorously that a formal solution is a classical solution, it is necessary to show the sum function defined by this series converges everywhere, and satisfies the PDE and also the initial and boundary conditions. The necessary proofs can be constructed for the examples and problems contained in this book, but the arguments involved can be difficult and tedious, so the proofs will be omitted. If instead of a classical solution a generalized solution is involved, the sum function will converge to the mid-point of any jump condition located on the boundary.

A similar argument leads to the solution of an IVBP for the one-dimensional form of (6.2), although this time we must set $T_m(t) = C_m \cos \lambda_m t + D_m \sin \lambda_m t$ to satisfy the second equation in (6.14). The coefficients C_m and D_m are found by using the orthogonality condition in the same way as above, though this time making use of the two initial conditions $u(x,0) = U(x)$ and $u_t(x,0) = V(x)$. Here, $u_t(x,0)$ must be found by differentiation of the expression for $u(x,t)$ in (6.19) with respect to t followed by setting $t = 0$.

To show how additional separation constants arise we will consider two particular cases of the multidimensional Sturm–Liouville problem arising from PDE (6.1). In case (a) we examine a two-dimensional Sturm–Liouville problem of a type that could arise from the study of the time-dependent heat equation in a circular region $0 \le r \le \rho$, subject to the boundary condition $\alpha R(\rho) + \beta(\partial R/\partial r)_{r=\rho} = 0$ on $r = \rho$. This will necessitate expressing the equation in terms of the plane polar coordinates (r, θ). Then, in case (b), we examine a three-dimensional Laplace equation in a spherical region of radius ρ, also subject to a homogeneous boundary condition of the same type as that in case (a), though applied to the spherical boundary $r = \rho$. This problem could describe the electrostatic potential inside a hollow metal sphere on which the potential is prescribed. This will involve expressing the equation in terms of the spherical polar coordinates (r, ϕ, θ).

Case (a): A Sturm–Liouville problem obtained from the heat equation in plane polar coordinates (r, θ), with $k = $ constant and $p = 0$

We will set $k = 1$, since this involves no loss of generality, because if $k \ne 1$ it is only necessary to replace w by w/k. The equation on the left of (6.13) simplifies to $\Delta_2 X + \lambda X = 0$, and the time variation is governed by the ODE on the right. The form of $\Delta_2 X$ in the plane polar coordinates (r, θ) follows from (1.12) by omitting the z term, to give the two-dimensional Sturm–Liouville equation

$$\frac{\partial^2 X}{\partial r^2} + \frac{1}{r}\frac{\partial X}{\partial r} + \frac{1}{r^2}\frac{\partial^2 X}{\partial \theta^2} + \lambda X = 0.$$

Separating the variables by setting $X(r,\theta) = R(r)\Theta(\theta)$, where R and Θ are functions of only r and θ, respectively, substituting for X, dividing the result by $R\Theta$, and rearranging terms gives

$$\frac{1}{R}\left(r^2\frac{d^2 R}{dr^2} + r\frac{dR}{dr} + r^2\lambda R\right) = -\frac{1}{\Theta}\frac{d^2\Theta}{d\theta^2}.$$

The expression on the left now depends only on the independent variable r, while that on the right depends only on the independent variable θ, so for this result to be true for all r and θ, each expression must equal a separation constant that we set equal to $\mu^2 > 0$. This being so, the equations can be uncoupled to give

$$\frac{d^2\Theta}{d\theta^2} + \mu^2\Theta = 0 \quad \text{and} \quad r^2\frac{d^2R}{dr^2} + r\frac{dR}{dr} + (\lambda r^2 - \mu^2)R = 0. \quad (6.24)$$

The choice $\mu^2 > 0$ is correct because, as would be expected, the $\Theta(\theta)$ variation must be periodic, and with this choice for μ^2 the solution becomes

$$\Theta_\mu(\theta) = A_\mu \cos\mu\theta + B_\mu \sin\mu\theta, \quad (6.25)$$

where A_μ and B_μ are arbitrary integration constants. If $\Theta_\mu(\theta)$ is to be defined for $0 \le \theta \le 2\pi$, which is usually the case, it must be periodic with period 2π, so we must set $\mu = n$, with $n = 1, 2, \ldots$. The $R(r)$ variation is more complicated, because from the second equation in (6.24) it is seen to be a solution of Bessel's equation of order μ. We know from (5.48) that the general solution of this form of Bessel's equation is

$$R(r) = CJ_n(r\sqrt{\lambda}) + DY_n(r\sqrt{\lambda}),$$

where C and D are arbitrary constants. The solution is required in the complete circular disc $0 \le r \le \rho$, and for it to be physically realistic it must remain bounded. Consequently we must set $D = 0$ to remove the infinite behavior of all Bessel functions $Y_n(r)$ of order n at the origin, and so

$$R(r) = CJ_n(r\sqrt{\lambda}).$$

This expression must satisfy the boundary condition for $R(r)$ on a circular boundary $r = \rho$. As the Bessel functions $J_n(r)$ of all orders have an infinite set of zeros, satisfying the boundary condition is only possible for some special set of values $k_{nm} = \rho\sqrt{\lambda}$, with $m = 0, 1, 2, \ldots$, where the k_{nm} are determined by the condition

$$\alpha J_n(\rho\sqrt{\lambda}) + \beta[dJ_n(r\sqrt{\lambda})/dr]_{r=\rho} = 0.$$

So, in this case, the two-dimensional eigenfunction $X_{nm}(r, \theta)$ will be the product of two one-dimensional eigenfunctions. One of these will be the $\Theta_n(\theta)$ obtained from (6.25) with $\mu = n$, while the others will be the Bessel functions

$$R_{nm}(r) = C_{nm}J_n(k_{nm}r/\rho), \quad (6.26)$$

with the C_{nm} constants. The constants in the function $\Theta_n(\theta)$ must be chosen to satisfy the boundary conditions for $\Theta(\theta)$, so the two-dimensional eigenfunctions become

$$X_{nm}(r, \theta) = (A_n \cos n\theta + B_n \sin n\theta)C_{nm}J_n(k_{nm}r/\rho). \quad (6.27)$$

If the solution is required in the annulus $\rho_1 \leq r \leq \rho_2$, a similar argument applies. In this case, as the solution is not required at the origin, the term $Y_n(r\sqrt{\lambda})$ must be retained so now

$$R(r) = CJ_n(r\sqrt{\lambda}) + DY_n(r\sqrt{\lambda}),$$

in which case (6.27) becomes

$$X_{nm}(r,\theta) = (A_n \cos n\theta + B_n \sin n\theta)(C_{nm}J_n(k_{nm}r/\rho) + D_{nm}Y_n(k_{nm}r/\rho)).$$

This example has shown how the introduction of the plane polar coordinates (r,θ) leads to functions $\Theta_n(\theta)$, each of which depends on the *single* separation constant $\mu = n$, and to functions $R_{nm}(r)$, each of which depends on the *two* separation constants λ and μ.

Note that the two Sturm–Liouville ODEs that arose can each be written in standard Sturm–Liouville form

$$\frac{d}{d\theta}\left(\frac{d\Theta}{d\theta}\right) + \mu^2 \Theta = 0 \quad \text{and} \quad \frac{d}{dr}\left(r\frac{dR}{dr}\right) + \frac{1}{r}(r^2\lambda - \mu^2)R = 0, \quad r \geq 0,$$

$$(6.28)$$

and that Bessel's equation for $R(r)$ is a *singular* Sturm–Liouville equation, because $r = 0$ at the origin.

Case (b): Spherical polar coordinates (r, θ, ϕ), $k = $ constant, $p = 0$, and $w = 0$

In this case, as $k = $ constant, the PDE reduces to the three-dimensional Laplace equation $\Delta_3 u = 0$, so from (1.14) the equation $\Delta_3 X = 0$ becomes

$$\frac{1}{r^2}\frac{\partial}{\partial r}\left(r^2 \frac{\partial X}{\partial r}\right) + \frac{1}{r^2 \sin\theta}\frac{\partial}{\partial \theta}\left(\sin\theta \frac{\partial X}{\partial \theta}\right) + \frac{1}{r^2 \sin^2\theta}\frac{\partial^2 X}{\partial \phi^2} = 0. \quad (6.29)$$

The variables must now be separated by setting $X(r,\theta,\phi) = R(r)\Theta(\theta)\Phi(\phi)$ and substituting into (6.29). When this is done, the result is multiplied by $R\Theta\Phi/r^2 \sin^2\theta$, and after terms have been rearranged, we find that

$$\frac{\sin^2\theta}{R}\frac{d}{dr}\left(r^2\frac{dR}{dr}\right) + \frac{\sin\theta}{\Theta}\frac{d}{d\theta}\left(\sin\theta\frac{d\Theta}{d\theta}\right) = -\frac{1}{\Phi}\frac{d^2\Phi}{d\phi^2}.$$

The expression on the left depends on the independent variables r and θ, while that on the right depends only on the independent variable ϕ, so each expression must equal the same separation constant, which will be denoted by m^2. The introduction of this separation constant now enables the equation to be uncoupled and written as two separate equations, the first of which is an

ODE with only ϕ as its independent variable

$$\frac{d^2\Phi}{d\phi^2} + m^2\Phi = 0 \quad \text{with } m = 0, 1, 2, \ldots, \tag{6.30}$$

while the second is the PDE

$$\frac{\sin^2\theta}{R}\frac{d}{dr}\left(r^2\frac{dR}{dr}\right) + \frac{\sin\theta}{\Theta}\frac{d}{d\theta}\left(\sin\theta\frac{d\Theta}{d\theta}\right) = m^2$$

with r and θ as independent variables.

A further separation of the variables r and θ in this last equation, followed by the introduction of another separation constant λ, leads to

$$\frac{1}{R}\frac{d}{dr}\left(r^2\frac{dR}{dr}\right) = -\left(\frac{1}{\Theta\sin\theta}\frac{d}{d\theta}\left(\sin\theta\frac{d\Theta}{d\theta}\right) - \frac{m^2}{\sin^2\theta}\right) = \lambda. \tag{6.31}$$

This result decouples once again to give the two separate ODEs with independent variables θ and r

$$\frac{1}{\sin\theta}\frac{d}{d\theta}\left(\sin\theta\frac{d\Theta}{d\theta}\right) + \left(\lambda - \frac{m^2}{\sin^2\theta}\right)\Theta = 0, \tag{6.32}$$

and

$$\frac{d}{dr}\left(r^2\frac{dR}{dr}\right) - \lambda R = 0. \tag{6.33}$$

In this case, separation of variables has reduced the three-dimensional Laplace equation to the three ODEs (6.30), (6.32), and (6.33), where the separation constants λ and m must be chosen to make the solutions of the ODEs satisfy the boundary conditions.

Expanding (6.33) shows that $R(r)$ satisfies the ODE

$$r^2\frac{d^2R}{dr^2} + 2r\frac{dR}{dr} - \lambda R = 0. \tag{6.34}$$

This is a *Cauchy–Euler* equation, and from the elementary theory of ODEs it is known its solution is of the form $R(r) = r^k$, where the constant k is to be found by substituting this expression for $R(r)$ into the ODE. Making this substitution leads to the following equation for k in terms of λ

$$(k^2 + k - \lambda)r^k = 0.$$

As $r \neq 0$, this shows the exponent k in the solution $R(r) = r^k$ must be such that $k^2 + k = \lambda$. When we assume for simplicity that k is an integer, then

$\lambda = n(n+1), n = 0, 1, \ldots$, and $k = n$ or $k = -(n+1)$, so

$$R(r) = A_n r^n + B_n \frac{1}{r^{n+1}}. \tag{6.35}$$

Setting $\lambda = n(n+1)$ in (6.32) gives the following ODE for $\Theta(\theta)$:

$$\frac{1}{\sin\theta} \frac{d}{d\theta}\left(\sin\theta \frac{d\Theta}{d\theta}\right) + \left(n(n+1) - \frac{m^2}{\sin^2\theta}\right)\Theta = 0. \tag{6.36}$$

This ODE can be simplified by making the substitution $x = \cos\theta$ and then setting $\Theta(\theta) = y(x)$, when it becomes

$$\frac{d}{dx}\left((1-x^2)\frac{dy}{dx}\right) + \left(n(n+1) - \frac{m^2}{1-x^2}\right)y = 0, \quad \text{for } -1 \le x \le 1. \tag{6.37}$$

This equation is called **Legendre's associated differential equation**, and it is written in standard Sturm–Liouville form. The equation has two linearly independent solutions called **Legendre functions of the first and second kind** and denoted, respectively, by $P_n^m(x)$ and $Q_n^m(x)$, so its general solution is

$$y_{nm}(x) = C_{nm}P_n^m(x) + D_{nm}Q_n^m(x), \tag{6.38}$$

with C_{nm} and D_{nm} being arbitrary constants. The functions $P_n^m(x)$ and $Q_n^m(x)$ are defined in terms of infinite series with useful and interesting properties, although they will not be used here.

In applications it often happens that a solution $u(r, \theta, \phi)$ has *axial symmetry* about the z axis, and so is independent of ϕ, in which case $m = 0$, causing (6.37) to simplify still further to the *Legendre differential equation*

$$(1 - x^2)\frac{d^2 y}{dx^2} - 2x\frac{dy}{dx} + n(n+1)y = 0, \quad \text{for } -1 \le x \le 1. \tag{6.39}$$

When written in standard Sturm–Liouville form this equation becomes

$$\frac{d}{dx}\left((1 - x^2)\frac{dy}{dx}\right) + n(n+1)y = 0, \quad \text{for } -1 \le x \le 1. \tag{6.40}$$

Examination of (6.40) shows Legendre's equation is a *singular* Sturm–Liouville equation because, like Legendre's associated differential equation, the factor $(1 - x^2)$ vanishes at the ends of the interval $-1 \le x \le 1$ over which the solution is defined.

A second more complicated linearly independent solution of Legendre's equation denoted by $Q_n(x)$ was encountered in Section 5.3 where, like $Q_n^m(x)$, it was seen to have singularities at $x = \pm 1$. However, the problems to be considered here will need neither of these functions.

When there is axial symmetry about the z axis, and $0 < \theta < \pi$, the multi-dimensional eigenfunctions of the Laplace equation become

$$X_n(r, \theta) = \left(A_n r^n + B_n r^{-(n+1)}\right)[C_n P_n(\cos \theta) + D_n Q_n(\cos \theta)], \quad (6.41)$$

where $n = 0, 1, 2, \ldots$. If in a problem it is necessary that θ should lie in the closed interval, $0 \le \theta \le \pi$, then in order to exclude the singularities of $Q_n(\cos \theta)$ at $\cos \theta = \pm 1$ the term $Q_n(\cos \theta)$ must be omitted from the general solution, so then

$$X_n(r, \theta) = \left(A_n r^n + B_n r^{-(n+1)}\right) P_n(\cos \theta), \quad n = 0, 1, 2, \ldots. \quad (6.42)$$

EXERCISES 6.1

1. Using the approach of Example 6.1, solve the boundary value problem for the Laplace equation

$$\frac{\partial^2 u}{\partial x^2} + \frac{\partial^2 u}{\partial y^2} = 0 \quad \text{for } 0 < x < a, \, 0 < y < b,$$

with $u(x, 0) = \frac{1}{2} \cos^2 \frac{\pi x}{a}$, $u(x, b) = 0$, $\frac{\partial u}{\partial x}\big|_{x=0} = 0$, and $\frac{\partial u}{\partial x}\big|_{x=a} = 0$. Explain why the boundary value problem for this same equation, in the same region, subject to the boundary conditions $u(x, 0) = \frac{1}{2} \cos^2 \frac{\pi x}{a}$, $u(x, b) = \sin^2 \frac{\pi x}{a}$, $\frac{\partial u}{\partial x}\big|_{x=0} = 0$, and $\frac{\partial u}{\partial x}\big|_{x=a} = 0$, is the sum of the solution of this problem and the solution of Example 6.1. Plot the solution when $a = b = 1$.

In Exercises 2 through 6 find the eigenvalues and eigenfunctions of the given regular one-dimensional Sturm–Liouville problems.

2. $X'' + \lambda X = 0$ with $X(0) = 0$, $X(a) = 0$.

3. $X'' + \lambda X = 0$ with $X(0) = 0$, $X'(a) = 0$.

4. $X'' + \lambda X = 0$ with $X'(0) = 0$, $X'(a) = 0$.

5. $X'' + \lambda X = 0$ with the periodic boundary conditions $X(-a) = X(a)$ and $X'(-a) = X'(a)$.

6. $X'' + \lambda X = 0$ with $X(0) + a X'(0) = 0$ and $X(a) = 0$.

7. By making the substitution $x = \cos \theta$ and $\Theta(\theta) = y(x)$ in (6.36), confirm that it reduces to Legendre's associated differential equation in (6.37).

8. Use separation of variables to show that when the wave equation

$$\Delta_3 X = \frac{1}{c^2} \frac{\partial^2 X}{\partial t^2}$$

is written in terms of cylindrical polar coordinates with $X(r, \theta, z, t) = R(r)\Theta(\theta) Z(z) T(t)$, the separated equations become

$$\frac{d^2 T}{dt^2} = -c^2 p^2 T, \qquad \frac{d^2 \Theta}{d\theta^2} = -m^2 \Theta, \qquad \frac{d^2 Z}{dz^2} = -q^2 Z,$$

$$\frac{d^2 R}{dr^2} + \frac{1}{r}\frac{dR}{dr} - \frac{m^2}{r^2} R + n^2 R = 0, \qquad n^2 = p^2 - q^2,$$

where p, q, m, and n are separation constants, with m being an integer.

9. Use separation of variables to show that when the wave equation

$$\Delta_3 X = \frac{1}{c^2}\frac{\partial^2 X}{\partial t^2}$$

is written in terms of spherical polar coordinates with $X(r, \theta, \phi, t) = R(r)\Theta(\theta)\Phi(\phi) T(t)$, the separated equations become

$$\frac{d^2 T}{dt^2} = -c^2 p^2 T, \qquad \frac{d^2 \Phi}{d\phi^2} = -m^2 \Phi,$$

$$\frac{1}{\sin\theta}\frac{d}{d\theta}\left(\sin\theta\frac{d\Theta}{d\theta}\right) + \left\{n(n+1) - \frac{m^2}{\sin^2\theta}\right\}\Theta = 0,$$

$$\frac{d^2 R}{dr^2} + \frac{2}{r}\frac{dR}{dr} + \left\{p^2 - \frac{n(n+1)}{r^2}\right\}R = 0,$$

where p, m, and n are separation constants, with m being an integer.

6.2 Properties of Eigenfunctions and Eigenvalues

The most important properties of eigenvalues and eigenvectors of Sturm–Liouville systems are given below, and all with the exception of some results in Property 5 are proved.

1. *Eigenfunctions can be scaled*: As eigenfunctions are solutions of a linear homogeneous ODE it is self-evident that if X_n is an eigenfunction, then so is cX_n, where $c \neq 0$ is an arbitrary constant.

2. *Orthogonality of eigenfunctions*: We start by establishing the orthogonality of the eigenfunctions $X_n(x)$ of the one-dimensional Sturm–Liouville system

$$\frac{d}{dx}\left(k\frac{dX}{dx}\right) + (w\lambda - p)X = 0, \quad \text{for } a \leq x \leq b, \qquad (6.43)$$

with $w = w(x) \geq 0$, $k = k(x) > 0$, subject to either the homogeneous boundary conditions

$$\alpha_1 X(a) + \beta_1 \left(\frac{dX}{dx} \right)_{x=a} = 0 \quad \text{and} \quad \alpha_2 X(b) + \beta_2 \left(\frac{dX}{dx} \right)_{x=b} = 0$$

$$(6.44)$$

or the periodic boundary conditions

$$X(a) = X(b) \quad \text{and} \quad \left(\frac{dX}{dx} \right)_{x=a} = \left(\frac{dX}{dx} \right)_{x=b}. \qquad (6.45)$$

The proof proceeds as follows. Let $X_m(x)$ and $X_n(x)$ be any two eigenfunctions of (6.43) and (6.44) or (6.45), corresponding to the *distinct* eigenvalues λ_m and λ_n, respectively, so

$$\frac{d}{dx} \left(k \frac{dX_m}{dx} \right) + (w\lambda_m - p) X_m = 0 \quad \text{and}$$

$$\frac{d}{dx} \left(k \frac{dX_n}{dx} \right) + (w\lambda_n - p) X_n = 0.$$

Multiplying the first ODE by X_n, the second by X_m, integrating each equation over the interval $a \leq x \leq b$ and subtracting the results we have

$$\int_a^b \left[X_n \frac{d}{dx} \left(k \frac{dX_m}{dx} \right) - X_m \left(k \frac{dX_n}{dx} \right) \right] dx = (\lambda_n - \lambda_m) \int_a^b w X_m X_n dx.$$

Applying integration by parts to the integral on the left reduces the above result to

$$\left(k X_n \frac{dX_m}{dx} \right)_{x=a}^{x=b} - \left(k X_m \frac{dX_n}{dx} \right)_{x=a}^{x=b} = (\lambda_n - \lambda_m) \int_a^b w X_m X_n \, dx.$$

The expression on the left vanishes if Dirichlet, Neumann, or periodic boundary conditions are imposed, and they also vanish if the homogeneous mixed boundary conditions in (6.44) are imposed, as can be seen by writing them in the form

$$\left(\frac{dX}{dx} \right)_{x=a} = -\frac{\alpha_1}{\beta_1} X(a) \quad \text{and} \quad \left(\frac{dX}{dx} \right)_{x=b} = -\frac{\alpha_2}{\beta_2} X(b).$$

Consequently, it follows that

$$(\lambda_n - \lambda_m) \int_a^b w X_m X_n dx = 0.$$

By supposition the eigenvalues λ_m and λ_n are distinct, so the factor $(\lambda_n - \lambda_m) \neq 0$, and we have proved the orthogonality condition

$$\int_a^b w\, X_m X_n\, dx = 0 \quad \text{for } m \neq n. \tag{6.46}$$

By hypothesis $w > 0$, so the square of the norm of X_n is given by

$$\| X_n \|^2 = \int_a^b w\, X_n^2 dx > 0. \tag{6.47}$$

Results (6.46) and (6.47) establish the claim made in Section 6.1 that the set of eigenfunctions $X_1, X_2, \ldots$ are mutually orthogonal over the interval $a \le x \le b$ with respect to the weight function w, which is simply the function w on the right of Eqs. (6.1) and (6.2).

It follows directly from the Sturm–Liouville forms of Bessel's equation in (6.28) and Legendre's equation in (6.40) that Bessel functions and Legendre polynomials are both sets of mutually orthogonal functions. From (5.52) it is seen that the Bessel functions $J_n(x)$ of the *same* order, but with *different* zeros in their arguments, are orthogonal over an interval $0 \le r \le R$ with respect to the weight function $w = x$. The situation with Legendre polynomials is quite different, because from (5.30) it is seen that Legendre polynomials of *different* orders are orthogonal over the interval $-1 \le x \le 1$ with respect to a weight function $w = 1$.

Multidimensional eigenfunctions involve a product of one-dimensional eigenfunctions, so their orthogonality follows directly from the orthogonality of these constituent eigenfunctions.

3. *The eigenvalues are all real:* Next we use Property 1 to prove all eigenvalues of Sturm–Liouville equation (6.43) subject to the boundary conditions in (6.44) or (6.45) are real. Suppose, if possible, that a complex eigenvalue $\lambda = \alpha + i\beta$ exists with $\beta \neq 0$, and that this eigenvalue corresponds to the complex eigenfunction $X = U + iV$. Substituting for X in (6.43) gives

$$\frac{d}{dx}\left(k\frac{d}{dx}(U + iV) \right) + (w(\alpha + i\beta) - p)(U + iV) = 0,$$

and after separating the real and imaginary parts we have

$$\frac{d}{dx}\left(k\frac{dU}{dx} \right) + (w\alpha - p)U - w\beta V = 0 \quad \text{and}$$

$$\frac{d}{dx}\left(k\frac{dV}{dx} \right) + (w\alpha - p)V + w\beta U = 0.$$

Multiplying the second equation by i, subtracting it from the first equation, and collecting terms gives

$$\frac{d}{dx}\left(k\frac{d}{dx}(U - iV)\right) + (w(\alpha - i\beta) - p)(U - iV) = 0.$$

This shows that $\lambda = \alpha - i\beta$ is an eigenvalue of the same equation corresponding to the complex conjugate eigenfunction $\overline{X} = U - iV$.

Using the orthogonality property established in (6.46), and the fact that X and $\overline{X}$ correspond to *distinct* eigenvalues, it follows that they must be orthogonal with respect to the weight function w, so

$$\int_a^b w(U + iV)(U - iV)dx = 0.$$

However, expanding the integrand gives

$$\int_a^b w(U + iV)(U - iV)dx = \int_a^b (U^2 + V^2)dx = 0,$$

which is impossible since $U^2 + V^2$ is nonnegative. Thus the assumption that an eigenvalue can be complex is false, and we have established that all eigenvalues must be real, and have associated with them real eigenfunctions.

4. *The eigenvalues are nonnegative.* To show the eigenvalues of the Sturm–Liouville equation (6.43) subject to the boundary conditions in (6.44) or (6.45) are all nonnegative, let λ_n be the eigenvalue corresponding to eigenfunction X_n. Then

$$\frac{d}{dx}\left(k\frac{dX_n}{dx}\right) + (\lambda_n w - p)X_n = 0.$$

Multiplying this ODE by X_n and integrating over the interval $a \le x \le b$ gives

$$\int_a^b X_n\frac{d}{dx}\left(k\frac{dX_n}{dx}\right)dx + \lambda_n \int_a^b w X_n^2 dx - \int_a^b p X_n^2 dx = 0.$$

Simplifying the first integral using integration by parts and solving for λ_n gives

$$\lambda_n = \frac{1}{\|X_n\|^2}\left[\int_a^b k\left(\frac{dX_n}{dx}\right)^2 dx + \int_a^b p X_n^2 dx - \left(kX_n\frac{dX_n}{dx}\right)_{x=a}^{x=b}\right].$$

The first two terms on the right are positive since $k(x) > 0$, $p(x) \ge 0$, and the third term vanishes if Dirichlet, Neumann, or periodic boundary conditions are imposed. These same terms also vanish if the homogeneous mixed boundary conditions in (6.44) are imposed, as was shown previously. Hence, as $\|X_n\|^2 > 0$, it follows immediately that the λ_n are positive.

A special case arises when $p \equiv 0$ and Neumann conditions are imposed at both ends of the interval, because it can be seen from this last result that $X = $ constant (usually taken to be $X = 1$) is then an eigenfunction, and the associated eigenvalue is $\lambda = 0$. An example of this type was encountered in Exercise 4 of Exercise Set 6.1.

5. *The spectrum of eigenvalues*: The eigenvalues $\lambda_1, \lambda_2, \ldots$ of all regular and all periodic Sturm–Liouville systems are infinite in number and can be ordered so that $\lambda_n < \lambda_{n+1}$ for $n = 1, 2, \ldots$. The infinite set of eigenvalues is called the **spectrum** of the Sturm–Liouville system. Furthermore, for regular and periodic Sturm–Liouville systems, the spectrum of eigenvalues contains no cluster points, in the sense that no points λ^* exist where there are infinitely many values of λ_n in every neighborhood. Finally, the spectrum of the eigenvalues is such that $\lim_{n\to\infty} \lambda_n = \infty$.

Part of this result was established in Property 3, where the eigenvalues of all Sturm–Liouville systems were shown to be real, so provided they are all distinct it follows directly that they can be ordered, with $\lambda_n < \lambda_{n+1}$ for $n = 1, 2, \ldots$. The proof that the spectrum contains no cluster points, and that $\lim_{n\to\infty} \lambda_n = \infty$, requires arguments that go beyond this first account of the subject, so they will be omitted.

6. *Two eigenfunctions corresponding to the same eigenvalue differ only by a constant multiplicative factor*: Let X and Y be two eigenfunctions of a *regular* Sturm–Liouville system, both of which correspond to the *same* eigenvalue λ. Then, by definition,

$$\frac{d}{dx}\left(k\frac{dX}{dx}\right) + (w\lambda - p)X = 0 \quad \text{and} \quad \frac{d}{dx}\left(k\frac{dY}{dx}\right) + (w\lambda - p)Y = 0,$$

subject to the boundary conditions

$$\alpha_1 X(a) + \beta_1\left(\frac{dX}{dx}\right)_{x=a} = 0 \quad \text{and} \quad \alpha_2 X(b) + \beta_2\left(\frac{dX}{dx}\right)_{x=b} = 0$$

and

$$\alpha_1 Y(a) + \beta_1\left(\frac{dY}{dx}\right)_{x=a} = 0 \quad \text{and} \quad \alpha_2 Y(b) + \beta_2\left(\frac{dY}{dx}\right)_{x=b} = 0.$$

At $x = a$ we have

$$\alpha_1 X(a) + \beta_1 \left(\frac{dX}{dx} \right)_{x=a} = 0$$

$$\alpha_1 Y(a) + \beta_1 \left(\frac{dY}{dx} \right)_{x=a} = 0,$$

which form two homogeneous algebraic equations for α_1 and β_1, both of which cannot be zero, but this can only be possible if the determinant

$$\begin{vmatrix} X(a) & (X')_{x=a} \\ Y(a) & (Y')_{x=a} \end{vmatrix} = 0.$$

This determinant is the Wronskian of two solutions at $x = a$, and its vanishing implies the linear dependence of X on Y. A similar argument applies at $x = b$, so the result is proved for mixed homogeneous boundary conditions. The result is also true for periodic boundary conditions, because they can be combined to give mixed homogeneous boundary conditions. A modification of this argument that is left as an exercise shows the eigenfunctions are proportional for $a < x < b$, so the proof is complete.

When establishing Property 4 the following expression was obtained

$$\lambda = \frac{1}{\|X\|^2} \left[\int_a^b k \left(\frac{dX}{dx} \right)^2 dx + \int_a^b p X^2 dx - \left(k X \frac{dX}{dx} \right)_{x=a}^{x=b} \right], \quad (6.48)$$

where λ is the eigenvalue corresponding to the eigenfunction X of a Sturm–Liouville system. This expression is called the **Rayleigh quotient** for the eigenvalue λ corresponding to the eigenfunction X, and it has various uses besides that in Property 4. Of these we mention only that the Rayleigh quotient can be used to provide a numerical approximation for the *smallest* eigenvalue of a Sturm–Liouville system.

To see how this information is obtained, let φ be any twice differentiable satisfying the boundary conditions of a Sturm–Liouville system. Then, by means of the calculus of variations, it can be shown that when X in (6.48) is replaced by φ, the Rayleigh quotient provides an *upper bound* for the *smallest* eigenvalue λ. The closer φ approximates the eigenfunction X, the sharper will be the upper bound.

This result is useful, because the smallest eigenvalue often has an important physical significance. Thus, although in principle an eigenfunction can only be found when its associated eigenvalue is known, the Rayleigh quotient enables the smallest eigenvalue to be estimated using *any* function that satisfies the boundary conditions. Often, on physical grounds, the approximate form of an eigenfunction can be guessed, in which case this approximation will yield an upper bound to the smallest eigenvalue. So, even when the smallest eigenvalue cannot be found analytically, it is still possible to deduce an upper bound. For example, in a mechanical system, the eigenvalue λ might represent the lowest natural frequency of vibration of an irregularly shaped metal panel, with corresponding applications in other disciplines.

6.3 Applications of Separation of Variables

This section presents various examples showing the details of some typical applications of the method of separation of variables. It was seen in Section 6.1 that the method of separation of variables applies to linear PDEs, irrespective of their type, provided the physical boundaries of regions involved coincide with coordinate lines from a system of orthogonal coordinates. The only significant difference between the treatment of hyperbolic or parabolic equations and that of elliptic equations is that in the first two types of PDE a time dependence that must be taken into account by using initial conditions is present, whereas in elliptic equations only space variables are involved.

Example 6.2. The temperature distribution $u(x, t)$ in a slab of metal in the interval $0 \leq x \leq L$ for $t > 0$ is determined by $\kappa u_{xx} = u_t$ with $\kappa > 0$, subject to the boundary conditions $u(x, t) = 0$ and $u(L, t) = 0$, and the initial condition $u(x, 0) = U_0 x(L - x)^2$ for $0 \leq x \leq L$. Find the $u(x, t)$, and with $\kappa = 0.1$ plot the result as a function of x/L and t.

> *Solution:* The PDE is a heat equation that is parabolic, the boundary conditions are of Dirichlet type, and the region is open, because although x is restricted to $0 \leq x \leq L$, the time t is unrestricted. Thus the conditions given in Section 3.3 which ensure the problem is well posed are satisfied.
>
> In terms of heat conduction, this problem could be considered to describe the transient temperature distribution in a slab of metal with parallel plane faces a distance L apart, when each face is maintained at zero temperature, and at time $t = 0$ the initial temperature distribution across the slab is $u(x, 0) = U_0 x(L - x)^2$. The solution $u(x, t)$ is then the temperature on any plane $x = $ constant in the slab at time t. Equivalently, if diffusion is involved, this represents a parallel slab of porous material through which diffusion can take place, when the plane faces of the slab are maintained at zero concentration, and the initial concentration across the slab is given by $u(x, 0) = U_0 x(L - x)^2$. The solution gives the concentration on any plane $x = $ constant in the slab of porous material at any time $t > 0$.
>
> Setting $u(x, t) = X(x)T(t)$ in the PDE and separating variables gives
>
> $$\frac{X''}{X} = \frac{T'}{\kappa T} = -\lambda^2,$$
>
> where λ^2 is a separation constant. Thus the space variation is determined by $X'' + \lambda^2 X = 0$ subject to the space boundary conditions $X(0) = 0$ and

$X(L) = 0$. The general solution for $X(x)$ is

$$X(x) = A \cos \lambda x + B \sin \lambda x.$$

Imposing the boundary condition $X(0) = 0$ shows $A = 0$, while imposing the boundary condition $X(L) = 0$ shows that $\lambda_n = n\pi/L$, with $n = 1, 2, \ldots$, so the space eigenfunction $X_n(x) = B_n \sin \frac{n\pi x}{L}$.

The time variation is determined by the equation $T' + \kappa \lambda^2 T = 0$, so $T(t) = C \exp(-\kappa \lambda^2 t)$, and hence $T_n(t) = C_n \exp\left(-\frac{\kappa n^2 \pi^2 t}{L^2}\right)$. A partial solution is $u_n(x, t) = X_n(x) T_n(t)$, so we seek a complete solution of the form

$$u(x, t) = \sum_{n=1}^{\infty} D_n X_n(x) T_n(t) = \sum_{n=1}^{\infty} D_n \sin \frac{n\pi x}{L} \exp\left(-\frac{\kappa n^2 \pi^2 t}{L^2}\right),$$

where as B_n and C_n are arbitrary constants their product, which is simply another arbitrary constant, is denoted by D_n.

To find the D_n, and hence to complete the solution of the problem, we now set $t = 0$ in the expression on the right and replace $u(x, 0)$ by the initial condition $u(x, 0) = U_0 x(L - x)^2$, when we obtain

$$U_0 x(L - x)^2 = \sum_{n=1}^{\infty} D_n \sin \frac{n\pi x}{L}.$$

This shows the D_n are the coefficients in the half-range Fourier sine series expansion of $U_0 x(L - x)^2$ over the interval $0 \le x \le L$. Thus the D_n are given by

$$D_n = \frac{2}{L} \int_0^L U_0 x(L - x)^2 \sin \frac{n\pi x}{L} \quad \text{for } n = 1, 2, \ldots,$$

and so

$$D_n = \frac{4 L^3 U_0}{\pi^3} \left(\frac{2 + (-1)^n}{n^3}\right) \quad \text{for } n = 1, 2, \ldots.$$

Substituting for D_n in the expression for $u(x, t)$ now gives the required solution

$$u(x, t) = \frac{4 L^3 U_0}{\pi^3} \sum_{n=1}^{\infty} \left(\frac{2 + (-1)^n}{n^3}\right) \sin \frac{n\pi x}{L} \exp\left(-\frac{\kappa n^2 \pi^2 t}{L^2}\right),$$

$$\text{for } 0 \le x \le L, \ t \ge 0.$$

A plot of $\pi^3 u(x, t)/U_0$ as a function of x and t, with $\kappa = 0.1$ and $L = 1$, is shown in Fig. 6.2, from which the rapid decay of the solution to 0 can be seen as $t \to \infty$. ∎

Example 6.3. Solve the heat equation $\kappa u_{xx} = u_t$ with $k > 0$ and $0 \le x \le L$, subject to the Dirichlet boundary condition $u(0, t) = 0$ at $x = 0$, the Newton cooling condition $(u_x + Hu)|_{x=L} = 0$ with $H > 0$, representing heat loss into a surrounding medium at zero temperature, and the initial condition $u(x, 0) = x$ for $0 \le x \le L$.

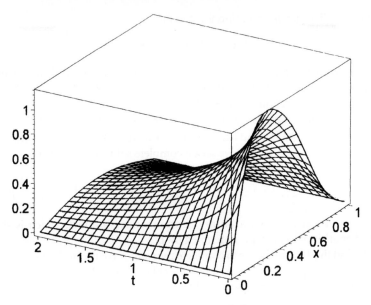

Figure 6.2   The temperature distribution $\frac{u(x,t)}{U_0}$ with $\kappa = 0.1$ as a function of $\frac{x}{L}$ and t as $t \to \infty$.

Solution: The conditions are suitable for a parabolic equation, so a unique and stable solution will exist. Separating variables as in the previous example gives

$$\frac{X''}{X} = \frac{T'}{\kappa T} = -\lambda^2,$$

where λ^2 is a separation constant, so

$$X'' + \lambda^2 X = 0 \quad \text{and} \quad T' + \lambda^2 \kappa T = 0.$$

The general solution for $X(x)$ is $X(x) = A \cos \lambda x + B \sin \lambda x$. As in the previous example, the condition $u(0, t) = 0$ implies $X(0) = 0$, showing that $A = 0$, but now the condition at $x = L$ implies that $X'(L) + HX(L) = 0$, from which it follows that

$$\lambda B \cos(\lambda L) + HB \sin(\lambda L) = 0, \quad \text{and as a result } \tan(\lambda L) = -\lambda/H.$$

Setting $\mu = \lambda L$ and $p = HL > 0$, we see that the eigenvalues μ (equivalently λ) are the zeros of the transcendental equation

$$\tan \mu = -\mu/p.$$

So the eigenfunctions of the problem become $X_n(x) = B_n \sin(\mu_n x/L)$ for $n = 1, 2, \ldots$.

For any given p, the approximate values of μ can be found by plotting the graphs of $\tan \mu$ and $-\mu/p$ and finding their points of intersection. More accurate values can be found by using these as starting approximations with an iterative numerical procedure like the Newton–Raphson method. For example, when $p = 1$, the first five positive values rounded to four decimal places are $\mu_1 = 2.0288$, $\mu_2 = 4.9132$, $\mu_3 = 7.9787$, $\mu_4 = 11.0855$, $\mu_5 = 14.2074$, and inspection of the graphs of $\tan \mu$ and $-\mu/p$ shows that as $n \to \infty$, so $\mu_{n+1} \to (2n+1)\pi/2$.

The time variation $T(t)$ is determined by $T' + \kappa \lambda^2 T = 0$, so

$$T_n(t) = C_n \exp\left[-\left(\frac{\kappa \mu_n^2}{L^2}\right)t\right].$$

Thus a partial solution can be written

$$X_n(x) T_n(t) = \sin\left(\frac{\mu_n x}{L}\right) \exp\left[-\left(\frac{\kappa \mu_n^2}{L^2}\right)t\right],$$

so we now seek a full solution of the form

$$u(x, t) = \sum_{n=1}^{\infty} D_n \sin\left(\frac{\mu_n x}{L}\right) \exp\left[-\left(\frac{\kappa \mu_n^2}{L^2}\right)t\right].$$

To determine the constants $D_n = B_n C_n$, we must make use of the initial condition $u(x, 0) = x$ and the orthogonality of the eigenfunctions $X_n(x)$ ensured by the Sturm–Liouville equation satisfied by $X_n(x)$ (see Property 2 in Section 6.2). Setting $t = 0$ we have

$$x = \sum_{n-1}^{\infty} D_n \sin\left(\frac{\mu_n x}{L}\right),$$

so multiplying by $\sin(\mu_m x/L)$, integrating over $0 \le x \le L$, and changing m to n gives

$$D_n = \frac{2L(\mu_n \cos \mu_n - \sin \mu_n)}{\mu_n(\cos \mu_n \sin \mu_n - \mu_n)}.$$

When arriving at this expression for D_n we have used the fact that

$$\int_0^L x \sin(\mu_n x/L)\,dx = \frac{L^2(\sin \mu_n - \mu_n \cos \mu_n)}{\mu_n^2}$$

and the orthogonality condition

$$\int_0^L \sin(\mu_m x/L) \sin(\mu_n x/L)\,dx = \begin{cases} 0, & m \ne n \\ \frac{L}{2}\left(\frac{p(p+1)+\mu_n^2}{p^2+\mu_n^2}\right), & m = n. \end{cases}$$

Thus the solution is

$$u(x, t) = \sum_{n=1}^{\infty} \frac{2L(\mu_n \cos \mu_n - \sin \mu_n)}{\mu_n(\cos \mu_n \sin \mu_n - \mu_n)} \sin\left(\frac{\mu_n x}{L}\right) \exp\left[-\left(\frac{\kappa \mu_n^2}{L^2}\right)t\right],$$

$$0 \le x \le L, \ t > 0.$$

This example has demonstrated how when there is a homogeneous mixed boundary condition, which in this case can be interpreted as Newton's law of cooling, the same method of solution applies, although the eigenvalues become solutions of a transcendental equation, and so must be found numerically. ■

Example 6.4. A uniform string of length L, constant line density ρ, and tension T_0 has its equilibrium position along the x axis, with the end at $x = 0$ fixed, and the end at $x = L$ free to move in a plane perpendicular to the x axis in such a way that the end of the string remains parallel to the x axis. If $u(x, t)$ is the transverse displacement of the string at point x and time $t > 0$, and the string is released from rest while in the displaced position $u(x, 0) = hx(L - x)^2$, where $h > 0$ is small, find the subsequent motion of the string. Make a 3d plot of the solution, interpret the eigenfunctions and eigenvalues of the problem, and use them to determine the energy distribution in the string.

Solution: Let $u(x, t)$ be the transverse displacement of the string at position x and time t. Then, as no external forces act on the string, its subsequent motion will be determined by the homogeneous wave equation $u_{tt} = c^2 u_{xx}$, where c is the speed with which a transverse vibration moves along the string and $c^2 = T_0/\rho$. The boundary conditions will be $u(0, t) = 0$, corresponding to the fixed end, and $u_x(L, t) = 0$, corresponding to the end at $x = L$ constrained to remain parallel to the x axis. The initial conditions are $u(x, 0) = hx(L - x)^2$, corresponding to the initial shape of the string, and $u_t(x, 0) = 0$, corresponding to the fact that the string starts from rest while in the displaced position. The wave equation is hyperbolic, the initial conditions are Cauchy conditions, and the region is open, because although the interval $0 \le x \le L$ is closed, the time variable $t > 0$ has no upper bound, so the region in space–time in which the solution is defined is open. These conditions satisfy those described in Section 3.3, so the problem is well posed. Note also that the initial condition for $u(x, 0)$ is correctly formulated, because it satisfies both boundary conditions. Setting $u(x, t) = X(x)T(t)$ to separate the variables leads to the result

$$\frac{X''}{X} = \frac{T''}{c^2 T} = -\lambda^2,$$

where λ^2 is a separation constant. This leads to the two separated equations

$$X'' + \lambda^2 X = 0 \quad \text{and} \quad T'' + c^2 \lambda^2 T = 0.$$

The general solution for $X(x)$ is $X(x) = A \cos \lambda x + B \sin \lambda x$, so an application of the boundary condition $X(0) = 0$, which follows from the condition

$u(0, t) = 0$, shows that $A = 0$. The corresponding boundary condition at $x = L$ is $X'(L) = 0$, and this leads to the condition $0 = \lambda \cos \lambda L$, so either $\lambda = 0$ or $0 = \cos \lambda L$. If $\lambda = 0$ then $X'' = 0$, so $X_0(x) = a + bx$. However, the boundary conditions show $a = b = 0$, so the eigenvalue $\lambda = 0$ must be rejected. Thus the eigenvalues λ must be solutions of $\cos \lambda L = 0$, showing that $\lambda_n = (2n - 1)\pi/2L$, with $n = 1, 2, \ldots$. The corresponding eigenfunctions are $X_n(x) = \sin[\frac{(2n-1)\pi x}{2L}]$, with $n = 1, 2, \ldots$, where for convenience the arbitrary multiplicative constant B has been set equal to 1. This involves no loss of generality, because the eigenfunction is a solution of a homogeneous equation, so any multiple of it will also be an eigenfunction.

Solving for the corresponding time variation gives

$$T_n(t) = C_n \cos\left[\frac{(2n - 1)\pi c t}{2L}\right] + D_n \sin\left[\frac{(2n - 1)\pi c t}{2L}\right],$$

so the complete solution will be of the form

$$u(x, t) = \sum_{n=1}^{\infty} \sin\left[\frac{(2n - 1)\pi x}{2L}\right]\left(C_n \cos\left[\frac{(2n - 1)\pi c t}{2L}\right]\right.$$
$$\left. + D_n \sin\left[\frac{(2n - 1)\pi c t}{2L}\right]\right).$$

Setting $t = 0$ and using the first initial condition we have

$$hx(L - x)^2 = \sum_{n=1}^{\infty} C_n \sin\left[\frac{(2n - 1)\pi x}{2L}\right],$$

showing the C_n are the coefficients in the half-range Fourier sine series expansion of $hx(L - x)^2$ over the interval $0 \le x \le L$.

Multiplying this result by $\sin[\frac{(2m - 1)\pi x}{2L}]$, integrating over the interval $0 \le x \le L$, and solving for C_m gives

$$C_m = \frac{32hL^3}{\pi^4}\left[\frac{6(-1)^m + (4m - 2)\pi}{(2m - 1)^4}\right], \quad \text{for } m = 1, 2, \ldots,$$

from which C_n follows when m is replaced by n.

To find the coefficients D_n we must differentiate the series solution for $u(x, t)$ with respect to t, set $t = 0$, and use the second initial condition $u_t(x, 0) = 0$. From this it follows at once that $D_n = 0$ for $n = 1, 2, \ldots$, so the complete solution is

$$u(x, t) = \sum_{n=1}^{\infty} C_n \sin\left[\frac{(2n - 1)\pi x}{2L}\right] \cos\left[\frac{(2n - 1)\pi c t}{2L}\right],$$

with

$$C_n = \frac{32hL^3}{\pi^4}\left[\frac{6(-1)^n + (4n - 2)\pi}{(2n - 1)^4}\right] \quad \text{for } 0 \le x \le L, \ t > 0.$$

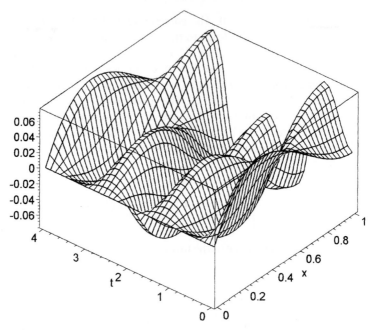

Figure 6.3   A 3d plot of the vibrations of the string showing how when $h = 0.5$, $L = 1$, and $c = 1$ the pattern of vibrations repeats with period $t = 4$.

Figure 6.3 shows a 3d plot of this solution when $h = 0.5$, $L = 1$, and $c = 1$. It can be seen from this that while the end of the string at $x = 0$ remains fixed, the end at $x = 1$ vibrates about the equilibrium position in a plane perpendicular to the x axis with period $t = 4$.

An examination of the form of this solution is instructive, because it shows how the solution of the problem comprises the sum of different vibrational frequencies, each characterized by a term $\sin\left[\frac{(2n-1)\pi x}{2L}\right] \cos\left[\frac{(2n-1)\pi ct}{2L}\right]$. The first factor in such a term, which is simply the nth *eigenfunction*, describes the space variation, and it is called the **nth normal mode** of vibration. The second factor describes the way in which this mode is modulated by an oscillatory time factor, with the frequency of vibration being determined by the nth *eigenvalue* λ_n. The rapid decrease of the coefficient C_n as n increases shows the form of the resulting wave on the string is largely determined by the lowest modes of vibration.

To examine the energy distribution between normal modes we see from $u(x, t)$ that the kinetic energy K_n of the nth mode is

$$K_n = \frac{1}{2}\rho \int_0^L (u_t)^2 dx = C_n^2 \frac{\rho \pi^2 c^2 (2n-1)^2}{16L} \sin^2\left[\frac{(2n-1)\pi ct}{2L}\right],$$

while it follows from Section 4.6 that within the accuracy of the wave equation for a string the potential energy $V = \frac{1}{2}T_0 \int_0^L u_x^2 dx$. Thus, from the expression for $u(x, t)$, the potential energy of the nth mode is

$$V_n = \frac{1}{2}\rho \int_0^L (u_t)^2 \, dx = C_n^2 \frac{\rho\pi^2 c^2 (2n-1)^2}{16L} \cos^2\left[\frac{(2n-1)\pi ct}{2L}\right].$$

Thus the total energy of the nth mode is $E_n = K_n + V_n$ where

$$E_n = C_n^2 \frac{\rho\pi^2 c^2 (2n-1)^2}{16L}, \quad \text{with } C_n = \frac{32hL^3}{\pi^4}\left[\frac{6(-1)^n + (4n-2)\pi}{(2n-1)^4}\right],$$

and so it is proportional to the square of the amplitude C_n of the mode. In this case as C_n decreases rapidly as n increases, most energy of the energy in the string is contained in the lowest modes. ∎

Example 6.5. Adapt the method of separation of variables to solve the nonhomogeneous wave equation $u_{tt} = c^2 u_{xx} + f(x, t)$ on the interval $0 \le x \le L$ subject to the homogeneous boundary conditions $u(0, t) = u(L, t) = 0$, and the initial conditions $u(x, 0) = F(x)$ and $u_t(x, 0) = G(x)$. Apply the method to the case when $f(x, t) = \sin t \sin(2\pi x/L)$, $F(x) = \sin(3\pi x/L)$, and $G(x) \equiv 0$.

Solution: Separation of variables cannot be used directly because of the nonhomogeneous term, so an attempt must be made to transform the problem in such a way that the method is applicable. The approach will be to seek a solution of the form $u(x, t) = U(x, t) + v(x, t)$, where $U(x, t)$ is a solution of the homogeneous wave equation $U_{tt} = c^2 U_{xx}$ subject to the homogeneous boundary conditions $U(0, t) = U(L, t) = 0$ and the initial conditions $U(x, 0) = F(x)$ and $U_t(x, 0) = G(x)$, and the function $v(x, t)$ must be determined.

Substituting $u(x, t)$ into the nonhomogeneous equation gives $U_{tt} + v_{tt} = c^2 U_{xx} + c^2 v_{xx} + f(x, t)$, so to make U a solution of the homogeneous wave equation we must set $v_{tt} = c^2 v_{xx} + f(x, t)$. Consideration of the *homogeneous* form of the equation for $v(x, t)$, and separating variables in the usual way by setting $v(x, t) = X(x)C(t)$ and introducing a separation constant λ^2, it follows that the general solution for $X(x)$ is $X(x) = A\cos(\lambda x/L) + B\sin(\lambda x/L)$. Imposing the boundary conditions $X(0) = X(L) = 0$, which follow from the homogeneous boundary conditions that must be satisfied by $v(x, t)$, it is found that the coefficient $B = 0$ and the eigenvalues are $\lambda_n = n\pi x/L$, so the associated eigenfunction is $X_n(x) = \sin(n\pi x/L)$.

The next step is to determine the function $v(x, t)$, and to do this it is first necessary to expand the nonhomogeneous term $f(x, t)$ in terms of the eigenfunctions $X_n(x)$, while allowing for the fact that the coefficients in the expansion must be functions of t by setting

$$f(x, t) = \sum_{n=1}^{\infty} B_n(t) \sin(n\pi x/L), \quad \text{so that}$$

$$B_n(t) = \frac{2}{L}\int_0^L f(x, t) \sin(n\pi x/L) \, dx.$$

The appropriate form for the function $v(x, t)$ will now be written in terms of the eigenfunctions $X_n(x)$ by writing

$$v(x, t) = \sum_{n=1}^{\infty} C_n(t) \sin(n\pi x/L),$$

where the functions $C_n(t)$ are to be found, and the factors $\sin(n\pi x/L)$ ensure $v(x, t)$ satisfies the necessary homogeneous boundary conditions.

As this expression for $v(x, t)$ must satisfy the original PDE, substituting both $v(x, t)$ and $f(x, t)$ into the equation gives

$$\sum_{n=1}^{\infty} \left[C_n''(t) + \left(\frac{n\pi c}{L} \right)^2 C_n(t) - B_n(t) \right] \sin(n\pi x/L) = 0.$$

This must be true for all $t \geq 0$ and $0 \leq x \leq L$, which is only possible if

$$C_n''(t) + \left(\frac{n\pi c}{L} \right)^2 C_n(t) - B_n(t) = 0 \quad \text{for } n = 1, 2, \ldots.$$

The functions $C_n(t)$ now follow by solving this linear ODE after using the initial conditions $C_n(0) = 0$ and $C_n'(0) = 0$, which follow from the initial conditions of the problem.

It is now possible to construct the solution $u = U + v$, because

$$v(x, t) = \sum_{n=1}^{\infty} C_n(t) \sin(n\pi x/L),$$

and the solution for $U(x, t)$ satisfying

$$U_{tt} = c^2 U_{xx}, \ U(0, t) = U(L, t) = 0 \text{ and } U(x, 0) = F(x), \ U_t(x, 0) = G(x)$$

is easily seen to be given by

$$U(x, t) = \sum_{n=1}^{\infty} (a_n \cos(n\pi c t/L) + b_n \sin(n\pi c t/L)) \sin(n\pi x/L),$$

with

$$a_n = \frac{2}{L} \int_0^L F(x) \sin(n\pi x/L) \, dx \quad \text{and} \quad b_n = \frac{2}{L} \int_0^L G(x) \sin(n\pi x/L) \, dx.$$

Thus the problem has been solved provided the operations performed are justified, and this can be shown to be the case provided: (a) $f(0, t) = f(L, t) = 0$; (b) $f(x, t)$ is continuous; and (c) $f_x(x, t)$ and $f_{xx}(x, t)$ are continuous.

To apply this method to the stated problem we set $f(x, t) = \sin t \, \sin(2\pi x/L)$, $F(x) = \sin(3\pi x/L)$, $G(x) \equiv 0$ and note that conditions (a) to (c) are satisfied.

The functions $B_n(t)$ are determined by

$$B_n(t) = \frac{2}{L} \int_0^L \sin t \sin(2\pi x/L) \sin(n\pi x/L)\, dx, \quad \text{for } n = 1, 2, \ldots,$$

so $B_2(t) = \sin t$ and $B_n(t) = 0$ for all $n \neq 2$.

Consequently, all but $C_2(t)$ must vanish, while $C_2(t)$ is the solution of

$$C_2''(t) + (2\pi c/L)^2 C_2(t) = \sin t \quad \text{with } C_2(0) = 0 \text{ and } C_2'(0) = 0.$$

This has the solution

$$C_2(t) = \frac{L^2[L \sin(2\pi ct/L) - 2\pi c \sin t]}{2\pi c(L^2 - 4\pi^2 c^2)},$$

and as $F(x) = \sin(3\pi x/L)$ and $G(x) \equiv 0$ it follows that all the b_n must vanish and all the a_n vanish with the exception of $a_3 = 1$, so $U(x,t) = \cos(3\pi ct/L)\sin(3\pi x/L)$.

The required solution $u(x,t) = U(x,t) + v(x,t)$ of the nonhomogeneous equation is thus

$$u(x,t) = \cos\left(\frac{3\pi ct}{L}\right) \sin\left(\frac{3\pi x}{L}\right)$$
$$+ \frac{L^2[L \sin(2\pi ct/L) - 2\pi c \sin t]}{2\pi c[L^2 - 4\pi^2 c^2]} \sin\left(\frac{2\pi x}{L}\right).$$

This is an exact closed form solution because of the simple nature of the functions $f(x,t)$ and $F(x)$. ∎

Example 6.6. Solve the diffusion equation $ku_{xx} = u_t$ on the interval $0 \leq x \leq L$ subject to the time-varying boundary conditions $u(0,t) = F(t)$, $u(L,t) = G(t)$, $t \geq 0$ and the initial condition $u(x,0) = f(x)$.

Solution: The method of separation of variables does not apply directly to the situation where time-varying boundary conditions arise. However, we show how by reformulating the problem it can be reduced to a nonhomogeneous diffusion equation.

Set $u(x,t) = U(x,t) + V(x,t)$, substitute this into $ku_{xx} = u_t$, and rearrange terms to obtain

$$kU_{xx} - U_t = -(kV_{xx} - V_t).$$

The appropriate boundary conditions are then

$$V(0,t) = F(t) - U(0,t) \quad \text{and} \quad V(L,t) = G(t) - U(L,t),$$

while the initial condition becomes

$$U(x,0) = f(x) - V(x,0) \quad \text{for } 0 \leq x \leq L.$$

The idea now is to make the boundary conditions for $U(x, t)$ homogeneous by making a suitable choice for $V(x, t)$. This is easily accomplished by setting

$$V(x, t) = F(t) + (x/L)[G(t) - F(t)].$$

This choice for $V(x, t)$ converts the equation for $U(x, t)$

$$kU_{xx} - U_t = [F'(t) + (x/L)(G'(t) - F'(t))],$$

which is a nonhomogeneous diffusion equation, although now it is subject to the homogeneous boundary conditions $U(0, t) = U(L, t) = 0$ and the initial condition $U(x, 0) = f(x) - V(x, 0)$.

A straightforward modification of the method of Example 6.5 allows this nonhomogeneous equation to be solved, although the details are left as an exercise. ∎

Example 6.7. Solve by separation of variables the nonhomogeneous wave equation $u_{tt} = c^2 u_{xx} + f(x)$ on the interval $0 \leq x \leq L$, subject to the homogeneous boundary and initial conditions $u(0, t) = 0$ and $u(L, t) = 0$ and $u(x, 0) = 0$ and $u_t(x, 0) = 0$. Apply the method to the case when $f(x) = hx(L - x)$ with $0 \leq x \leq L$ and h a constant.

Solution: Separation of variables cannot be applied directly to this problem because of the nonhomogeneous term, so instead we will attempt to find a solution of the form $u(x, t) = U(x, t) + g(x)$ with $U(x, t)$ being a solution of the *homogeneous* wave equation. Substituting for $u(x, t)$ in the PDE gives $U_{tt} = c^2 U_{xx} + c^2 g''(x) + f(x)$, so $U(x, t)$ will be a solution of the homogeneous wave equation if $c^2 g''(x) + f(x) = 0$. The boundary conditions for $U(x, t)$ follow from $u(x, t) = U(x, t) + g(x)$ and the boundary conditions for $u(x, t)$, so $u(0, t) = U(0, t) + g(x)$ and $u(L, t) = U(L, t) + g(L)$. However, $u(0, t) = 0$ and $u(L, t) = 0$, and so $U(0, t) = -g(0)$ and $U(L, t) = -g(L)$. Thus $U(x, t)$ will satisfy *homogeneous* boundary conditions if $g(0) = g(L) = 0$. The unknown function $g(x)$ now follows by solving $d^2 g/dx^2 = -(1/c^2) f(x)$ with $g(0) = g(L) = 0$.

The boundary conditions for $U(x, t)$ have already been chosen to be the conditions $U(0, t) = U(L, t) = 0$. One initial condition for $U(x, t)$ follows from $u(x, t) = U(x, t) + g(x)$ by setting $t = 0$ and using the fact that $u(x, 0) = 0$, when we find that $U(x, 0) = -g(x)$. The second initial condition follows from $u(x, t) = U(x, t) + g(x)$ by differentiating it with respect to t and using the fact that $u_t(x, 0) = 0$, when we find that $U_t(x, 0) = 0$.

Hence $U(x, t)$ is the solution of $U_{tt} = c^2 U_{xx}$, subject to the boundary conditions $U(0, t) = U(L, t) = 0$ and the initial conditions $U(x, 0) = -g(x)$ and $U_t(x, 0) = 0$. The general problem is now solved, because $g(x)$ can be obtained by direct integration, and once it is known, the function $U(x, t)$ can be found in the usual manner by separation of variables.

We now use this approach to solve the above problem when $f(x) = hx(L - x)$, with h being a constant. The function $g(x)$ follows by integrating

$d^2g/dx^2 = -(h/c^2)x(L - x)$ and then imposing the conditions $g(0) = 0$ and $g(L) = 0$. The result is that $g(x) = \frac{h}{12c^2}(x^4 - 2Lx^3 + L^3x)$, so $U(x, t)$ is the solution of $U_{tt} = c^2 U_{xx}$ subject to the boundary conditions $U(0, t) = 0$ and $U(L, t) = 0$, and the initial conditions $U(x, 0) = \frac{-h}{12c^2}(x^4 - 2Lx^3 + L^3x)$ and $U_t(x, 0) = 0$.

Separating variables in the usual manner by setting $U(x, t) = X(x)T(t)$ we find that $\frac{X''}{X} = \frac{T''}{c^2 T} = \lambda^2$, where λ^2 is a separation constant. Thus $X'' + \lambda^2 X = 0$ and $T'' + \lambda^2 c^2 T = 0$, and so $X(x) = A\cos\lambda x + B\sin\lambda x$ and $T(t) = C\cos(\lambda ct) + D\sin(\lambda ct)$.

The boundary conditions for $X(x)$ follow from those for $U(x, 0)$ and are seen to be $X(0) = 0$ and $X(L) = 0$, so that $A = 0$ and $\lambda_n = n\pi/L$, from which it follows that $X_n(x) = \sin(n\pi x/L)$. Thus the solution will be of the form

$$U(x, t) = \sum_{n=1}^{\infty} \left(C_n \cos\left(\frac{n\pi ct}{L}\right) + D_n \sin\left(\frac{n\pi ct}{L}\right) \right) \sin\left(\frac{n\pi x}{L}\right).$$

The initial condition $U(x, 0) = -g(x)$ implies that

$$\frac{-h}{12c^2}(x^4 - 2Lx^3 + L^3x) = \sum_{n=1}^{\infty} C_n \sin\left(\frac{n\pi x}{L}\right),$$

while the condition $U_t(x, 0)$ implies that $D_n = 0$ for all n. Multiplying the previous result by $\sin(m\pi x/L)$, integrating over the interval $0 \le x \le L$, using the orthogonality of the sine functions, and then changing m to n gives

$$C_n = \frac{-4hL^4}{c^2\pi^5 n^5}[1 - (-1)^n].$$

Substituting C_n and $g(x)$ into the expression for $u(x, t)$ gives

$$u(x, t) = \frac{h}{12c^2}(x^4 - 2Lx^3 + L^3x)$$
$$- \frac{8hL^4}{c^2\pi^5} \sum_{n=0}^{\infty} \frac{\cos[(2n + 1)\pi ct/L]\sin[(2n + 1)\pi x/L]}{(2n + 1)^5}.$$

The series converges very rapidly, so it can be approximated by its first term. ∎

Example 6.8. An empty cylindrical cavity of radius ρ and height h has its curved walls and flat base maintained at zero potential, while its flat top is maintained at the constant potential U. Find the potential at a general point inside the cavity, and the potential along the axis of the cavity.

Solution: The electrostatic potential in empty space satisfies the Laplace equation, so as the boundary conditions are Dirichlet conditions, and the region of space in the cylindrical cavity is closed, the problem is well posed. The natural system of coordinates to be used is cylindrical polar coordinates (r, θ, z) with the z axis taken along the axis of the cavity and the origin located such that $z = 0$ on the base of the cavity. If the potential is u, then as the surface distribution of potential is symmetrical about the z axis there can be no variation with θ.

Thus if the potential is $u(r, \theta)$ the appropriate form of the Laplace equation will be $\frac{\partial^2 u}{\partial r^2} + \frac{1}{r}\frac{\partial u}{\partial r} + \frac{\partial^2 u}{\partial z^2} = 0$.

To separate variables we now set $u(r, z) = R(r)Z(z)$ and substitute into the Laplace equation when we find that $\frac{R''}{R} + \frac{1}{r}\frac{R'}{R} + \frac{Z''}{Z} = 0$. Introducing a separation constant λ^2 in the usual way gives

$$\frac{R''}{R} + \frac{1}{r}\frac{R'}{R} = -\frac{Z''}{Z} = -\lambda^2,$$

so after separation of the variables we find that

$$r R'' + R' + \lambda^2 r R = 0 \quad \text{and} \quad Z'' - \lambda^2 Z = 0.$$

The first equation is Bessel's equation of order zero, so from (5.48) its general solution is seen to be $R(r) = A J_0(\lambda r) ++ B Y_0(\lambda r)$. As the potential must be finite everywhere inside the cavity, including along its axis, to exclude the infinite value of the Bessel function Y_0 of the second kind we must set $B = 0$, and so $R(r) = A J_0(\lambda r)$.

The boundary condition $u(\rho, z) = 0$ on the curved wall of the cavity requires that $R(\rho) = 0$, and so $\lambda_n \rho = j_{0,n}$, where $j_{0,n}$ is the nth zero of $J_0(x)$, values of which are to be found in Table 5.1. Thus the eigenvalue $\lambda_n = j_{0,n}/\rho$, and the corresponding eigenfunction, is $R_n(r) = J_0(r j_{0,n}/\rho)$.

Using these values of λ_n when solving for $Z(z)$ we obtain $Z_n(z) = C \cosh(j_{0,n}z/\rho) + D \sinh(j_{0,n}z/\rho)$ so the complete solution will be of the form

$$u(r, z) = \sum_{n=1}^{\infty} J_0(j_{0,n}r/\rho)[C_n \cosh(j_{0,n}z/\rho) + D_n \sinh(j_{0,n}z/\rho)].$$

To find the coefficients C_n and D_n we must now use the boundary conditions on the plane ends of the cavity at $z = 0$ and $z = h$. When $z = 0$ the potential is $u(r, 0) = 0$, so $0 = \sum_{n=1}^{\infty} C_n J_0(j_{0,n}r/\rho)$ from which we see that $C_n = 0$ for $n = 1, 2, \ldots$, so the solution simplifies to

$$u(r, z) = \sum_{n=1}^{\infty} D_n J_0(j_{0,n}r/\rho) \sinh(j_{0,n}z/\rho).$$

Imposing the boundary condition $u(r, h) = U$ this becomes

$$U = \sum_{n=1}^{\infty} D_n \sinh(j_{0,n}h/\rho) J_0(j_{0,n}r/\rho).$$

Thus the coefficients $D_n \sinh(j_{0,n}h/\rho)$ are the coefficients in the expansion of the function $U = $ constant over the interval $0 \le r \le \rho$, in terms of the Bessel function J_0. Multiplying the expression for U by $r J_0(j_{0,m}r/\rho)$, integrating over the interval $0 \le r \le \rho$, and using the orthogonality of Bessel functions in (5.52) gives

$$U \int_0^{\rho} r J_0(j_{0,m}r/\rho)\, dr = \frac{1}{2}\rho^2 D_m [J_1(j_{0,m})]^2 \sinh(j_{0,m}h/\rho).$$

From Section 4.6 $\int r J_0(\alpha r)\, dr = (r/\alpha) J_1(\alpha r) + \text{constant}$, so

$$U \int_0^{\rho} r J_0(j_{0,m} r/\rho)\, dr = \frac{U\rho^2}{j_{0,m}} J_1(j_{0,m}), \text{ because } J_1(0) = 0.$$

Thus

$$\frac{U\rho^2}{j_{0,m}} J_1(j_{0,m}) = \frac{1}{2}\rho^2 D_m [J_1(j_{0,m})]^2 \sinh(j_{0,m}h/\rho),$$

so solving for D_m and replacing m by n we obtain

$$D_n = \frac{2U}{j_{0,n} J_1(j_{0,n}) \sinh(j_{0,n}h/\rho)}.$$

Finally, substituting for D_n in the series for $u(r, z)$ gives the required solution

$$u(r, z) = 2U \sum_{n=1}^{\infty} \frac{J_0(j_{0,n}r/\rho) \sinh(j_{0,n}z/\rho)}{j_{0,n} J_1(j_{0,n}) \sinh(j_{0,n}h/\rho)} \quad \text{for } 0 \le r \le \rho,\ 0 \le z \le h.$$

The solution along the z axis follows by setting $r = 0$, and using the fact that $J_0(0) = 1$ we obtain

$$u(0, z) = 2U \sum_{n=1}^{\infty} \frac{\sinh(j_{0,n}z/\rho)}{j_{0,n} J_1(j_{0,n}) \sinh(j_{0,n}h/\rho)}.$$

A 3d plot of $u(r, z)/U$ with $\rho = 1$ and $h = 1$ is shown in Fig. 6.4, using only the first 10 terms of the expansion. The oscillations about $z = 1$ are due to

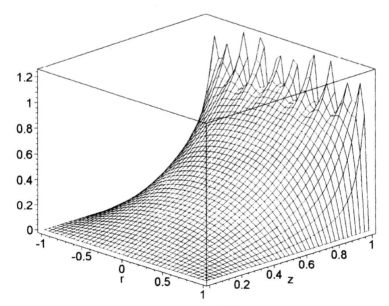

Figure 6.4 A 3d plot of the potential $\frac{u(r, z)}{U}$ using a 10-term approximation with $\rho = 1$ and $h = 1$.

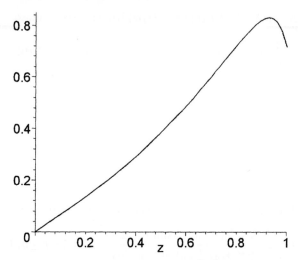

Figure 6.5 A plot of the potential $\frac{u(0,\,z)}{U}$ along the z axis using a 10-term approximation with $\rho = 1$ and $h = 1$.

the Gibbs-type phenomenon caused by the discontinuous change of potential between the top of the cylindrical cavity at $z = h$ and the curved walls. Figure 6.5 shows a plot of the potential along the z axis using the same 10-term approximation. ∎

Example 6.9. Find the fundamental modes of oscillation of a uniform circular membrane of unit radius when its rim is rigidly clamped.

Solution: The boundary shape suggests it will be natural to use the polar coordinate system (r, θ, t). Accordingly, the origin will be located at the center of the membrane when in its equilibrium position, and $u(r, \theta, t)$ will denote the displacement of the membrane at position (r, θ) and time t in a direction perpendicular to its equilibrium position. It follows from Chapter 1, and the Laplacian in polar coordinates, that the displacement u must be a solution of the wave equation

$$u_{tt} = c^2 \left(u_{rr} + \frac{1}{r} u_r + \frac{1}{r^2} u_{\theta\theta} \right).$$

The boundary condition will be $u(1, \theta, t) = 0$, corresponding to the fact that the rim of the membrane $r = 1$ is clamped rigidly for all time. Although we will not make use of them, appropriate initial conditions would be the requirement that

$$u(r,\ \theta,\ 0) = f(r,\ \theta) \quad \text{and} \quad u_t(r,\ \theta,\ 0) = g(r, \theta),$$

where f specifies the initial shape of the membrane and g the speed with which each element of the displaced membrane moves at time $t = 0$. The wave equation is hyperbolic and the initial conditions are of the Cauchy type, so as the region of space–time in which the solution is required is open since the solution is required for $t > 0$, it follows that the problem is well posed.

The variables will be separated in two stages, starting by setting $u(r, \theta, t) = H(r, \theta) T(t)$. After proceeding in the usual manner and introducing a separation constant $-\lambda^2$ we arrive at the result

$$\frac{1}{c^2} \frac{T''}{T} = \frac{\Delta H}{H} = -\lambda^2.$$

Thus

$$T'' + \lambda^2 c^2 T = 0 \quad \text{and} \quad \Delta H + \lambda^2 H = 0,$$

where the second equation, which is a PDE, is called the **Helmholtz equation**.

We now separate the variables in $H(r, \theta)$ by writing $H(r, \theta) = R(r)\Theta(\theta)$, as a result of which the equation for H becomes

$$\Theta \left(R'' + \frac{1}{r} R' \right) + \frac{R}{r^2} \Theta'' + \lambda^2 R\Theta = 0.$$

After division by $R\Theta$, and the introduction of another separation constant m, this becomes

$$\frac{r^2}{R} \left(R'' + \frac{1}{r} R' \right) + \lambda^2 r^2 = -\frac{\Theta''}{\Theta} = m,$$

as a result of which we arrive at the ODEs for R and Θ

$$r^2 R'' + r R' + (\lambda^2 r^2 - m) R = 0 \quad \text{and} \quad \Theta'' + m\Theta = 0.$$

The equation for R is seen to be Bessel's equation, but before solving it the equation for Θ must be considered. It has the general solution $\Theta(\theta) = A\cos(\sqrt{m}\theta + \phi)$, with ϕ being an arbitrary constant. As θ is the polar angle, the solution must be periodic in θ with period 2π, because we must have $u(r, \theta, t) = u(r, \theta + 2n\pi, t)$, for $n = 0, 1, \ldots$. Thus we must set $m = n^2$ with $n = 0, 1, \ldots$, when $\Theta(\theta) = A\cos(n\theta + \phi)$. After setting $m = n^2$ in the Bessel equation its general solution is seen to be given by $R(r) = B J_n(\lambda r) + C Y_n(\lambda r)$.

However, the displacement of the membrane must always remain finite, so this will only be possible for $0 \le r \le 1$ if $C = 0$, so the solution for $R(r)$ reduces to $R(r) = B J_n(\lambda r)$. To proceed further it is necessary to determine the separation constant λ. This follows from the boundary condition $u(1, \theta, t) = 0$, which implies that $R(1) = 0$, so λ must be a solution of $J_n(\lambda) = 0$. Thus the separation constant λ must be a zero of the Bessel function $J_n(r)$ so, if $j_{n,m}$ is the mth zero of $J_n(r)$, it follows that $\lambda = j_{n,m}$ for $n = 0, 1, 2, \ldots$ and $m = 1, 2, \ldots$.

Combining $R(r)$ and $\Theta(\theta)$, a partial solution of $H(r, \theta)$ becomes $H_{nm}(r, \theta) = J_n(j_{n,m}r) \cos(n\theta)$, where the arbitrary constant ϕ has been set equal to 0 by choosing a suitable reference line from which the polar angle θ is measured. The function $H_{nm}(r, \theta)$ is the two-dimensional eigenfunction for the membrane problem.

Solving for the time variation gives $T_{nm}(t) = D_{nm} \cos(j_{n,m}ct) + E_{nm} \sin(j_{n,m}ct)$, so the solution of an initial boundary value problem (IBVP) will be of the form

$$u(r, \theta, t) = \sum_{n=0, m=1}^{\infty} J_n(j_{n,m}r) \cos(n\theta)[D_{nm} \cos(j_{n,m}ct) + E_{nm} \sin(j_{n,m}ct)].$$

Although the solution of an IVBP is not needed, we mention that the coefficients D_{nm} and E_{nm} can be found by requiring $u(r, \theta, t)$ to satisfy the initial conditions $u(r, \theta, 0) = f(r, \theta)$ and $u_t(r, \theta, 0) = g(r, \theta)$.

The fundamental modes of vibration of the membrane are simply the functions $H_{nm}(r, \theta) = J_n(j_{n,m}r) \cos(n\theta)$, and these are often called the **eigenmodes** of the problem. Three typical examples of these eigenmodes are shown in Fig. 6.6, corresponding, respectively, to (a) $n = 0$, $m = 2$, (b) $n = 1$, $m = 2$, and (c) $n = 2$, $m = 2$, and associated with each is a contour plot of the eigenmode. When the time variation is imposed on an eigenmode it simply causes its amplitude to change sign periodically, and in the contour plots associated with these eigenmodes the straight lines represent points on the membrane that always remain at rest. These lines are called **nodal lines**, and the displacement of the membrane is in the opposite direction across each nodal line. ■

Example 6.10. Let $u(r, \theta, \phi)$ be a potential in spherical polar coordinates, and the potential on a spherical surface of unit radius be such that $u(1, \phi, \theta) = U/\sqrt{2}$ for $0 \leq \theta \leq \frac{1}{4}\pi$ and $u(1, \theta, \phi) = U \sin \theta$ for $\frac{1}{4}\pi \leq \theta \leq \pi$. Find the potential inside the spherical surface.

Solution: The potential u inside the sphere satisfies the Laplace equation, and it satisfies Dirichlet boundary conditions on the surface of the sphere of radius 1, so the conditions for a well-posed problem are satisfied.

As the potential is independent of the azimuthal angle ϕ, the spherical polar form of the Laplace equation to be satisfied by u becomes

$$r^2 u_{rr} + 2r u_r + \cot\theta\, u_\theta + u_{\theta\theta} = 0.$$

Before separating variables we will make a change of variables by setting $\xi = \cos\theta$, when it follows from the chain rule that $u_\theta = -\sin\theta\, u_\xi$ and $u_{\theta\theta} = \sin^2\theta\, u_{\xi\xi} - \cos\theta\, u_\xi$, so the Laplace equation becomes $r^2 u_{rr} + 2r u_r - 2\xi u_\xi + (1 - \xi^2)u_{\xi\xi} = 0$.

To separate variables we now set $u(r, \xi) = R(r)Q(\xi)$, and after introducing a separation constant k we find that

$$\frac{r^2 R'' + 2r R'}{R} = \frac{2\xi Q' - (1 - \xi^2)Q''}{Q} = k.$$

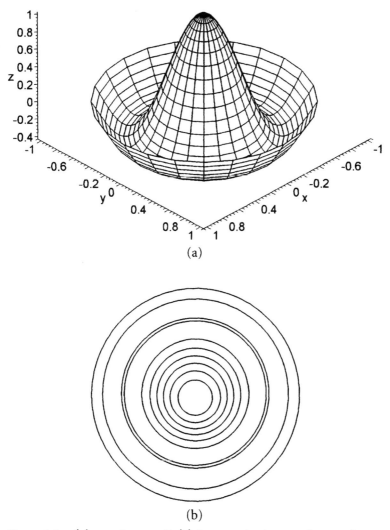

(a)

(b)

Figure 6.6 (a) $n = 0$, $m = 2$. (b) Contour plot for $n = 0$, $m = 2$.
(c) $n = 1$, $m = 2$. (d) Contour plot for $n = 1$, $m = 2$. (e) $n = 2$,
$m = 2$. (f) Contour plot for $n = 2$, $m = 2$.

There now follow from this the two ODEs

$$r^2 R'' + 2r R' - kR = 0 \quad \text{and} \quad (1 - \xi^2) Q'' - 2\xi Q' + kQ = 0.$$

Setting the separation constant $k = n(n + 1)$, with $n = 0, 1, \ldots$, the equation
for Q becomes the Legendre equation $(1 - \xi^2) Q'' - 2\xi Q' + n(n + 1) Q = 0$,
with the solution $Q(\xi) = P_n(\xi)$, where $P_n(\xi)$ is the Legendre polynomial of
degree n. The equation for R becomes the Cauchy–Euler equation $r^2 R'' + 2r R' - n(n+1) R = 0$. This has solutions of the form $R(r) = r^\alpha$, where the permissible

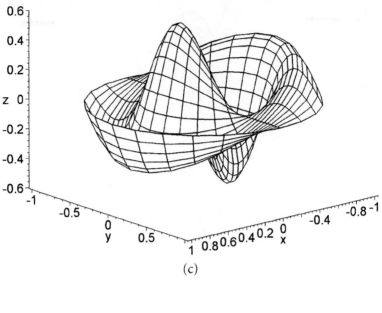

(c)

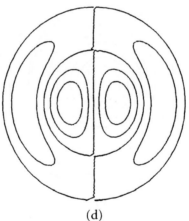

(d)

Figure 6.6   (*Continued*).

values of α are found by substituting into the ODE, when we find that $\alpha = n$ and $\alpha = -(n+1)$, so the general solution for $R(r)$ is $R(r) = Ar^n + Br^{-(n+1)}$. Combining $R(r)$ and $Q(\xi)$ shows the general solution to be

$$u(r,\theta,\phi) = \sum_{n=0}^{\infty} \left(A_n r^n + B_n \frac{1}{r^{n+1}} \right) P_n(\xi),$$

with $-1 \le \xi \le 1$ and $\xi = \cos\theta$.

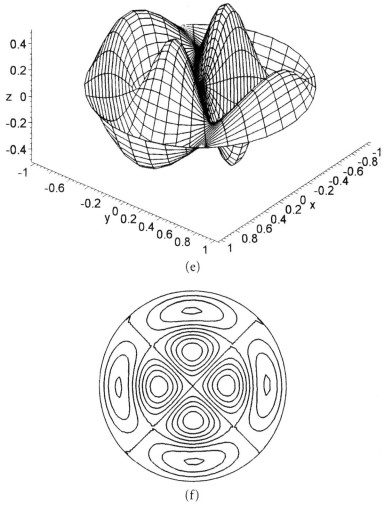

(e)

(f)

Figure 6.6   (*Continued*).

For this solution to remain bounded inside the sphere we must set $B_n = 0$ with $n = 0, 1, \ldots$, causing the solution to simplify to

$$u(r, \xi, \phi) = \sum_{n=0}^{\infty} A_n r^n P_n(\xi), \quad \text{with } -1 \leq \xi \leq 1 \text{ and } 0 \leq r \leq 1.$$

As $\sin \theta = (1 - \xi^2)^{1/2}$, the boundary condition becomes

$$u(1, \xi, \phi) = \begin{cases} U(1 - \xi^2)^{1/2}, & -1 \leq \xi \leq 1/\sqrt{2} \\ U/\sqrt{2}, & 1/\sqrt{2} \leq \xi \leq 1. \end{cases}$$

The coefficients A_n now follow by requiring $u(r, \xi, \phi)$ satisfy this boundary condition when $r = 1$, and using the orthogonality properties of $P_n(\xi)$ given in (5.30).

Setting $r = 1$ in $u(r, \xi, \phi)$, using the boundary condition, multiplying the series by $P_m(\xi)$, and integrating over the interval $-1 \leq \xi \leq 1$ gives

$$\int_{-1}^{1} u(1, \xi, \phi) P_m(\xi) \, d\xi = \sum_{n=0}^{\infty} A_n \int_{-1}^{1} P_m(\xi) P_n(\xi) \, d\xi.$$

Substituting for $u(1, \phi, \xi)$ and using the orthogonality condition (5.30) this becomes

$$\int_{-1}^{1/\sqrt{2}} U(1 - \xi^2)^{1/2} P_m(\xi) \, d\xi + U/\sqrt{2} \int_{1/\sqrt{2}}^{1} P_m(\xi) \, d\xi = \frac{2 A_m}{2m + 1},$$

$$\text{for } m = 0, 1, \ldots.$$

Thus

$$A_0 = \left(\frac{1}{\sqrt{2}} - \frac{1}{4} + \frac{3\pi}{8} \right) U, \quad A_1 = \frac{U}{12\sqrt{2}}, \quad A_2 = -\frac{3\pi U}{64},$$

$$A_3 = \frac{5U}{96\sqrt{2}}, \ldots,$$

and so

$$u(r, \xi, \phi) = U(1.6352 \, P_0(\xi) + 0.0589r \, P_1(\xi) - 0.1473r^2 \, P_2(\xi)$$
$$+ 0.0368r^3 \, P_3(\xi) + \cdots). \qquad \blacksquare$$

So far, the Examples illustrating the application of the method of separation of variables have been applied to second-order partial differential equations. The method can also be applied to certain types of higher-order equation that are not of Sturm–Liouville type, and the next example shows how it can be used to determine the transverse vibrations of a beam. Whereas the *longitudinal* vibrations of an elastic rod are governed by the wave equation, the *transverse* vibrations of an elastic beam are governed by a linear fourth-order partial differential equation.

Let E be Young's modulus of elasticity for the material of a uniform beam, I be the moment of inertia of a cross section of the beam about a line along the beam through the centroid of its cross section, and m be the mass per unit length of the beam (its line density). Then if x is the distance along the beam measured from its left end, t is the time, and $u(x, t)$ is the transverse deflection of the beam as shown in Fig. 6.7, its transverse vibrations are governed by the fourth-order equation

$$E I \frac{\partial^4 u}{\partial x^4} = -m \frac{\partial^2 u}{\partial t^2},$$

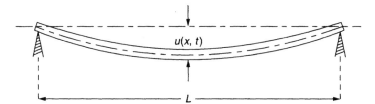

Figure 6.7   A uniform beam of length L hinged at its ends with a transverse deflection $u(x, t)$.

which for convenience will be written

$$\frac{\partial^2 u}{\partial t^2} + a^4 \frac{\partial^4 u}{\partial x^4} = 0,$$

where $a^4 = EI/m$.

The following Example shows how the transverse vibrations of a beam of length L, hinged at its ends, can be determined by separation of variables when the vibrations are started by forcing the beam into a displaced shape $u(x, 0) = f(x)$, and then at time $t = 0$ releasing it from rest.

Example 6.11. Solve the beam equation

$$u_{tt} + a^4 u_{xxxx} = 0 \quad \text{for } 0 \leq x \leq L \text{ and } t > 0,$$

subject to the boundary conditions

$$u(0, t) = 0, \ u(L, t) = 0, \quad u_{xx}(0, t) = 0, \quad u_{xx}(L, t) = 0,$$

and the initial conditions

$$u(x, 0) = f(x), \qquad u_t(x, 0) = 0.$$

Find the ratio of the energy in the second and third modes of vibration to that in the first mode, when (a) $f(x) = x(L - x)$, (b) $f(x) = x^2(L - x)$, and (c) $f(x) = x^3(L - x)$.

Solution: As usual we start by setting $u(x, t) = X(x)T(t)$, and after substituting this into the PDE and dividing by XT we find that

$$\frac{X^{(4)}}{X} = -\frac{T''}{a^4 T} = \lambda,$$

and so

$$X^{(4)} - \lambda X = 0 \quad \text{and} \quad T'' + \lambda a^4 T = 0.$$

It is a familiar fact that walking across a plank supported at its ends induces vertical (transverse) vibrations, so for these to be described by the equation for $T(t)$ it is necessary that $\lambda > 0$, so for convenience when solving the fourth-order

ODE for $X(x)$ we will set $\lambda = \mu^4$. With this choice for λ the general solution for $X(x)$ is found to be

$$X(x) = A\cos\mu x + B\sin\mu x + C\cosh\mu x + D\sinh\mu x.$$

The boundary conditions for $u(x, t)$ are appropriate for a beam with fixed but hinged ends, and these imply that $X(0) = 0$, $X''(0) = 0$, $X(L) = 0$ and $X''(L) = 0$. Imposing the conditions $X(0) = 0$ and $X''(0) = 0$ at $x = 0$ on $X(x)$ shows that $A = C = 0$, and so $X(x) = B\sin\mu x + D\sinh\mu x$. Imposing the conditions $X(L) = 0$ and $X''(L) = 0$ at the other end $x = L$ on $X(x)$ shows that $B\sin\mu L = 0$ and $D\sinh\mu L = 0$. However, $\mu \neq 0$ so the second result is only possible if $D = 0$, and as $u(x, t)$ cannot be identically zero we must have $B \neq 0$, and so $\sin\mu L = 0$, showing that the eigenvalue $\mu_n = n\pi/L$, for $n = 1, 2, \ldots$. The eigenfunction is thus $X_n(x) = \sin(n\pi x/L)$ and the corresponding eigenvalue is $\lambda_n = n^4\pi^4/L^4$.

The general solution for $T(t)$ is

$$T(t) = P\cos\left(\frac{n^2\pi^2 a^2 t}{L^2}\right) + Q\sin\left(\frac{n^2\pi^2 a^2 t}{L^2}\right),$$

so from the initial condition $u_t(x, 0) = 0$, it follows that

$$T_n(t) = P_n\cos\left(\frac{n^2\pi^2 a^2 t}{L^2}\right).$$

Thus the partial solutions are

$$u_n(x, t) = \sin\left(\frac{n\pi x}{L}\right)\cos\left(\frac{n^2\pi^2 a^2 t}{L^2}\right),$$

so the required solution $u(x, t)$ must be of the form

$$u(x, t) = \sum_{n=1}^{\infty} P_n\sin\left(\frac{n\pi x}{L}\right)\cos\left(\frac{n^2\pi^2 a^2 t}{L^2}\right).$$

To determine the coefficients P_n we now use the remaining initial condition $u(x, 0) = f(x)$, when after setting $t = 0$ we find that $f(x) = \sum_{n=1}^{\infty} P_n\sin\left(\frac{n\pi x}{L}\right)$. As $f(x)$ is expressed as the half-range Fourier sine series, it follows in the usual way that

$$P_n = \frac{2}{L}\int_0^L f(x)\sin\left(\frac{n\pi x}{L}\right)dx, \quad \text{for } n = 1, 2, \ldots.$$

Once the Fourier coefficients P_n have been determined, the series solution for the vibrations induced by the initial shape $u(x, 0) = f(x)$ with $u_t(x, 0) = 0$ follow by substituting c_n into

$$u(x, t) = \sum_{n=1}^{\infty} P_n\sin\left(\frac{n\pi x}{L}\right)\cos\left(\frac{n^2\pi^2 a^2 t}{L^2}\right).$$

Note that each partial solution has a characteristic frequency $\omega_n = n^2\pi^2 a^2/L^2$, and the lowest frequency, called the **fundamental frequency** of the beam, is $\omega_1 = \pi^2 a^2/L^2$. Multiples of this fundamental frequency are called harmonics. In the context of musical instruments, the tone of a note (vibration) depends on the proportion of energy it contains from each of its harmonics, and it is this that determines the quality of a musical instrument. The *tone* of a note must not be confused with its **pitch**, which is simply the fundamental frequency of the note.

Thus a musical instrument like the xylophone, which produces its sound by the vibration of metal bars, produces a purer, but less rich sound than a stringed instrument, because its higher frequencies are spaced according to n^2, where in a stringed instrument they are spaced according to n.

By analogy with Example 6.4, the energy in the nth mode of vibration is proportional to C_n^2, where $C_n = P_n$, so the ratio of the energy in the nth mode to that of the fundamental mode is C_n^2/C_1^2. Thus the required ratios for cases (a), (b), and (c) are 1, C_2^2/C_1^2 and C_3^2/C_1^2.

Case (a): $C_n = \dfrac{4L^2[1-(-1)^n]}{n^3\pi^3}$, so $C_2^2/C_1^2 = 0$, $C_3^2/C_1^2 \approx 0.0014$.

Case (b): $C_n = \dfrac{4L^3[2(-1)^n - 1]}{n^3\pi^3}$, so $C_2^2/C_1^2 \approx 0.1406$, $C_3^2/C_1^2 \approx 0.0014$.

Case (c): $C_n = \dfrac{12L^4[(4 - n^2\pi^2)(-1)^n - 4]}{n^5\pi^5}$, so $C_2^2/C_1^2 \approx 0.4354$,

$$C_3^2/C_1^2 \approx 0.0317.$$

Examination of these results indicates that the more the initial displacement departs from the symmetrical displacement in Case (a), the more is the energy that enters into the higher-order modes. If the vibrations of the beam due to these excitations $f(x)$ are interpreted in terms of sound waves, it follows that the excitation due to the symmetrical displacement in Case (a) produces the purest tone, since its higher harmonics contain the least energy. Thus, in a xylophone, the purest tone can be expected to be produced when a vibrating bar is struck in its center. ∎

So far no example involving heat flow in a material where it is necessary to use the internal boundary conditions given in (1.38) and (1.39) to take account of the continuity of temperature and heat flux across an internal boundary has been considered. These problems are difficult and are best solved purely numerically. However, it is desirable to provide at least one example of this type to show how the eigenfunction method must be adapted, and to discover the effect of a change of material on the temperature distribution. The last example is sufficiently simple for the method to be applied, and it shows how the unsteady temperature distribution can be found in a two-layer composite slab, part of which is aluminum and part is steel.

Before formulating the problem mathematically we will describe the general nature of the problem. Two long slabs of different heat-conducting materials, one of thickness h_1 and the other of thickness h_2, have their common plane faces in perfect contact to form a composite slab of thickness $h_1 + h_2$. The specific heat, density, and thermal conductivity of the material in the first slab are c_1, ρ_1, and k_1, respectively, and the corresponding quantities in the second slab are c_2, ρ_2, and k_2. Let the first slab occupy the region $0 \leq x < h_1$ and the second slab the region $h_1 < x \leq h_1 + h_2$, and suppose the composite slab is heated in such a way that at time $t = 0$ the temperature distribution across the slab is $f(x)$. The problem will be to find the temperature distribution $V(x, t)$ in the composite slab as a function of x and the time t if, at time $t = 0$, the outer sides $x = 0$ and $x = h_1 + h_2$ of the composite slab are suddenly cooled to and then maintained at zero temperature.

The difficulty in this problem comes from the fact that, at first sight, because of the composite nature of the slab, there seems no obvious way of constructing a set of eigenfunctions $X_n^*(x)$ that are orthogonal over the width $0 \leq x \leq h_1 + h_2$ of the slab. The way the difficulty is resolved will be seen to be by relating the x variation in the second slab $h_1 < x \leq h_1 + h_2$ to the x variation in the first slab $0 \leq x < h_1$, finding the eigenvalues of the problem, and then weighting the variations across each slab in such a way that for each eigenvalue a set of composite functions $X_n^*(x)$ form an orthogonal system over the interval $0 \leq x \leq h_1 + h_2$. When arriving at this result it will be seen that as the eigenvalues are no longer the zeros of elementary functions they must be determined numerically, and also that suitable weighting functions must be found to arrive at the composite set of orthogonal eigenfunctions $X_n^*(x)$.

Example 6.12. Find the temperature distribution $V(x, t)$ in a composite slab of two different heat-conducting materials, one occupying the region $0 \leq x < h_1$ and the other the region $h_1 < x \leq h_1 + h_2$ with their common faces at $x = h_1$ in perfect contact, and at time $t = 0$ when the slab has an initial temperature distribution $f(x)$, the outside faces of the slab at $x = 0$ and $x = h_1 + h_2$ are suddenly cooled to and maintained at zero temperature. Describe the conditions to be satisfied by an initial temperature distribution $f(x)$ for such a composite slab of material. Let the specific heat, density, and thermal conductivity of the material in the first slab be c_1, ρ_1, and k_1, respectively, and the corresponding values in the second slab be c_2, ρ_2, and k_2. Apply the general solution to a block of metal formed by joining a slab of aluminum and a slab of steel, each 5 cm thick, when the initial steady temperature distribution is $f(x) = T_0 \sin(\pi x/10)$. Take the material constants for aluminum to be $c_1 = 0.21, \rho_1 = 2.7$ gm cm^{-3}, $k_1 = 0.51$, and those for steel to be $c_2 = 0.11$, $\rho_2 = 7.8$ gm cm^{-3}, $k_2 = 0.12$ (see Fig. 6.8a).

Solution: It is convenient to set $\beta_1 = c_1\rho_1/k_1$ and $\beta_2 = c_2\rho_2/k_2$ so the unsteady heat equations to be solved become

$$\frac{\partial^2 V}{\partial x^2} = \beta_1 \frac{\partial V}{\partial t} \quad \text{for } 0 \leq x < h_1 \quad \text{and} \quad \frac{\partial^2 V}{\partial x^2} = \beta_2 \frac{\partial V}{\partial t} \quad \text{for } h_1 \leq x < h_1 + h_2.$$

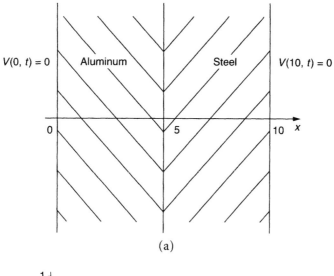

(a)

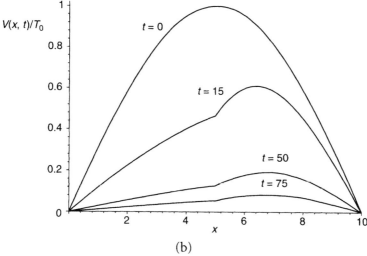

(b)

Figure 6.8 (a) The geometry of the composite slab. (b) The approximate temperature distribution $V(x,t)/T_0$ for $t = 0, 15, 50,$ and 75 s.

Consider the region $0 \leq x < h_1$ and introduce a dimensionless length variable $\xi = x/h_1$, which causes the two unsteady heat equations to become

$$\frac{1}{h_1^2} \frac{\partial^2 V}{\partial \xi^2} = \beta_1 \frac{\partial V}{\partial t} \quad \text{for } 0 \leq x < h_1 \quad \text{and}$$

$$\frac{1}{h_1^2} \frac{\partial^2 V}{\partial \xi^2} = \beta_2 \frac{\partial V}{\partial t} \quad \text{for } h_1 \leq x < h_1 + h_2.$$

Writing $V(\xi, t) = X(\xi)T(t)$ and substituting this into the first equation gives, after separating variables,

$$\frac{1}{X}\frac{d^2 X}{d\xi^2} = h_1^2\beta_1\frac{1}{T}\frac{dT}{dt} = -\lambda^2,$$

where λ^2 is a separation constant. This leads to the two ODEs

$$\frac{d^2 X}{d\xi^2} + \lambda^2 X = 0 \quad \text{and} \quad \frac{dT}{dt} = -\frac{\lambda^2}{h_1^2\beta_1} T.$$

Solving the first equation and substituting for ξ gives

$$X(x) = A\cos\frac{\lambda x}{h_1} + B\sin\frac{\lambda x}{h_1},$$

while solving the second equation gives

$$T(t) = \overline{T}\exp\left(-\lambda^2 / h_1^2\beta_1\right).$$

As $V(0, t) = 0$ it follows that $X(0) = 0$, showing that $A = 0$, so $X(x)$ simplifies to

$$X(x) = B\sin\frac{\lambda x}{h_1},$$

where as yet the eigenvalues λ are unknown.

To proceed further we must solve for $V(x, t)$ in the second slab by setting $V(x, t) = \tilde{X}(\xi)\tilde{T}(t)$, where again $\xi = x/h_1$, and using the fact that after separating variables the unsteady heat equation becomes

$$\frac{d^2\tilde{X}}{d\xi^2} = h_1^2\beta_2\frac{1}{T}\frac{d\tilde{T}}{dt}.$$

To obtain a generalization of the usual eigenfunction expansion it is necessary that the time variation is the same in both slabs, so to achieve this we must multiply this last equation by β_1/β_2 before introducing the separation constant $-\lambda^2$. This gives

$$\frac{\beta_1}{\beta_2}\frac{1}{\tilde{X}}\frac{d^2\tilde{X}}{dx^2} = h_1^2\beta_1\frac{1}{T}\frac{dT}{dt} = -\lambda^2,$$

showing that

$$\frac{d^2\tilde{X}}{dx^2} + \frac{\beta_2}{\beta_1}\lambda^2\tilde{X} = 0 \text{ and, once again, } T(t) = \overline{T}\exp\left(-\lambda^2 / h_1^2\beta_1\right).$$

After substituting for ξ the function $\tilde{X}(x)$ must be such that it satisfies the boundary condition $\tilde{X}(h_1 + h_2) = 0$, and consideration of the equation for

$\tilde{X}(x)$ shows this condition is satisfied by setting

$$\tilde{X}(x) = P \sin\left(\lambda\sqrt{\frac{\beta_2}{\beta_1}}\frac{(h_1 + h_2 - x)}{h_1}\right).$$

Thus far we have succeeded in satisfying the homogeneous boundary conditions $V(0, t) = V(h_1 + h_2, t) = 0$ at the outside edges of the slab. It now remains for us to satisfy the internal boundary conditions (1.38) and (1.39) at $x = h_1$, namely $X(h_1) = \tilde{X}(h_1)$, representing continuity of the temperature at the internal boundary, and $k_1(dX/dx)_{x=h_1} = k_2(d\tilde{X}/dx)_{x=h_1}$, representing continuity of heat flux at this internal boundary. These respective conditions lead to the results

$$B \sin\lambda = P \sin\left(\lambda\sqrt{\frac{\beta_2}{\beta_1}\frac{h_2}{h_1}}\right),$$

and

$$Bk_1\frac{\lambda}{h_1}\cos\lambda = -P\lambda k_2\sqrt{\frac{\beta_2}{\beta_1}}\frac{1}{h_1}\cos\left(\lambda\sqrt{\frac{\beta_2}{\beta_1}\frac{h_2}{h_1}}\right).$$

Dividing these two equations shows the eigenvalues $\lambda_1, \lambda_2, \ldots$ must be the sequential positive roots of

$$\frac{k_2}{k_1}\tan\lambda + \sqrt{\frac{\beta_1}{\beta_2}}\tan\left(\lambda\sqrt{\frac{\beta_2}{\beta_1}\frac{h_2}{h_1}}\right) = 0.$$

This is a transcendental equation for the eigenvalues λ_n, and because of its complexity the λ_n must be found numerically. It also follows directly from the equations relating the scale constants B and P that the nth partial eigenfunction

$$X_n(x) = \sin\left(\lambda_n\sqrt{\frac{\beta_2}{\beta_1}\frac{h_2}{h_1}}\right)\sin\left(\frac{\lambda_n x}{h_1}\right) \quad \text{for } 0 \le x \le h_1$$

and the nth partial eigenfunction

$$\tilde{X}_n(x) = \sin\lambda_n \sin\left(\lambda_n\sqrt{\frac{\beta_2}{\beta_1}}\frac{(h_1 + h_2 - x)}{h_1}\right) \quad \text{for } h_1 \le x \le h_1 + h_2.$$

Assembling the pieces of information obtained so far shows the required solution will be of the form

$$V(x, t) = \sum_{n=1}^{\infty} R_n \tilde{X}_n(x) \exp\left(-\lambda_n^2 t / h_1^2 \beta_1\right),$$

where the nth eigenfunction

$$\bar{X}_n(x) = \begin{cases} \sin\left(\lambda_n\sqrt{\frac{\beta_2}{\beta_1}\frac{h_2}{h_1}}\right)\sin\left(\frac{\lambda_n x}{h_1}\right), & 0 \le x \le h_1 \\ \sin\lambda_n\sin\left(\lambda_n\sqrt{\frac{\beta_2}{\beta_1}}\frac{(h_1+h_2-x)}{h_1}\right), & h_1 \le x \le h_1 + h_2, \end{cases}$$

and the coefficients R_n must be found by using the initial condition $V(x,0) = f(x)$.

The functions $X_n(x)$ are not orthogonal over the interval $0 \le x \le h_1 + h_2$, but consideration of the arguments concerning Sturm–Liouville problems given at the start of this chapter shows that by weighting $X_n(x)$ by $\sqrt{c_1\rho_1}$ and $\bar{X}_n(x)$ by $\sqrt{c_2\rho_2}$ the composite sequence of functions

$$X_n^*(x) = \begin{cases} \sqrt{c_1\rho_1}\sin\left(\lambda_n\sqrt{\frac{\beta_2}{\beta_1}\frac{h_2}{h_1}}\right)\sin\left(\frac{\lambda_n x}{h_1}\right), & 0 \le x \le h_1 \\ \sqrt{c_2\rho_2}\sin\lambda_n\sin\left(\lambda_n\sqrt{\frac{\beta_2}{\beta_1}}\frac{(h_1+h_2-x)}{h_1}\right), & h_1 \le x \le h_1 + h_2 \end{cases}$$

becomes orthogonal over the interval $0 \le x \le h_1 + h_2$.

The coefficients R_n must be found by using the initial condition $V(x,0) = f(x)$ in the series solution for $V(x,t)$

$$f(x) = \sum_{n=1}^{\infty} R_n\bar{X}_n(x).$$

However, the initial condition cannot be entirely arbitrary, because it must be such that at $x = h_1$ the function $f(x)$ satisfies the internal boundary conditions (1.38) and (1.39). The simplest type of initial condition will be one for which $f(x)$ is continuous for $0 \le x \le h_1 + h_2$ and $f'(h_1) = 0$, because then condition (1.38) will certainly be satisfied at the internal boundary $x = h_1$, while the internal boundary condition (1.39) will be satisfied trivially.

Multiplying the series for $f(x)$ by $X_n^*(x)$ and integrating over the interval $0 \le x \le h_1$, multiplying the series for $f(x)$ by $X_n^*(x)$ and integrating over the interval $h_1 \le x \le h_1 + h_2$, followed by adding the results and using the orthogonality of the eigenfunctions $X_n^*(x)$ over the complete interval $0 \le x \le h_1 + h_2$ shows the coefficients R_n are given by the formula

$$R_n = \frac{c_1\rho_1\int_0^{h_1} f(u)X_n(u)\,du + c_2\rho_2\int_{h_1}^{h_1+h_2} f(u)\bar{X}_n(u)\,du}{c_1\rho_1\int_0^{h_1}[X_n(u)]^2\,du + c_2\rho_2\int_{h_1}^{h_1+h_2}[\bar{X}_n(u)]^2\,du}.$$

Now that the coefficients R_n have been found the unsteady temperature distribution in the composite slab has been determined.

Even when the initial temperature distribution $f(x)$ is sufficiently simple for the coefficients R_n to be found analytically the calculations are tedious, and when this is not the case the use of numerical methods becomes essential. Consequently, as the eigenvalues must be found numerically, it is convenient for the coefficients R_n also to be found numerically. When technology is used, the determination of the eigenvalues λ_n and the coefficients R_n becomes straightforward.

Using the material constants for aluminum and steel, setting $h_1 = h_2 = 5$, and noting that the initial condition is appropriate since $f'(5) = 0$, some straightforward but tedious calculations show that

$$\lambda_1 = 0.9603, \quad \lambda_2 = 1.7451, \quad \lambda_3 = 2.6105, \quad \lambda_4 = 3.6028, \quad \lambda_5 = 4.4829, \ldots;$$

$$R_1 = 1.2499\,T_0, \quad R_2 = -0.3294\,T_0, \quad R_3 = 0.2228\,T_0,$$
$$R_4 = 0.0756\,T_0, \quad R_5 = -0.0050\,T_0, \ldots.$$

Thus the solution obtained using the first five terms of the series expansion with the time t measured in seconds is

$$V(x, t)/T_0 = 1.2449\,\tilde{X}_1(x)e^{-0.0332t} - 0.3294\,\tilde{X}_2(x)e^{-0.1096t}$$
$$+ 0.2228\,\tilde{X}_3(x)e^{-0.2452t} + 0.0756\,\tilde{X}_4(x)e^{-0.4670t}$$
$$- 0.0050\,\tilde{X}_5(x)e^{-0.7230t} + \cdots.$$

Figure 6.8a shows the geometry of the composite slab, while Fig. 6.8b shows plots of $V(x, t)/T_0$ for $t = 0, 15, 50$, and 75 s. The plots in Fig. 6.8b illustrate the fact that although the temperature remains continuous across the interface at $x = 5$, after time $t = 0$ a discontinuity in the slope of the temperature plot occurs across $x = 5$, reflecting the need to satisfy the continuity of heat flux across the interface. ∎

EXERCISES 6.3

Exercises 1 through 4 involve the derivation of useful formulas for the solution of some simple standard problems involving parabolic and hyperbolic PDEs in one space variable and time, and elliptic PDEs in rectangular Cartesian coordinates. All four results follow by making a direct application of the method of separation of variables to the given PDE and then using its stated auxiliary conditions.

1. Show the solution $u(x, t)$ of the homogeneous diffusion (heat) equation $u_t = k u_{xx}$ with $0 \leq x \leq L, t > 0$, and $k > 0$, subject to the homogeneous boundary conditions $u(0, t) = u(L, t) = 0$ and the initial condition $u(x, 0) = f(x)$ for $0 \leq x \leq L$, is given by

$$u(x, t) = \sum_{n=1}^{\infty} A_n \sin\frac{n\pi x}{L} \exp\left(-\frac{n^2\pi^2 kt}{L^2}\right),$$

with

$$A_n = \frac{2}{L} \int_0^L f(x) \sin \frac{n\pi x}{L} \, dx.$$

2. Show the solution $u(x, t)$ of the homogeneous wave equation $u_{tt} = c^2 u_{xx}$, with $0 \le x \le L$ and $t > 0$, subject to the homogeneous boundary conditions $u(0, t) = u(L, t) = 0$ and the initial conditions $u(x, 0) = f(x)$ and $u_t(x, 0) = g(x)$ for $0 \le x \le L$, is given by

$$u(x, t) = \sum_{n=1}^{\infty} \left(A_n \cos \frac{n\pi ct}{L} + B_n \sin \frac{n\pi ct}{L} \right) \sin \frac{n\pi x}{L},$$

where

$$A_n = \frac{2}{L} \int_0^L f(x) \sin \frac{n\pi x}{L} \, dx, \qquad B_n = \frac{2}{n\pi c} \int_0^L g(x) \sin \frac{n\pi x}{L} \, dx.$$

3. Show the solution $u(x, y)$ of the Laplace equation $u_{xx} + u_{yy} = 0$ in the region $0 \le x \le a, 0 \le y \le b$, subject to the three homogeneous Dirichlet conditions $u(0, y) = u(a, y) = u(x, b) = 0$ and the nonhomogeneous Dirichlet condition $u(x, 0) = f(x)$, is

$$u(x, y) = \sum_{n=1}^{\infty} A_n \sin \frac{n\pi x}{a} \sinh \frac{n\pi (b - y)}{a},$$

where

$$A_n = \frac{2}{a \sinh(n\pi b/a)} \int_0^a f(x) \sin \frac{n\pi x}{a} \, dx.$$

4. Show the solution $u(x, y)$ of the Laplace equation $u_{xx} + u_{yy} = 0$ in the region $0 \le x \le a, 0 \le y \le b$, subject to the two homogeneous Neumann conditions $u_x(0, y) = u_x(a, y) = 0$, the homogeneous Dirichlet condition $u(x, 0) = 0$, and the nonhomogeneous Dirichlet condition $u(x, b) = f(x)$, is

$$u(x, y) = A_0 y + \sum_{n=1}^{\infty} A_n \cos \frac{n\pi x}{a} \sinh \frac{n\pi y}{a},$$

with

$$A_0 = \frac{1}{ab} \int_0^a f(x) \, dx, \qquad A_n = \frac{2}{a \sinh(n\pi b/a)} \int_0^a f(x) \cos \frac{n\pi x}{a} \, dx.$$

5. Solve the diffusion equation $u_t = k u_{xx}$ for $0 \le x \le \pi, t > 0$, given that $u(0, t) = u(\pi, t) = 0$ for $t > 0$ and $u(x, 0) = x^2(\pi - x)$.

6. Solve the diffusion equation $u_t = ku_{xx}$ for $0 \le x \le L, t > 0$, given that $u(0, t) = u(L, t) = 0$ and

$$u(x, 0) = \begin{cases} x, & 0 \le x \le L/2 \\ L - x, & L/2 \le x \le L. \end{cases}$$

7. Solve the nonhomogeneous wave equation $u_{tt} = u_{xx} + f(x, t)$ for $0 \le x \le \pi$, subject to the respective boundary and initial conditions $u(0, t) = u(\pi, t) = 0$ and $u(x, 0) = 0 = u_t(x, 0) = 0$ when (a) $f(x, t) = t \sin x$ and (b) $f(x, t) = x \sin x$.

8. Generalize the method of solution for the nonhomogeneous wave equation to solve the nonhomogeneous diffusion equation $u_t = ku_{xx} + f(x, t)$ over the interval $0 \le x \le L$ subject to the homogeneous boundary conditions $u(0, t) = u(L, t) = 0, t \ge 0$ and the initial condition $u(x, 0) = F(x)$ for $0 \le x \le L$.

9. A uniform string of length L has its ends at $x = 0$ and $x = L$ clamped, and its transverse vibrations are governed by the equation $u_{tt} = c^2 u_{xx}$, where $u(x, t)$ is the displacement of the string perpendicular to the x axis at position x and time t. Find $u(x, t)$ if a point $x = a$ on the string is moved perpendicular to the x axis through a distance $h \ll L$ and then released from rest at time $t = 0$. The solution of this problem corresponds to the motion of a *plucked string*.

10. Solve the nonhomogeneous wave equation $u_{tt} = u_{xx} + t \sin(2\pi x/L)$ with $0 \le x \le L$, given that $u(0, t) = 0, u(L, t) = 0, u(x, 0) = 0$ and $u_t(x, 0) = 0$.

11. A uniform rectangular membrane occupying the region $0 \le x \le \alpha$, $0 \le y \le \beta$ is clamped at its edges. Its small oscillations are governed by the equation $u_{tt} = c^2(u_{xx} + u_{yy})$, where $u(x, y, t)$ is the displacement of the membrane from its equilibrium position at position (x, y) and time $t > 0$. Derive a series solution for $u(x, y, t)$, using the boundary conditions corresponding to the clamped edges, and the initial conditions

$$u(x, y, 0) = f(x, y) \quad \text{and} \quad u_t(x, y, 0) = g(x, y),$$

with f and g being arbitrary continuous functions subject only to the requirement that they vanish on the boundary of the membrane. Use the result to find the solution when $f(x, y) = 2 \sin(3\pi x/\alpha)\sin(\pi y/\beta)$, $g(x, y) \equiv 0$, and explain why the solution is so simple.

12. The solution of the Laplace equation $u_{xx} + u_{yy} = 0$ is required subject to the two nonhomogeneous Dirichlet conditions $u(x, 0) = 3 \sin(2\pi x/a)$ for $0 \le x \le a$ and $u(0, y) = 2y(b - y)$ for $0 \le y \le b$, and the two homogeneous Dirichlet conditions $u(a, y) = 0$ for $0 \le y \le b$ and $u(x, b) = 0$ for

$0 \leq x \leq a$. Explain why the solution of this problem can be expressed as the sum of two simpler problems, each with homogeneous Dirichlet conditions on three sides of the rectangle and a nonhomogeneous Dirichlet condition on the fourth side. Solve one of these problems using the form of solution derived in Exercise 3 and the other by an adaptation of this same method, and hence solve the stated boundary value problem. Explain why the part of the solution due to one of the nonhomogeneous Dirichlet conditions is so simple.

13. Radially symmetric vibrations of a uniform circular membrane of radius R, with its rim clamped, are described by the equation $u_{tt} = c^2[u_{rr} + (1/r)u_r]$, where the displacement $u(r, t)$ at radius r and time t is measured perpendicular to the equilibrium position of the membrane. Find $u(r, t)$ if, initially, the membrane is at rest, and at time $t = 0$ it is set in motion in such a way that its radially symmetric speed is $u_t(r, 0) = h(1 - r^2/R^2)$, where $h \ll R$.

14. Solve the Laplace equation $u_{rr} + (1/r)u_r + (1/r^2)u_{\theta\theta} = 0$ for $u(r, \theta)$ in the semi-circle $0 \leq r \leq \rho, 0 \leq \theta \leq \pi$, subject to the boundary conditions $u = U$ (constant) on the semi-circular boundary $r = \rho, 0 \leq \theta \leq \pi$, and $u = 0$ on the diameter forming the lower boundary of the semi-circle.

15. Solve the Laplace equation in Exercise 14 in a circle of radius ρ, given that the solution is finite and on the boundary of the circle

$$u(\rho, \theta) = \begin{cases} U, & 0 < \theta < \pi \\ -U, & \pi < \theta < 2\pi. \end{cases}$$

16. A uniform square membrane has its corners at the points $(0, 0), (a, 0), (a, a)$ and $(0, a)$, with its edges clamped, so $u = \partial u/\partial n = 0$ on the boundary. If $u(x, y, t)$ is the displacement of a point (x, y) of the membrane at time t, its vibrations are determined by the equation $u_{tt} = c^2(u_{xx} + u_{yy})$. Find $u(x, y, t)$ if the membrane starts from rest while in the displaced position $u(x, y, 0) = hxy(a - x)(a - y)$, where $h \ll a$.

17. Let $u(r, \theta, \phi)$ be the potential outside a sphere of unit radius due to a potential distribution $u(1, \theta, \phi) = U \sin\theta$ on the surface of the sphere. By setting $\xi = \cos\theta$ and using spherical polar coordinates, show the potential outside the sphere up to and including the term in $P_4(\xi)$ is given by

$$u(r, \theta, \phi) = \frac{\pi U}{4}\left(\frac{1}{r}P_0(\xi) - \frac{5}{8r^3}P_2(\xi) - \frac{9}{64r^5}P_4(\xi)\cdots\right).$$

18. A cavity is formed by the space between two concentric spheres of radii r_1 and r_2, with $r_1 < r_2$. Let the surface of the inner sphere be at a constant potential U and the surface of the outer sphere be at a constant potential V.

By expanding the potential in the cavity in terms of spherical polar coordinates, show that

$$u(r,\theta,\phi) = U + \frac{r_2(U-V)}{(r_2-r_1)}\left(\frac{r_1}{r}-1\right), \quad r_1 \le r \le r_2.$$

Use the fact that by symmetry $u(r,\phi,\theta)$ can only depend on the radius r to derive this result in a simpler manner.

19. A cavity is formed by the space between two concentric spheres of radii $\frac{1}{2}$ and 1. Find $u(r,\phi,\theta)$ inside the cavity if the surface of the inner sphere is maintained at a constant potential U, and the surface of the outer sphere is at a potential

$$u(1,\phi,\theta) = \begin{cases} U, & -1 \le \xi \le 0 \\ U(1-\xi^2)^{1/2}, & 0 \le \xi \le 1 \end{cases} \quad \text{with} \quad \xi = \cos\theta.$$

20. A long circular cylinder of radius ρ is raised to a uniform temperature U. At time $t = 0$ the cylinder is placed in a room at zero temperature when cooling takes place from its cylindrical sides according to Newton's law of cooling $(u_r + Hu)|_{r=\rho} = 0$. Find the temperature distribution in the cylinder as a function of r and t, on the assumption that the cylinder is sufficiently long for the temperature variation along its length to be ignored, so the temperature $u(r, t)$ is determined by the heat equation $u_t = \kappa[u_{rr} + (1/r)u_r]$.

21. The temperature in a circular cylinder of radius ρ and height H obeys the steady-state heat equation $u_{rr} + (1/r)u_r + u_{zz} = 0$. Find the temperature distribution $u(r, z)$ given that the circular base is maintained at zero temperature, heat is lost through its cylindrical surface to the surrounding air at zero temperature according to Newton's law of cooling $(u_r + hu)|_{r=\rho} = 0$, and the circular top is maintained at a constant temperature U.

22. Repeat Exercise 21, but this time use the condition that there is a parabolic temperature across the circular top given by $u(r, H) = U(1 - r^2/\rho^2)$.

23. Solve the modified unsteady heat equation $u_t = \kappa u_{xx} - \alpha u$ with $0 \le x \le L$ and $\alpha > 0$, given that $u_x(0, t) = u_x(L, t) = 0$ and at time $t = 0$ the initial temperature distribution along the rod is $u(x, 0) = hx(L - x)$. Suggest a physical problem in heat conduction that could be described by this initial boundary value problem.

24. Use separation of variables to find the eigenvalues and eigenfunctions of the time-dependent beam equation (unsteady biharmonic wave equation) $u_{tt} + c^2 u_{xxxx} = 0$ over the interval $0 \le x \le L$, subject to the boundary conditions $u(0,t) = u(L,t) = 0$, $u_x(0,t) = u_x(L,t) = 0$ for $t \ge 0$.

Explain why the eigenfunctions are *not* mutually orthogonal over the interval $0 \leq x \leq L$, and find approximations for the nth eigenvalue and eigenfunction when n is large.

25. A spherical metal shell of inner radius R_1 and outer radius R_2 has its inner and outer surfaces maintained at the respective constant temperatures T_0 and $T_0|\cos\theta|$. Find the first three nonzero terms of the Fourier–Legendre expansion of the temperature distribution $T(r,\theta)$ in terms of the radius r and the Legendre polynomials $P_n(\cos\theta)$, with $R_1 \leq r \leq R_2$ and $0 \leq \theta \leq \pi$.

26. In terms of the cylindrical polar coordinates (r,θ,z), a cylindrical metal cylinder occupying the region $0 \leq r \leq 1$ and $0 \leq z \leq 4$ has its base and cylindrical sides maintained at zero temperature, while the temperature distribution on the top surface $z = 4$ is $T(r,\theta,4) = 100(1 - r^2)$. Find the temperature distribution $T(r,\theta,z)$ at a general point in the cylinder, and also along its axis.

27. Using the data and results of Example 6.12, find the first five terms in the series solution for the temperature distribution in the composite slab considered in that example when the initial temperature distribution is $f(x) = T_0$ (constant). Plot the temperature distribution $V(x,t)/T_0$ at times $t = 0, 15, 50,$ and 75 s. Hence show that when $t = 0$ the Gibbs phenomena at $x = 0$ and $x = 10$ reflect the difference in each material, but that after 15 s the phenomena are no longer apparent and the subsequent temperature distributions are similar to those in Fig. 6.8b, though with greater magnitudes at corresponding times.

General Results for Linear Elliptic and Parabolic Equations

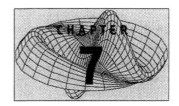

CHAPTER 7

7.1 General Results for Elliptic and Parabolic Equations

In previous chapters the method of separation of variables was used to solve elliptic, parabolic, and hyperbolic equations, where in each case the method of approach was essentially the same. This method provides solutions to specific problems, but it offers no general insight into any of the characteristic features of specific types of PDE. A number of general results were derived for the linear wave equation, though not by using separation of variables, and these included the existence of solutions, their continuous dependence on the initial data, the domain of dependence of a solution, and the difference in the form and property of solutions in even and odd numbers of space dimensions. However, so far no equivalent general results have been derived for the Laplace equation or the heat equation, both of which occur repeatedly in applications, and each of which is as important as the wave equation. It is the purpose of this chapter to address this situation by deriving some of the most fundamental properties of these two equations.

Included among the topics to be examined are the important questions of the existence of solutions, their uniqueness, the derivation of maximum/ minimum properties, which can provide useful qualitative information about solutions, and the Poisson integral formula, which will be used to derive several general results. The last topic to be introduced is in the form of an introduction to what are called *self-similar* solutions. This important class of solutions is very different from those obtained from equations where the length, mass, and time variables can be made nondimensional by dividing each by conveniently chosen units of length, mass, and time suggested by the physical problem involved. This approach causes solutions to scale according to the reference units used when making the solution nondimensional, but if the reference units are changed to ones more appropriate for another problem, the solution will also be changed. Self-similar solutions are different, and they occur in problems where there are no natural units of length, mass, and time with which to scale solutions. Instead, they are found when it is possible to introduce a special nondimensional combination of variables called a *similarity variable*, when it turns out that a solution can then be made to scale on itself. A single solution then represents all possibilities, simply by assigning suitable values to the similarity variable.

To pursue the topic of self-similar solutions in detail would take us far beyond the scope of this first account of partial differential equations, although when they exist self-similar solutions greatly simplify the task of solving PDEs, so it is important to see by example how the method works. The approach will be applied to an unsteady heat conduction problem in a semi-infinite plane slab of metal, where the variables involved are the perpendicular distance x into the slab and the time t. It will be seen that by the introduction of a similarity variable the solution of the PDE is reduced to the solution of an ODE that can then be solved. In fact, in general, when a PDE involves n independent variables, if $m < n$ similarity variables can be found, the problem reduces to the solution of a simpler PDE involving only $n - m$ independent variables. This powerful approach is applicable to a variety of physical problems, typical of which is the boundary layer flow of a viscous fluid layer adjacent to a flat plate.

7.2 Laplace Equation

The first result to be proved will be the uniqueness of the solution of the Laplace equation subject to Dirichlet conditions. This can be established very quickly by using Green's first integral theorem.

Theorem 7.1. (Uniqueness of the Solution of the Laplace Equation Subject to Dirichlet Conditions)
Let D be a closed region with a smooth external boundary Γ, at each point of which an outward drawn unit normal $\mathbf{n}$ is defined. Then the solution of the Laplace equation $\Delta u = 0$ subject to a continuous Dirichlet condition $u(\mathbf{r}) = u(\Gamma)$ on the boundary Γ of D is unique.

Proof: Green's first integral theorem asserts that if f and g are two twice differentiable functions in a closed region D and an outward drawn unit normal $\mathbf{n}$ on each point of its boundary Γ, then

$$\iiint_D (f\Delta g + \nabla f \cdot \nabla g)\, dV = \iint_\Gamma (f\nabla g) \cdot d\mathbf{A},$$

where dV is the volume element in D and $d\mathbf{A} = \mathbf{n}\, dA$ is the vector element of area on Γ. Set $u = \phi - \psi$, and let ϕ and ψ be two different solutions of the Laplace equation (harmonic functions) in D, each of which satisfies the *same* boundary condition on Γ. Then u is a harmonic function in D and $u = 0$ on Γ. Setting $f = g = u$ in Green's first integral theorem it becomes

$$\iiint_D (\nabla u)^2 \, dV = \iint_\Gamma (u\nabla u) \cdot d\mathbf{A}.$$

However, $u = 0$ on Γ so this reduces to

$$\iiint_D (\nabla u)^2 \, dv = 0,$$

but the integrand is nonnegative, so this result is only possible if $\nabla u = \mathbf{0}$, showing that $u = $ constant. As $u = 0$ on the boundary Γ, it follows at once that $u \equiv 0$ in D and on its boundary Γ, so $\phi \equiv \psi$, and the theorem is proved. ∎

Although the justification will not be given here, this result remains true under conditions far more general than the restrictive ones used in this proof. For example, the condition of continuity of the Dirichlet condition can be relaxed and the smooth boundary Γ can be replaced by a piecewise smooth one.

Theorem 7.2. (Uniqueness of the Solution of the Laplace Equation Subject to Neumann Conditions)
Under the conditions of Theorem 7.1, let u be a solution of the Laplace equation $\Delta u = 0$ in a region D with boundary Γ subject to the Neumann condition $\mathbf{n} \cdot \nabla u = f(\mathbf{r})$, where $\mathbf{r}$ is the position vector of a general point on Γ. Then the solution is unique apart from an arbitrary additive constant and the solution will only exist if

$$\iint_\Gamma f(\mathbf{r}) \, dA = 0,$$

where dA is the area element on Γ.

Proof: Let ϕ and ψ be two different solutions of the Neumann problem, then $u = \phi - \psi$ must be harmonic in D subject to a Neumann condition corresponding to $f(\mathbf{r}) = 0$ on Γ. An application of Green's first integral theorem with $f = g = u$ gives

$$\iiint_D (\nabla u)^2 \, dV = 0,$$

showing that $u = $ constant in D. So $\phi = \psi + $ constant, and the first result is established. The condition to be satisfied for a solution to exist follows by writing the Laplacian $\Delta u = 0$ as $\Delta u = \nabla \cdot (\nabla u) = 0$, and then applying the Gauss divergence theorem to the result when we obtain

$$0 = \iiint_D \Delta u \, dV = \iiint_D \nabla \cdot (\nabla u) \, dV = \iint_\Gamma \nabla u \cdot \mathbf{n} \, dA = \iint_\Gamma f(\mathbf{r}) \, dA.$$

∎

A physical reason for this condition is easily found if u is considered to be the steady-state temperature distribution in a region D through the boundary Γ of which passes a heat flux given by the Neumann condition $\mathbf{n} \cdot \nabla u = f(\mathbf{r})$. Clearly, unless the total heat flux through Γ is zero, there will either be an increase or a decrease in the heat contained in D, causing the temperature to change with time. However, this is impossible, since a steady state has been postulated in D, and no mechanism exists for heat to be either added or removed from D other than through Γ, so the condition $\iint_\Gamma f(\mathbf{r}) \, dA = 0$ is necessary.

Theorem 7.3. (The Poisson Integral Formula for a Circle)
Let $f(\theta)$ be a piecewise continuous function with period 2π. Then the solution of the Laplace equation $\Delta u = 0$ in the open disk $r < r_0$ centered on the origin, and subject to the Dirichlet boundary condition $u(r_0, \theta) = f(\theta)$ on $r = r_0$, is

$$u(r, \theta) = \frac{1}{2\pi} \int_0^{2\pi} f(\psi) \frac{r_0^2 - r^2}{r_0^2 - 2rr_0 \cos(\psi - \theta) + r^2} \, d\psi \quad (r < r_0),$$

where (r, θ) are plane polar coordinates.
The function $u(r, \theta)$, which is harmonic in the open disk $r < r_0$, is piecewise continuous on $r = r_0$ where it satisfies the Dirichlet condition everywhere except at points θ_j where $f(\theta)$ is discontinuous, in which case it assumes the value $\bar{u}(r_0, \theta_j) = \frac{1}{2}(f(\theta_j^{(-)}) + f(\theta_j^{(+)}))$, where $f(\theta_j^{(-)})$ and $f(\theta_j^{(+)})$ are the values of $f(\theta)$ to the immediate left and right of θ_j.

Proof: We prove the theorem first for the unit circle $r_0 = 1$, and then extend the result to the case of general r_0. The Laplace equation in plane polar coordinates is

$$u_{rr} + (1/r)u_r + (1/r^2)u_{\theta\theta} = 0,$$

so separating variables in the usual manner by setting $u(r, \theta) = R(r)\Theta(\theta)$ leads to the results

$$\Theta''(\theta) + \lambda^2 \Theta = 0 \quad \text{and} \quad r^2 R'' + rR' - \lambda^2 R = 0,$$

where λ^2 is a separation constant.
As the solution in the disk must be periodic with period 2π, it follows that we must set $\lambda^2 = n^2$ with n being an integer, so $\Theta_n(\theta) = a_n \cos n\theta + b_n \sin n\theta$, and $R(r)$ must be a solution of

$$r^2 R'' + rR' - n^2 R = 0.$$

This has the solutions $R_n(r) = C_1 r^n$ and $R_n(r) = C_2 r^{-n}$, but the second of these must be rejected since u must be finite at the origin, so the partial solutions are

$$u_n(r,\theta) = r^n(a_n \cos n\theta + b_n \sin n\theta).$$

The separation constant n can also assume the value $n = 0$, so there will also be a solution $u_0(r,\theta) = a_0$ (a constant). Thus we will seek a general solution of the form

$$u(r,\theta) = a_0 + \sum_{n=1}^{\infty} r^n(a_n \cos n\theta + b_n \sin n\theta).$$

Setting $r = 1$ and using the Dirichlet condition this becomes

$$f(\theta) = a_0 + \sum_{n=1}^{\infty} (a_n \cos n\theta + b_n \sin n\theta),$$

showing that the expression on the right is simply the Fourier series representation of the Dirichlet condition, so

$$a_0 = \frac{1}{2\pi} \int_0^{2\pi} f(\theta)\, d\theta, \qquad a_n = \frac{1}{\pi} \int_0^{2\pi} f(\theta) \cos n\theta\, d\theta,$$

$$\text{and} \quad b_n = \frac{1}{\pi} \int_0^{2\pi} f(\theta) \sin n\theta\, d\theta.$$

Substituting for a_n and b_n in the expression for $u(r,\theta)$, and replacing the variable of integration by the dummy variable ψ to avoid confusion with the variable θ gives

$$u(r,\theta) = \frac{1}{2\pi} \int_0^{2\pi} f(\psi)\, d\psi + \frac{1}{\pi} \sum_{n=1}^{\infty} r^n$$

$$\times \left\{ \int_0^{2\pi} f(\psi)(\cos n\psi \cos n\theta + \sin n\psi \sin n\theta)\, d\psi \right\}.$$

The integrals can be combined and the result written

$$u(r,\theta) = \frac{1}{\pi} \int_0^{2\pi} f(\psi) \left(\frac{1}{2} + \sum_{n=1}^{\infty} r^n \cos n(\psi - \theta) \right) d\psi.$$

Next we use the trigonometric identity

$$\frac{1}{2} + \sum_{n=1}^{n} r^n \cos n(\psi - \theta) = \frac{1 - r}{2(1 - 2r \cos(\psi - \theta) + r^2)},$$

the proof of which forms Exercise 1 in Exercise Set 7.2. When this result is used, the expression for $u(r,\theta)$ becomes the Poisson integral formula in a unit disk

$$u(r,\theta) = \frac{1}{2\pi} \int_0^{2\pi} f(\psi) \frac{1 - r^2}{1 - 2r\cos(\psi - \theta) + r^2} \, d\psi.$$

The more general result given in the theorem follows by replacing r by r/r_0, and the convergence property of the solution on the boundary of the circle $r = r_0$ is an immediate consequence of the convergence property of the Fourier series.   ∎

Sometimes, to emphasize the relationship between $u(r,\theta)$ and its value $f(\psi)$ on the boundary, the Poisson integral formula is written

$$u(r,\theta) = \frac{1}{2\pi} \int_0^{2\pi} u(r_0,\psi) \frac{r_0^2 - r^2}{r_0^2 - 2rr_0\cos(\psi - \theta) + r^2} \, d\psi. \qquad (7.1)$$

The function

$$P(r, r_0, \chi) = \frac{r_0^2 - r^2}{r_0^2 - 2rr_0\cos\chi + r^2}, \quad \text{with } \chi = \psi - \theta, \qquad (7.2)$$

is called the **Poisson kernel**, so the Poisson integral formula can be written

$$u(r,\theta) = \frac{1}{2\pi} \int_0^{2\pi} f(\psi) P(r_0, r, \psi - \theta) \, d\psi. \qquad (7.3)$$

The Poisson kernel has the following useful properties:

(i) $P(r_0, r, \chi) > 0$ for $r < r_0$, as can be seen from the fact that both its numerator and denominator are positive, because $(r_0 - r)^2 < r_0^2 - 2rr_0\cos\chi + r^2 < (r_0 + r)^2$.

(ii) An antiderivative of $P(r_0, r, \chi)$ is

$$\int P(r_0, r, \chi) \, d\chi = 2\arctan\left[\left(\frac{r_0 + r}{r_0 - r}\right)\tan\left(\frac{1}{2}\chi\right)\right], \qquad (7.4)$$

as may be easily verified by differentiation.

(iii) As $0 < \theta \le 2\pi$ and $0 < \psi \le 2\pi$, the argument $\chi = \psi - \theta$ must lie in the interval $-2\pi < \chi \le 2\pi$. The arctangent function has more than one branch, and the argument $\chi = \psi - \theta$ in integral (7.3) must be a monotonic increasing function, so it is necessary to take the branch that assigns the value $-\pi$ when $\frac{1}{2}\chi = -\frac{1}{2}\pi$ and the value π when $\frac{1}{2}\chi = \frac{1}{2}\pi$. Then, with this choice of branch, it follows that

$$\int_0^{2\pi} P(r_0, r, \psi - \theta) \, d\psi = 2\pi. \qquad (7.5)$$

The Poisson integral formula in Theorem 7.1 is difficult to evaluate unless the function $f(\theta)$ is very simple, although the formula can be used with a numerical integration routine to provide a numerical approximation to $u(r, \theta)$. A more practical way of obtaining a numerical solution of the Laplace equation in a circle $r = r_0$ subject to the Dirichlet condition $u(r_0, \theta) = f(\theta)$ on its boundary is to use the representation obtained when deriving the integral formula, namely

$$u(r, \theta) = a_0 + \sum_{n=1}^{\infty} r^n (a_n \cos n\theta + b_n \sin n\theta), \qquad (7.6)$$

where

$$a_0 = \frac{1}{2\pi} \int_0^{2\pi} f(\theta) \, d\theta, \qquad a_n = \frac{1}{\pi} \int_0^{2\pi} f(\theta) \cos n\theta \, d\theta, \quad \text{and}$$

$$b_n = \frac{1}{\pi} \int_0^{2\pi} f(\theta) \sin n\theta \, d\theta. \qquad (7.7)$$

The success of this approach when the boundary condition is a simple trigonometric function is illustrated by the following example.

Example 7.1. Find the function $u(r, \theta)$, which is harmonic in the unit circle and assumes the value $u(1, \theta) = \sin^3 \theta$ on its boundary.

Solution: To attempt a solution by using Theorem 7.3 directly requires evaluating the complicated definite integral

$$u(r, \theta) = \frac{1}{2\pi} \int_0^{2\pi} \frac{(1 - r^2) \sin^3 \psi}{1 - 2r \cos(\psi - \theta) + r^2} \, d\psi.$$

However, by using the trigonometric identity $\sin^3 \theta = \frac{3}{4} \sin \theta - \frac{1}{4} \sin 3\theta$ the task of finding the solution is simplified if use is made of result (7.6), which on $r = 1$ becomes

$$\frac{3}{4} \sin \theta - \frac{1}{4} \sin 3\theta = a_0 + \sum_{n=1}^{\infty} (a_n \cos n\theta + b_n \sin n\theta).$$

The orthogonality of the sine and cosine functions over the interval $0 \le \theta \le 2\pi$ could be used to evaluate the coefficients a_n and b_n, but this is unnecessary because the expression on the left is its own Fourier series. Thus in this case the coefficients can be found by equating the coefficients of corresponding trigonometric functions, and when this is done we find that

$$a_n = 0 \quad \text{for } n = 0, 1, \ldots,$$

$$b_1 = \frac{3}{4}, \quad b_2 = 0, \quad b_3 = -\frac{1}{4}, \quad \text{and} \quad b_n = 0 \quad \text{for } n = 4, 5, \ldots.$$

As the series for $u(r, \theta)$ only contains a finite number of terms for the solution, namely

$$u(r,\theta) = \frac{3}{4}r \sin \theta - \frac{1}{4}r^3 \sin 3\theta \quad \text{for } 0 \leq r \leq 1 \quad \text{and} \quad 0 \leq \theta \leq 2\pi,$$

is exact, because it satisfies both the boundary condition and the Laplace equation.

An immediate consequence of this result is that the Poisson integral determining the solution has been evaluated by an indirect method, so we have also shown that

$$\int_0^{2\pi} \frac{(1-r^2) \sin^3 \psi}{1 - 2r \cos(\psi - \theta) + r^2} \, d\psi = \frac{1}{2}\pi(3r \sin \theta - r^3 \sin 3\theta), \quad 0 \leq r \leq 1. \quad \blacksquare$$

The main use of the Poisson integral formula is to establish general results concerning harmonic functions. To illustrate this we will prove the **Gauss mean value theorem** for harmonic functions. This involves showing a harmonic function possesses the *mean value property*, meaning that at every point P of the region D where the u is harmonic, its value $u(P)$ at P is equal to the average of u around any circle centered on P that lies entirely in D.

Theorem 7.4. (The Gauss Mean Value Theorem for Harmonic Functions)
Let u be harmonic in a region D, and let P at (x_0, y_0) be any point in D about which a circle Γ of radius $r = r_0$ that lies entirely inside D can be constructed. Then u has the mean value property throughout D

$$u(x_0, y_0) = \frac{1}{2\pi} \int_0^{2\pi} u(x_0 + \rho \cos \theta, y_0 + \rho \sin \theta) \, d\theta,$$

where ρ is the radial distance from the center of the circle at P to its boundary Γ, and ds is the element of length around Γ.

Proof: Let P be any point in a region D where the function u is harmonic. Shifting the origin to the point P leaves the function harmonic, so the Poisson integral formula can be applied to the point P provided the circle $r = r_0$ centered on P lies strictly inside D. The result follows immediately by setting $r = 0$ in Theorem 7.3, so the theorem is proved. $\blacksquare$

It is a simple matter to use the Poisson integral formula for a circle to prove the continuous dependence of the solution inside the circle on the Dirichlet boundary data. However, this property will be established later for a region of arbitrary shape, and not simply for a circle, by using the maximum/minimum principle to be proved later, so the proof for a circle based on Theorem 7.3 is left as an exercise.

For many practical situations it is desirable to replace the circle in Theorem 7.3 by the upper half of the (x, y) plane. However, before doing so, it is necessary to ask if this is a reasonable requirement, because a boundary value problem for the Laplace equation is only well posed in a bounded region. Fortunately, the imposition of the additional requirement that the solution remains bounded is sufficient to enable a unique solution to be found in the half-plane $y > 0$ when a bounded Dirichlet condition is imposed on the line $y = 0$. Before deriving the half-plane version of the Poisson integral theorem it will be helpful if we first solve two simple problems, the first of which is illustrated in Fig. 7.1, where a bounded function u such that

$$\Delta u = 0 \quad \text{for } -\infty < x < \infty, \quad y > 0, \quad \text{with } u(x,0) = \begin{cases} k_1, & x < 0 \\ k_2, & x > 0, \end{cases} \quad (7.8)$$

is required.

As u must remain bounded, and the geometry of the problem shows the solution can only depend on the angle θ, we will try a solution of the form $u(r,\theta) = A + B\theta$. Using the boundary conditions $u(r,\pi) = k_1$ and $u(r,0) = k_2$ we find that $k_1 = A + B\pi$ and $k_2 = A$, which shows that

$$u(r,\theta) = k_2 + \left(\frac{k_1 - k_2}{\pi}\right)\theta. \quad (7.9)$$

This simple function, which is linear in θ, is harmonic and satisfies the boundary conditions, so as the solution is unique this is the required bounded solution for the boundary value problem shown in Fig. 7.1.

The second problem, a generalization of the first one, is shown in Fig. 7.2a, where a bounded solution is required in the half-plane $y > 0$, satisfying the boundary conditions

$$\Delta u = 0, \quad \text{for } -\infty < x < \infty, \quad y > 0,$$

$$\text{with } u(x,y) = \begin{cases} 0, & x < a \\ f, & a < x < b \\ 0, & x > b, \end{cases} \quad f = \text{constant.}$$

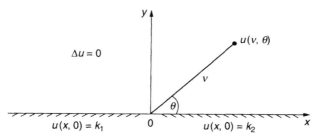

Figure 7.1   A simple half-plane problem with the boundary condition assuming two different constant values.

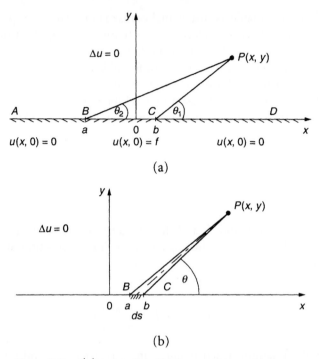

Figure 7.2   (a) A simple half-plane problem with the boundary condition assuming three different constant values. (b) The equivalent situation when segment BD in (a) is an element of length ds.

Let us consider the function $u(x, y) = \frac{f}{\pi}(\theta_1 - \theta_2)$. This satisfies the stated boundary conditions, because when P lies on the line segment CD angles $\theta_1 = \theta_2 = 0$, so $u = 0$; when P lies on the line segment BC angle $\theta_1 = \pi$ and $\theta_2 = 0$, so $u = f$; and when P lies on the line segment AB angles $\theta_1 = \theta_2 = \pi$, so again $u = 0$. Suffice it to say that in the half-plane $y > 0$, by using an argument unnecessary for what is to follow, the function $u(x, y)$ can be shown to be harmonic in the half-plane $y > 0$. Thus, as it satisfies the boundary conditions on $y = 0$, it is the solution of the problem. The significance of this will become apparent when the next theorem is proved.

Theorem 7.5. (The Poisson Integral Formula for a Half-Plane)
Let $f(x)$ be a piecewise continuous and bounded function for $-\infty < x < \infty$. Then the function

$$u(x, y) = \frac{y}{\pi} \int_{-\infty}^{\infty} \frac{f(s)}{(x - s)^2 + y^2} \, ds$$

is bounded and harmonic in the half-plane $y > 0$, and on the x axis it equals $f(x)$ wherever $f(x)$ is continuous, and the value $\tilde{u}(x_j, 0) = \frac{1}{2}\{f(x_j^{(-)}) + f(x_j^{(+)})\}$ at any point x_j where $f(x)$ is discontinuous, with $f(x_j^{(-)})$ and $f(x_j^{(+)})$ being the values of $f(x)$ to the immediate left and right of x_j.

Proof: To motivate ideas we first derive the result in an intuitive manner, after which we give a formal proof based on the Fourier transform. Thus a reader unfamiliar with the Fourier transform will still have an idea of why the theorem is true.

Consider Fig. 7.2b, and suppose this time that the interval BD in Fig. 7.2a is now a small element of length ds on the x axis with its midpoint at $x = s$, and that the boundary condition is now $u(x, 0) = f(x)$ with $f(x)$ being a bounded function of x, so the value of $f(x)$ at $x = s$ is $f(s)$. Then, if P is the point (x, y) in the half-plane $y > 0$, from the reasoning used with Fig. 7.2a, the contribution du to the value of u at P, namely $u(x, y)$, due to the boundary condition on the element ds will be approximately $du = f(s)(d\theta/ds)\, ds$. Thus, integrating over $-\infty < x < \infty$ we have

$$u(x, y) = \frac{1}{\pi} \int_{-\infty}^{\infty} f(s) \left(\frac{d\theta}{ds} \right) ds.$$

To proceed further we must now find the angle θ by using $\tan\theta = y/(x - s)$. When $y/(x - s) > 0$ we have $\theta = \arctan[y/(x - s)] > 0$, but when $y/(x - s) < 0$, because the tangent function has the domain of definition $-\pi/2 < x < \pi/2$ and the range $-\infty < x < \infty$, we see that $\theta = \pi + \arctan[y/(x - s)]$. Furthermore, when $y/(x - s) = \pm\infty$ it follows from Fig. 7.2b that $\theta = \frac{1}{2}\pi$. So, we define a *new* **Arctangent** function as

$$\theta = \text{Arctan}\,[y/(x - s)] = \begin{cases} \arctan[y/(x - s)], & y/(x - s) > 0 \\ \pi + \arctan[y/(x - s)], & y/(x - s) < 0 \\ \frac{1}{2}\pi, & y/(x - s) = \pm\infty. \end{cases}$$

Differentiation then shows that

$$\frac{d\theta}{ds} = \frac{y}{(x - s)^2 + y^2}, \quad \text{irrespective of the sign of } y/(x - s).$$

Using this result in the previous expression for $u(x, y)$ we arrive at the Poisson integral formula for the half-plane $y > 0$,

$$u(x, y) = \frac{y}{\pi} \int_{-\infty}^{\infty} \frac{f(s)}{(x - s)^2 + y^2}\, ds. \qquad \blacksquare$$

With care this intuitive proof can be made rigorous. To do this $f(x)$ must be approximated by a piecewise constant function with n segments, and the result $u(x, y) = \frac{1}{\pi}(\psi_1 - \psi_2)$ obtained previously must then be generalized to take account of n segments. Then the number n of segments where $f(x)$ is approximated by a constant value must be allowed to increase to infinity, as the length of each segment tends to 0. As a result, it turns out that the Poisson integral formula for the half-plane $y > 0$ will be defined in terms of the limiting sum used when defining a Riemann integral. Proceeding in this manner, and letting $n \to \infty$, produces the required formula. This form of proof is lengthy, so instead our formal proof will be based on the Fourier transform.

However, the result

$$d\theta = \frac{y}{(x - s)^2 + y^2} \, ds$$

obtained rigorously in the intuitive proof given above will prove useful when applying the Poisson integral formula to examples.

Proof using the Fourier transform: Let $F(\omega) = \frac{1}{\sqrt{2\pi}} \int_{-\infty}^{\infty} f(x) \exp\{-i\omega x\} dx$. Then $F(\omega)$ is called the **Fourier transform** of $f(x)$, and the integral will exist if, for example, $f(x)$ is piecewise continuous over every finite interval and $\int_{-\infty}^{\infty} |f(x)| \, dx$ is finite. We now consider the boundary value problem $\Delta u = 0$ for $-\infty < x < \infty$ and $y > 0$, where u is bounded and $u(x, 0) = f(x)$, with $f(x)$ satisfying the conditions for the existence of its Fourier transform. Taking the Fourier transform of $u_{xx} + u_{yy} = 0$ with respect to x gives

$$(i\omega)^2 U(\omega, y) + \frac{d^2}{dy^2} U(\omega, y) = 0,$$

where $U(\omega, y) = \mathcal{F}\{u(x, y)\}$ is the Fourier transform of $u(x, y)$ with respect to x, with y treated as a constant. The general solution of this ODE for $U(\omega, y)$ is

$$U(\omega, y) = A(\omega) \exp(\omega y) + B(\omega) \exp(-\omega y),$$

where $A(\omega)$ and $B(\omega)$ are arbitrary functions to be determined, while y is to be regarded as a parameter. As the solution must be bounded in the half-plane $y > 0$ for both positive and negative ω, this is only possible if $A(\omega) = 0$ when $\omega < 0$ and $B(\omega) = 0$ when $\omega < 0$. So, defining $C(\omega) = A(\omega) + B(\omega)$, we can write

$$U(\omega, y) = C(\omega) \exp(-y \, |\omega|), \quad \text{for } -\infty < \omega < \infty \quad \text{and} \quad y > 0.$$

By hypothesis, the initial condition $u(x, 0) = f(x)$ has the Fourier transform $\mathcal{F}\{f(x)\} = F(\omega)$, so setting $y = 0$ in $U(\omega, y)$ and using this last result gives

$$U(\omega, y) = F(\omega) \exp(-y \, |\omega|).$$

From a table of Fourier transform pairs we have

$$\mathcal{F}\left\{\sqrt{\frac{2}{\pi}}\left(\frac{y}{x^2+y^2}\right)\right\} = \exp(-y\,|\omega|),$$

so an application of the Fourier transform convolution theorem to $U(\omega, y)$ gives the result

$$u(x, y) = \frac{y}{\pi}\int_{-\infty}^{\infty}\frac{f(s)}{(x-s)^2+y^2}\,ds,$$

which is the half-plane Poisson integral formula.   ∎

The Poisson integral formula in Theorem 7.5 is easier to evaluate than the formula for the circle in Theorem 7.3, as can be seen in the following examples.

Example 7.2. Find the function $u(x, y)$ bounded and harmonic in the half-plane $y > 0$, and on the x axis satisfies the boundary condition $u(x, y) = f(x)$, where

$$f(x) = \begin{cases} 0, & x < -1 \\ -1, & -1 < x < 0 \\ 1, & 0 < x < 1 \\ 0, & x > 1. \end{cases}$$

Solution: The solution follows by making a direct application of Theorem 7.5. Substituting for $f(x)$ we have

$$u(x, y) = \frac{y}{\pi}\int_{-1}^{1}\frac{f(s)}{(x-s)^2+y^2}\,ds$$

$$= \frac{y}{\pi}\int_{-1}^{0-}\frac{-1}{(x-s)^2+y^2}\,ds + \frac{y}{\pi}\int_{0+}^{1}\frac{1}{(x-s)^2+y^2}\,ds.$$

An application of the results $\theta = \text{Arctan}\left(\frac{y}{x-s}\right)$ and $d\theta = \frac{y\,ds}{(x-s)^2+y^2}$ gives

$$u(x, y) = -\frac{1}{\pi}\text{Arctan}\,[y/(x-s)]\,\big|_{s=-1}^{0-} + \frac{1}{\pi}\text{Arctan}\,[y/(x-s)]\,\big|_{s=0+}^{1}$$

and so

$$u(x, y) = \frac{1}{\pi}\{\text{Arctan}[y/(x+1)] + \text{Arctan}[y/(x-1)] - 2\,\text{Arctan}(y/x)\},$$

for $-\infty < x < \infty$ and $y > 0$.   ∎

Example 7.3. Find the function $u(x, y)$ bounded and harmonic in the half-plane $y > 0$, and on the x axis satisfies the boundary condition $u(x, y) = f(x)$, where

$$f(x) = \begin{cases} 0, & x < 0 \\ x, & 0 \le x \le 1 \\ 0, & x > 1. \end{cases}$$

Solution: From Theorem 7.5 we have

$$
\begin{aligned}
u(x, y) &= \frac{y}{\pi} \int_0^1 \frac{s}{(x - s)^2 + y^2} \, ds \\
&= -\frac{y}{\pi} \int_0^1 \frac{(x - s)}{(x - s)^2 + y^2} \, ds + \frac{x}{\pi} \int_0^1 \frac{y \, ds}{(x - s)^2 + y^2} \\
&= \frac{y}{2\pi} \ln \left[\frac{(x - 1)^2 + y^2}{x^2 + y^2} \right] + \frac{x}{\pi} \operatorname{Arctan} \left[y/(x - s) \right] \big|_{s=0}^1 \\
&= \frac{y}{2\pi} \ln \left[\frac{(x - 1)^2 + y^2}{x^2 + y^2} \right] + \frac{x}{\pi} \operatorname{Arctan} \left(\frac{y}{x - 1} \right) - \frac{x}{\pi} \operatorname{Arctan} \left(\frac{y}{x} \right).
\end{aligned}
$$

∎

Example 7.4. Show, using Theorem 7.5, that the function $u(x, y)$, harmonic for $y > 0$ and equal to the constant value $u(x, 0) = k$ on the x axis, is $u(x, y) \equiv k$.

Solution: From Theorem 7.5 we have

$$u(x, y) = \frac{k}{\pi} \int_{-\infty}^{\infty} \frac{y}{(x - s)^2 + y^2} \, ds = \frac{k}{\pi} \operatorname{Arctan} \left[y/(x - s) \right] \big|_{s=-\infty}^{\infty}.$$

Care must now be taken when evaluating the limits to ensure the correct values of the Arctangent function are used. We have

$$u(x, y) = \frac{k}{\pi} \lim_{s \to \infty} \operatorname{Arctan} \left[y/(x - s) \right] - \frac{k}{\pi} \lim_{s \to -\infty} \operatorname{Arctan} \left[y/(x - s) \right]$$

$$= (k/\pi)[\pi + \arctan 0 - \arctan 0] = k.$$

∎

Example 7.5. Prove that the bounded solution of the Laplace equation in the half-plane $y > 0$ depends continuously on the Dirichlet condition on the x axis.

Solution: Let $u(x, y)$ and $v(x, y)$ be solutions of the Laplace equation corresponding to the respective bounded and continuous Dirichlet conditions $u(x, 0) = f(x)$ and $v(x, 0) = g(x)$. Furthermore, let $|u(x, y) - v(x, y)| < \varepsilon$ for some arbitrarily small $\varepsilon > 0$. Then from Theorem 7.5

$$u(x, y) - v(x, y) = \frac{y}{\pi} \int_{-\infty}^{\infty} \frac{f(s) - g(s)}{(x - s)^2 + y^2} \, ds,$$

so

$$|u(x, y) - v(x, y)| \leq \frac{y}{\pi} \int_{-\infty}^{\infty} \frac{|f(s) - g(s)|}{(x - s)^2 + y^2} ds < \frac{\varepsilon y}{\pi}$$

$$\int_{-\infty}^{\infty} \frac{ds}{(x - s)^2 + y^2} ds = \left(\frac{\varepsilon y}{\pi}\right)\left(\frac{\pi}{y}\right) = \varepsilon.$$

Thus a small change in the Dirichlet condition produces a correspondingly small change in the solution, and the result is established. ∎

Note that because the Dirichlet condition enters into the integrand of Theorem 7.5 (and also of Theorem 7.3), a change of Dirichlet data in even a smallest interval on the boundary will influence the solution throughout the entire half-plane (circle). This situation is very different from the case of the wave equation, where it was seen that a localized change in an initial condition only produces a localized change in the solution. The reason being, of course, that in the wave equation disturbances travel at a finite speed. When considering the Laplace equation, because a change of boundary data influences the entire solution, a disturbance caused by such a change can be considered to travel at an infinite speed throughout D.

The final result to be established, called the maximum/minimum principle for the Laplace equation, is a property with many uses. The result can be proved in a variety of different ways, and the one used here is based on a proof by contradiction due to Petrovsky.

Theorem 7.6. (The Maximum/Minimum Principle for Harmonic Functions) *Let D be a bounded region with boundary Γ, and let $u(x, y)$ be a nonconstant function defined and harmonic in D and continuous on Γ. Then $u(x, y)$ attains its maximum and minimum values on Γ. Furthermore, if $u(x, y) = k$ (constant) on Γ then $u(x, y) \equiv k$ in D.*

Proof: Let $u(x, y)$ be harmonic in D and continuous on the boundary Γ of D and suppose, contrary to the statement of the theorem, that $u(x, t)$ attains its maximum $\tilde{M}$ at some interior point $P(x_0, y_0)$ of D, so if M is the maximum of u on Γ we have $M < \tilde{M}$. Translate the origin of coordinates to the point P by making the linear transformation $X = x - x_0, Y = y - y_0$, when $u(x, y)$ becomes $U(X, Y)$ and because of the linearity of the transformation and of the Laplacian, $U(X, Y)$ is also harmonic in D.

Now construct an auxiliary function

$$v(X, Y) = U(X, Y) + \frac{\tilde{M} - M}{2d^2}(X^2 + Y^2),$$

where d is the length of the longest chord that can be constructed joining any two points on Γ, so in some sense it can be considered to be the diameter of D.

Then $X^2 + Y^2 < d^2$ for any point (X, Y) in D. If, however, a point (X, Y) lies on Γ we have

$$v(X, Y) \le M + \frac{1}{2}(\tilde{M} - M) = \frac{1}{2}(M + \tilde{M}) < \tilde{M},$$

which shows that, like $U(X, Y)$, the auxiliary function $v(X, Y)$ must also attain its maximum at an interior point of D. However, this result produces a contradiction, because a simple calculation shows that

$$v_{XX} + v_{YY} = U_{XX} + U_{YY} + (\tilde{M} - M)/d^2 > 0,$$

while from elementary calculus it is known that at a maximum of a function of two variables none of its second derivatives can be positive. Consequently this contradiction implies that the maximum of $u(x, y)$ cannot occur inside D, so the statement in the theorem concerning the maximum value is proved. The statement concerning the minimum of $u(x, y)$ follows by applying the above reasoning to the function $-u(x, y)$. Finally, if $u(x, y) = k$ (constant) on the boundary Γ, then because inside D the values of $u(x, y)$ must lie between its maximum and minimum values on Γ, it must follow that $u \equiv k$ at all points of D. The proof of the theorem is complete. ∎

This powerful and very useful theorem enables the uniqueness of solutions of the Laplace equation, and their continuous dependence on the Dirichlet boundary data, to be proved quickly and easily.

Theorem 7.7. (The Uniqueness and Continuous Dependence of Solutions of the Laplace Equation on Dirichlet Boundary Conditions)
Let $u(x, y)$ be the solution of the Laplace equation in a bounded region D, on the boundary Γ of which it satisfies the Dirichlet boundary condition $u(x, y)_\Gamma = f(s)$, where s is the distance along Γ measured from some convenient reference point. Then (a) the solution is unique, and (b) the solution depends continuously on $f(s)$.

Proof:

(a) If possible, let there be two different solutions $u(x, y)$ and $v(x, y)$, both satisfying the same boundary condition $f(s)$. Then $w(x, y) = u(x, y) - v(x, y)$ is a solution of the Laplace equation subject to a boundary condition corresponding to $f(s) = 0$. By Theorem 7.6 we must have $w(x, y) \equiv 0$ in D, so $u(x, y) \equiv v(x, y)$.

(b) Construct the function $w(x, y)$ used in (a), but let $u(x, y)$ depend on a Dirichlet condition $f(s)$ and $v(x, y)$ on the Dirichlet condition $g(s)$, where $|f(s) - g(s)| < \varepsilon$ for some arbitrarily small $\varepsilon > 0$. Then from Theorem 7.6, throughout D, it follows that $|u(x, y) - v(x, y)| < \varepsilon$, so the continuous dependence of the solution on the Dirichlet boundary condition is established. ∎

EXERCISES 7.2

1. Set $z = r(\cos\alpha + i\sin\alpha)$, and show using DeMoivre's theorem that $\sum_{n=0}^{\infty} r^n \cos n\alpha = \text{Re}\{1/(1 - z)\}$. Express $1/(z - 1)$ in terms of r and α to show that $\sum_{n=0}^{\infty} r^n \cos n\alpha = \frac{1-r\cos\alpha}{1-2r\cos\alpha+r^2}$, and hence prove that

$$\frac{1}{2} + \sum_{n=1}^{\infty} r^n \cos n\alpha = \frac{1}{2}\left(\frac{1 - r^2}{1 - 2r\cos\alpha + r^2}\right).$$

2. Place bounds on the Poisson kernel $P(r_0, r, \psi - \theta)$ in terms of r and r_0, and use them to estimate $u(r, \theta)$ in terms of $\int_0^{2\pi} f(\psi)\, d\psi$.

3. Use the Poisson integral formula for a circle to show that when the Dirichlet condition $u(r_0, \theta) = f(\theta) \equiv k$ (constant) is imposed on the circle $r = r_0$, then $u \equiv k$ for $0 \le r \le r_0$.

4. Use result (7.6) to find a series solution for the function $u(r, \theta)$ that is harmonic in the unit circle $0 \le r < 1$, and on its boundary $r = 1$ satisfies the Dirichlet conditions

$$u(1, \theta) = \begin{cases} 0, & 0 < \theta < 2\pi/3 \\ 1, & 2\pi/3 < \theta < 4\pi/3 \\ -1, & 4\pi/3 < \theta < 2\pi. \end{cases}$$

5. Find an exact solution $u(r, \theta)$ of the Laplace equation in the unit circle, given that it satisfies the boundary condition $u(1, \theta) = \sin^2\theta + 2\cos^3\theta$.

6. (a) Show that $u(r, \theta) = 2r^2 \cos 2\theta + 3r \cos\theta + 4$ is harmonic, and verify its mean value property at the origin.

   (b) Show that $u(x, y) = 3x^2 + x - 3y^2 + 4$ is harmonic. Verify its mean value property at the origin, and also at the point $(1, -2)$. (Hint: use cylindrical polar coordinates.)

7. Use the maximum/minimum principle to prove the uniqueness of the Dirichlet problem for the Laplace equation in a bounded region D with boundary curve Γ.

8. Let D be a bounded region with boundary curve Γ. Use the maximum/minimum principle to prove the continuous dependence of the solution of the Laplace equation in D on Dirichlet boundary data on Γ.

9. Find the bounded solution $u(x, y)$ of the Laplace equation in the half-plane $y > 0$, given that it satisfies the boundary condition $u(x, 0) = f(x)$, where

$$f(x) = \begin{cases} 0, & x < -2 \\ 1, & -2 < x < 2 \\ 0, & x > 2. \end{cases}$$

10. Use the Poisson formula for the half-plane $y > 0$ to find the bounded solution of $\Delta u = 0$, for $-\infty < x < \infty$ and $y > 0$, given that $u(x, 0) = f(x)$, where

$$f(x) = \begin{cases} k_1, & x < 0 \\ k_2, & x > 0. \end{cases}$$

11. Find the bounded solution of the boundary value problem $\Delta u = 0$, for $-\infty < x < \infty$ and $y > 0$, given that $u(x, 0) = f(x)$, where

$$f(x) = \begin{cases} 0, & x < -1 \\ 1 - |x|, & -1 \le x \le 1 \\ 0, & x > 1. \end{cases}$$

12. Let C_1 and C_2 be two concentric circles with the respective radii R_1 and R_2, where $R_1 > R_2 > 0$. Let the functions $u_1(r, \theta)$ and $u_2(r, \theta)$ be harmonic inside C_1 and C_2, respectively, with $u_1(R_1, \theta) = \sin \theta$ and $u_2(R_2, \theta) = \sin \theta$. Find u_1 and u_2 and show that

$$u_1(r, \theta) < u_2(r, \theta) \quad \text{for } 0 \le \theta \le 2\pi \quad \text{and} \quad 0 \le r \le R_2.$$

Is this result in agreement with Theorem 7.6 and, if so, why? Will the relationship remain true if the common boundary condition is changed from $\sin \theta$ to $\sin^3 \theta$? Give reasons for your answer.

7.3 The Heat Equation

The one-dimensional time-dependent heat equation $u_t = \kappa u_{xx}$ in a region with constant thermal diffusivity κ can always be reduced to an equation of the form

$$\frac{\partial u}{\partial t} = \frac{\partial^2 u}{\partial x^2} \tag{7.10}$$

by scaling the variable t.

Like the Laplace equation, the heat equation also satisfies a maximum/minimum principle, though in a somewhat different form from that of Theorem 7.6. The result to be derived here is somewhat weaker than the most general principle that can be formulated, but it will suffice for the purposes of this book.

The theorem simply states in mathematical terms a consequence suggested by the physical observation that heat flows from a higher to a lower temperature. This result is motivated by considering a rectangle D in the (x, t) plane, with its base on $0 \le x \le L$ on $t = 0$, and its sides the lines $x = 0$ and $x = L$ for

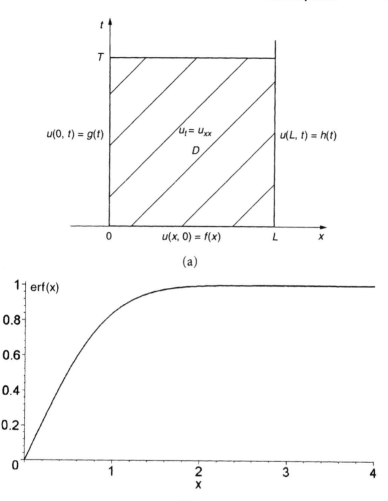

Figure 7.3 (a) The rectangle D in the maximum/minimum principle for the heat equation. (b) The error function erf (x) (see Eq. (7.24)).

$0 \le t \le T$, with arbitrary $T > 0$, as shown in Fig. 7.3. Then, as heat can only flow from a higher to a lower temperature, the maximum/minimum values of u can only occur on the base or sides of the rectangle D, for any value of $T > 0$.

Theorem 7.8. (The Maximum/Minimum Principle for the Heat Equation) *Let the solution $u(x, t)$ of the heat equation (7.10) be defined inside the closed rectangle D in the (x, t) plane $0 \le x \le L$ and $0 \le t \le T$. Then the solution $u(x, t)$ attains its maximum and minimum values on the base $t = 0$ of rectangle D, or on its sides $x = 0, x = L$. If $u(x, t)$ is constant on the base and sides of rectangle D it is identically constant throughout D.*

Proof: As with Theorem 7.6, this result will be established by using proof by contradiction. Suppose, if possible, that the maximum value $\tilde{M}$ of u occurs at an interior point $(\tilde{x}, \tilde{t})$ of D, and that the maximum value of u on the base $t = 0$ of D in Fig. 7.3 is M, where $\tilde{M} - M = \varepsilon$, for some arbitrary $\varepsilon > 0$. We now define a new function $v(x, t)$ in terms of $u(x, t)$ by

$$v(x, t) = u(x, t) + \frac{\varepsilon}{4L^2}(x - \tilde{x})^2.$$

Then $v \geq u$, and on the base $t = 0$ and sides $x = 0$ and $x = L$ of D; because $0 < \tilde{x} < L$, so we have

$$v(x, t) \leq \tilde{M} - \varepsilon + \frac{\varepsilon}{4} = \tilde{M} - \frac{3\varepsilon}{4}.$$

Thus, as $v \geq u$, the maximum of $v(x, t)$ in D cannot occur on the base of D, on $x = 0$, or on $x = L$. Now suppose $v(x, t)$ attains its maximum at a point $(\bar{x}, \bar{t})$ in D. Then, from elementary calculus, it follows that the continuously differentiable function of two variables $v(x, t)$ will have a maximum at $(\bar{x}, \bar{t})$ if $v_t(\bar{x}, \bar{t}) = 0$ and $v_{tt}(\bar{x}, \bar{t}) \geq 0$, in which case

$$v_t - v_{xx} \geq 0.$$

However, differentiation of $v(x, t)$ shows that

$$v_t - v_{xx} = u_t - u_{xx} - \varepsilon/(2L^2) < 0.$$

This contradicts the result $v_t - v_{xx} \geq 0$, so the assumption that the maximum of u occurs inside D is incorrect, and the statement about the maximum property of u is proved. The statement about the minimum property of u follows by applying the previous argument to the function $w(x, t) = -u(x, t)$.

Up to this point the statement in Theorem 7.8 is called the **weak maximum/minimum principle** for the heat equation. A **strong maximum/minimum principle** exists and asserts that if $u(x, t)$ is *continuous* in D and on its base and sides either the extrema of $u(x, t)$ occur on the base and sides of D or $u \equiv$ constant in D. The proof of this more general result is difficult, so we only establish the last assertion of Theorem 7.8. It follows because if $u(x, t) = C$ is constant on the base and sides of D then u is continuous there, so the maximum and minimum values attained by u on the base and sides of D must also be C. Thus the difference of the extrema of u in D must be zero, and this is only possible if $u(x, t) \equiv C$. The theorem is proved. ∎

Theorem 7.8 enables the uniqueness of the solution of an initial boundary value problem and the continuous dependence of the solution on its boundary data to be proved very quickly.

Theorem 7.9. (The Uniquenesss of an IVBP for the Heat Equation and the Continuous Dependence of the Solution on the Boundary Data)

(a) *There is at most one solution of the IVBP*

$$u_t - u_{xx} = 0 \quad for\ 0 < x < L \quad and \quad t > 0,$$

given that

$$u(x,0) = f(x),\ u(0,t) = g(t), \quad and \quad u(L,t) = h(t).$$

(b) *The solution u depends continuously on the initial and boundary data in (a).*

Proof:

(a) If possible let there be two solutions u and v in D, both satisfying the heat equation and each satisfying the same initial and boundary data. Then $w = u - v$ satisfies the heat equation subject to zero initial and boundary value data. From Theorem 7.8, for any $T > 0$, we have $w(x,t) = 0$ for $0 \leq x \leq L$ and $0 \leq t \leq T$. However, $T > 0$ was arbitrary, so $w \equiv 0$ for $0 \leq x \leq L$ and $t \geq 0$, and so $u \equiv v$.

(b) Let u and v be solutions of the heat equation corresponding to the initial boundary value data $f(x), g(t), h(t)$ and $F(x), G(t), H(t)$, respectively. For some time $T > 0$ let the data for u and v be close, in the sense that for $0 \leq x \leq L$ and $0 \leq t \leq T$ there is an arbitrarily small number $\varepsilon > 0$ such that

$$\max_{0 \leq x \leq L} |f(x) - F(x)| \leq \varepsilon, \qquad \max_{0 \leq t \leq T} |g(t) - G(t)| \leq \varepsilon,$$

$$and \quad \max_{0 \leq t \leq T} |h(t) - H(t)| \leq \varepsilon.$$

Then, as $w = u - v$ satisfies the heat equation and its initial boundary data are $w(x,0) = f(x) - F(x)$, $w(0,t) = g(t) - G(t)$, and $w(L,t) = h(t) - H(t)$, it follows from Theorem 7.8 that inside rectangle D in the (x,t) plane $0 \leq x \leq L, 0 \leq t \leq T$ we must have $|w| < \varepsilon$. Hence $|u - v| < \varepsilon$ throughout D, and the theorem is proved. ■

Example 7.6. Without solving the initial boundary value problem

$$u_t = 4u_{xx} \quad for\ 0 \leq x \leq 2, \quad t \geq 0,$$

subject to the initial and boundary conditions

$$u(0,t) = u(2,t) = 0 \quad and \quad u(x,0) = x(x-2)^3 e^{-x},$$

place numerical bounds on the solution $u(x,t)$.

Solution: Theorem 7.8 asserts that the maximum and minimum values of $u(x, t)$ must occur either on the line $0 \leq x \leq 2$ when $t = 0$, or on the lines $x = 0$ and $x = 2$ for all $t > 0$. The maximum and minimum of the solution on the lines $x = 0$ and $x = 2$ are zero, so it only remains to examine the extrema of the initial condition $u(x, 0) = x(x - 2)^3 e^{-x}$ for $0 \leq x \leq 2$. The function $u(x, 0)$ is negative in this interval and vanishes at $x = 0$ and $x = 2$, and a routine calculation shows it has an absolute minimum at $x^* = 3 - \sqrt{7}$ where $u(x^*, 0) = -1.108033$. Thus the required bounds for $u(x, t)$ are

$$-1.108033 < u(x, t) < 0 \quad \text{for } 0 \leq x \leq 2 \text{ and } t > 0. \qquad \blacksquare$$

EXERCISES 7.3

1. Find a change of variable that transforms the heat equation $u_t = \kappa u_{xx}$ into an equation of the form $v_t = v_{xx}$.

2. Show the transformation $v(x, t) = e^{-at} u(x, t)$ reduces the equation

$$v_t = \kappa v_{xx} - ae^{-at} v \quad \text{to } u_t = \kappa u_{xx}.$$

3. Show an equation of the form $u_t = u_{xx} - b(t)u_x$ can be transformed into the equation $U_t = U_{\xi\xi}$ by using the transformation $x = \xi + \int_0^t b(\tau)\, d\tau$ and then setting $U(\xi, t) = u(\xi + b(\tau)\, d\tau, t)$.

4. If u and v are continuous solutions of $u_t = u_{xx}$ in some region D and on its boundary Γ, prove that if $u \leq v$ on Γ, then $u \leq v$ in D.

5. Show that if $u(x, t)$ is a solution of the equation $u_t = ku_{xx}$ on the real line $-\infty < x < \infty$, then u_t, u_x, and u_{xx} are also solutions.

6. Let $u(x, t)$ be a solution of the equation $u_t = ku_{xx}$ for $0 \leq x \leq L$. By multiplying the equation by u and integrating by parts show that

$$\frac{d}{dt} \int_0^L \frac{1}{2} u^2 \, dx = k(uu_x)|_0^L - k \int_0^L (u_x)^2 \, dx.$$

This is the analogue of the energy equation for the wave equation, though without its physical meaning. Use the result to prove the uniqueness of the solution subject to the initial condition $u(x, 0) = f(x)$ for $0 \leq x \leq L$ and the boundary conditions $u(0, t) = g(t)$ and $u(L, t) = h(t)$.

7. Let $u(x, t)$ be the solution of $u_t = ku_{xx}$ subject to the initial condition $u(x, 0) = 4x^3 - 5x^2 + x$ for $0 \leq x \leq 1$, and the boundary conditions $u(0, t) = \tanh t$ and $u(1, t) = te^{-t/3}$ for $t > 0$. Use Theorem 7.8 to place bounds on the solution $u(x, t)$ for $0 \leq x \leq 1$ and $t > 0$.

8. Give a physical reason why the maximum/minimum principle for the heat equation is valid for the heat equation with radial symmetry $\frac{\partial u}{\partial r} = \frac{\kappa}{r} \frac{\partial}{\partial r}\left(r \frac{\partial u}{\partial r}\right)$, subject to the continuous initial condition $u(r, 0) = f(r)$ for $R_1 \leq r \leq R_2$

and boundary conditions $u(R_1, t) = g(t)$ and $u(R_2, t) = h(t)$ for $t > 0$. Use the principle to place bounds on the solution $u(r, t)$ in terms of $f(r), g(t)$, and $h(t)$.

7.4 Self-Similarity Solutions

The concept of a *self-similar* solution arises with certain important types of partial differential equation, although this very special form of solution is only possible for partial differential equations where no natural scales exist with which to nondimensionalize the problem. When the method is applicable, the introduction of what are called *similarity variables* enables the number of independent variables involved to be reduced, thereby simplifying the task of finding a solution. Thus, in certain situations, finding the solution of the heat equation in one space variable and time can be reduced to solving an ordinary differential equation. First though, before illustrating the approach by considering an unsteady temperature distribution problem, it is necessary to say something about the general process of nondimensionalization.

In most physical problems a characteristic length X_0, a characteristic mass m_0, and a characteristic time t_0 arise naturally and are suggested by the nature of the problem, its geometry, and its boundary and initial conditions. For example, if the problem involves a space variable X, a mass M, and a time T, by reformulating the problem in terms of $x = X/X_0, m = M/m_0$, and $t = T/t_0$, a nondimensional formulation is obtained. A typical example is provided by the wave equation determining the vibrations of a nonuniform stretched string of finite length with its ends clamped. A characteristic length could be taken to be the length X_0 of the string between the clamped ends, a characteristic line density ρ_0 could be taken to be the average line density of the string, and a characteristic time t_0 could be taken to be the period of the lowest mode of vibration of the string when the line density is assumed to be constant and equal to ρ_0. The transverse displacement U of the string at the nondimensional position x would then become the nondimensional quantity $u = U/X_0$.

The formulation of physical problems in nondimensional form is to be expected, since nature has no natural length, mass, and time scales. Consequently it is to be expected that every problem must be capable of being expressed either in terms of dimensionless variables or in terms of dimensionally consistent quantities that are combinations of length, mass, and time. This is just another way of saying that the solution of all physical problems must be independent of the system of units used to describe the problem.

In problems where characteristic units are available to make the equation nondimensional, a change of scaling will change the form of the solution. Consequently, different problems will only have the same nondimensional solution if in each case the dimensionless quantities involved have all been scaled in the same way.

Self-similarity is different, and the problem we will use to illustrate this is a time-dependent temperature distribution problem in a semi-infinite uniform block of metal, when at time $t = 0$ the uniform temperature U_0 throughout the block is caused to change due to a step change in temperature from U_0 to U_1 at its plane face. We will consider the heat equation for the temperature $u(x, t)$ in a semi-infinite block of metal with uniform density ρ, constant specific heat c, and constant conductivity k, where x is measured perpendicular to the plane face of the block that coincides with $x = 0$, so the heat equation becomes

$$\rho c u_t = k u_{xx}, \quad \text{with } \rho, c, \text{ and } k \text{ constants.} \tag{7.11}$$

The initial condition is

$$u(x, 0) = U_0 \quad \text{for } x > 0, \tag{7.12}$$

which says that at $t = 0$ the temperature in the block has the constant value U_0. The boundary condition is

$$u(0, t) = U_1 \quad \text{for } t \geq 0, \tag{7.13}$$

which says that for $t > 0$ the plane face of the block at $x = 0$ is maintained at a constant temperature U_1, which may be either greater or less than U_0.

However, as the equation is second order in x a second boundary condition is necessary, and this must be such that it takes account of the fact that the block is semi-infinite. The extra boundary condition is provided by the obvious requirement that far from the plane face of the block the temperature is U_0, so the second boundary condition takes the form of the limit

$$\lim_{x \to \infty} u(x, t) = U_0. \tag{7.14}$$

It is now necessary to consider the dimensions of the physical quantities involved. In dimensional terms the temperature u is in degrees, a mass of metal has the dimension M, a distance measured into the block has the length dimension L, and the time measured from the sudden change of temperature of the plane face has the dimension T. In terms of these quantities ρ, c, and k have the dimensions:

density ρ has the dimensions	M/L^3
the specific heat c has dimensions	calories$/(M \times \text{degrees})$
the thermal conductivity k has dimensions	calories$/(\text{degrees} \times L \times T)$

An examination of these physical quantities shows it is not possible to combine them to form either a physical constant with the dimension of a length or one with the dimension of a time, which could be used to nondimensionalize the length x or the time t in the heat equation. It turns out that this inability to nondimensionalize the length and the time is the key indicator of the fact that

the equation possesses a similarity solution, because the only other way solutions can be independent of the choice of characteristic quantities is if the variables involved are themselves all expressed in nondimensional combinations.
Setting the diffusivity $\kappa = k/(\rho c)$ brings (7.11) into the form

$$\frac{1}{\kappa}\frac{\partial u}{\partial t} = \frac{\partial^2 u}{\partial x^2},$$

from which an examination of the denominators shows that κt and x^2 must have the same dimensions, so any power of the nondimensional quantity $x^2/\kappa t$, like $x/\sqrt{\kappa t}$, will also be nondimensional. The approach to be adopted now is that instead of x and t we use a new dimensionless variable like $x/\sqrt{\kappa t}$. **Self-similar** solutions arise when, instead of characteristic scales being used to nondimensionalize a problem, nondimensional combinations of independent variables must be used instead. The temperature in the heat equation can be made nondimensional by forming the combination $(u-U_0)/(U_1-U_0)$, so writing $\alpha = x/\sqrt{\kappa t}$, which is a nondimensional quantity, we can take this to be what is called the **similarity variable** in the problem, and seek a solution of the form

$$\frac{u - U_0}{U_1 - U_0} = f(\alpha), \tag{7.15}$$

where f is a function to be determined. The way f is found is by substituting (7.15) into the heat equation $u_t = \kappa u_{xx}$ to arrive at an *ordinary differential equation* for f.

Then, with this new choice of variable, *all* solutions of the heat equation reduce to a single curve relating $(u - U_0)/(U_1 - U_0)$ to the similarity variable $\alpha = x/\sqrt{\kappa t}$. So, unlike the nondimensionalization discussed previously, in this case the scaling of the solution takes place relative to *itself*, because now u is only a function of the similarity variable α. Thus in this case the introduction of a similarity variable reduces the solution of a partial differential equation in two independent variables to the solution of an ordinary differential equation with the independent variable α.

Instead of proceeding with the similarity variable $\alpha = x/\sqrt{\kappa t}$ found by means of physical arguments, we now show how this variable can be deduced directly from the problem itself using only mathematical reasoning. To do so, ignoring the previous arguments, while recognizing that some dimensionless combination of x and t is necessary, let us assume a dimensionless similarity variable of the form

$$\beta = Cx/t^n, \tag{7.16}$$

where the constant C has the dimensions $(\text{time})^n/\text{length}$, and n is to be chosen to make the partial differential equation reduce to an ordinary differential equation.

Consequently, with this choice for β, we will seek a solution of the form

$$\frac{u - U_0}{U_1 - U_0} = f(\beta), \qquad (7.17)$$

and then substitute for u in the heat equation $u_t = \kappa u_{xx}$. To do this it is first necessary to find u_t and u_{xx}, and routine calculations show that

$$u_t = -C(U_1 - U_0)(df/d\beta)nx/t^{n+1} \text{ and } u_{xx} = C^2(U_1 - U_0)(df^2/d\beta^2)/t^{2n}.$$

After substitution into the heat equation and simplification, the equation for f becomes

$$\frac{d^2 f}{d\beta^2} + \frac{nxt^{n-1}}{\kappa C^2}\frac{df}{d\beta} = 0.$$

The variable x can be eliminated by using the result $\beta = Cx/t^n$, as a result of which the equation for f becomes

$$\frac{d^2 f}{d\beta^2} + \frac{nt^{2n-1}}{\kappa C}\beta\frac{df}{d\beta} = 0. \qquad (7.18)$$

For this to be an ordinary differential equation it is necessary that t vanishes, and this can be accomplished by setting $n = \frac{1}{2}$. With the choice $n = \frac{1}{2}$ (7.18) becomes

$$\frac{d^2 f}{d\beta^2} + \frac{1}{2\kappa C^2}\beta\frac{df}{d\beta} = 0, \quad \text{where now } \beta = Cx/\sqrt{t}. \qquad (7.19)$$

The constant C is still arbitrary, although because the equation must be dimensionless C must be such that the quantity $2\kappa C^2$ is dimensionless. A natural choice is to set $C = 1/\sqrt{2\kappa}$, because then f becomes the solution of the simple linear ordinary differential equation

$$\frac{d^2 f}{d\beta^2} + \beta\frac{df}{d\beta} = 0, \qquad (7.20)$$

where now $\beta = \frac{x}{\sqrt{2\kappa t}}$, which, apart from an unimportant numerical factor, is the same as the variable α obtained previously by physical reasoning.

To solve this equation for f we need conditions on f that take into account the fact that β can range from 0 to ∞. These can be deduced from the original initial and boundary conditions for u, because (7.13) and (7.14) require, respectively, that

$$f(0) = 1 \quad \text{and} \quad \lim_{\beta \to \infty} f(\beta) = 0. \qquad (7.21)$$

To find the general solution of Eq. (7.20) we first set $df/d\beta = g$, when the equation reduces to $g' + \beta g = 0$, with the general solution $g(\beta) = B\exp(-\beta^2/2)$, where B is an arbitrary constant. However, $g = df/d\beta$, so a further integration over the interval $0 \le s \le \beta$ shows the general solution for f to be

$$f(\beta) = A + B \int_0^\beta \exp\left(-\frac{1}{2}s^2\right) ds,$$

where A is another arbitrary constant.

The constants A and B must be found from the boundary conditions (7.21), the first of which shows that $A = 1$ and the second that $B = -1/\int_0^\infty \{\exp(-\frac{1}{2}\}s^2) ds$. Using the standard result $\int_0^\infty \exp(-\frac{1}{2}s^2) ds = \sqrt{\pi/2}$ allows $f(\beta)$ to be written

$$f(\beta) = 1 - \sqrt{\frac{2}{\pi}} \int_0^\beta \exp\left(-\frac{1}{2}s^2\right) ds. \tag{7.22}$$

Thus the solution of the time-dependent temperature distribution is seen to be

$$\frac{u - U_0}{U_1 - U_0} = 1 - \sqrt{\frac{2}{\pi}} \int_0^\beta \exp\left(-\frac{1}{2}s^2\right) ds. \tag{7.23}$$

In the study of heat transfer and diffusion expressions like this arise frequently, and they are usually written in a different form by making use of a special function called the **error function**, denoted by the symbol erf. The error function, taken from statistics, but used extensively throughout the study of heat conduction and diffusion, is defined as

$$\mathrm{erf}(x) = \frac{2}{\sqrt{\pi}} \int_0^x \exp(-\tau^2) \, d\tau. \tag{7.24}$$

A plot of erf (x) for $x > 0$ is given in Fig. 7.3, which shows how, starting from the value zero, erf (x) increases steadily and rapidly until it approaches its asymptotic value 1.

A simple change of variable shows that in terms of the error function the integral in (7.24) becomes

$$\sqrt{\frac{2}{\pi}} \int_0^\beta \exp\left(-\frac{1}{2}s^2\right) ds = \mathrm{erf}(\beta/\sqrt{2}),$$

so solution (7.23) becomes

$$\frac{u - U_0}{U_1 - U_0} = 1 - \mathrm{erf}(\beta/\sqrt{2}), \tag{7.25}$$

or, as $\beta = x/\sqrt{2\kappa t}$,

$$\frac{u - U_0}{U_1 - U_0} = 1 - \mathrm{erf}\left(\frac{x}{2\sqrt{\kappa t}}\right). \tag{7.26}$$

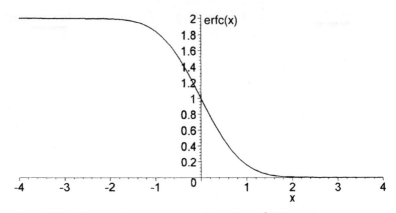

Figure 7.4   The complementary error function erfc(x).

If the **complementary error function** denoted by the symbol erfc is introduced, where

$$\text{erfc}(x) = 1 - \text{erf}(x),\qquad(7.27)$$

solution (7.26) can be written

$$\frac{u - U_0}{U_1 - U_0} = \text{erfc}\left(\frac{x}{2\sqrt{\kappa t}}\right).\qquad(7.28)$$

Like the error function, the complementary error function occurs frequently in the study of heat conduction and diffusion problems, and also throughout statistics. Figure 7.4 shows a plot of erfc (x), which, like erf (x), is defined for both positive and negative x.

Note that if instead of setting $\beta = Cx/t^n$ some other dimensionless combination had been chosen, like $\beta = Dt/x^n$ where D has the dimensions (length)^n/ time, and some other value had been taken for the constant C in (7.19) while ensuring that $2\kappa C^2$ remains dimensionless, the intermediate calculations would have been altered but the final result would remain unchanged. The only reason for favoring one choice of β over another in the initial stage of the analysis is if it is expected to simplify the intermediate calculations.

In general, finding similarity solutions for more complicated partial differential equations is difficult, and instead of the intuitive approach used here more sophisticated methods must be employed when finding a similarity variable and its associated simplified differential equation. Such methods, which use techniques from group theory and Lie algebra, are beyond the level of this introductory account, so the subject will not be developed any further.

EXERCISES 7.4

1. Show by solving the equation $f'' + kxf' = 0$ with k being a constant that the general solution is $f(x) = A + B \, \mathrm{erf}(x\sqrt{k/2})$.

2. Use the result of Exercise 1 to demonstrate that a different choice for constant C in (7.19), made so the coefficient of the term xf' remains dimensionless, will leave the final solution unchanged.

3. The purpose of this problem is to illustrate how a different choice of β from that used in (7.16) will lead to the same similarity variable and equation for the heat equation. Let $u(x, t)$ be a solution of the heat equation $u_t = \kappa u_{xx}$. Set $\frac{u - U_0}{U_1 - U_0} = f(\beta)$, with $\beta = Dt^m/x^n$, and find the equation satisfied by $f(\beta)$. Determine the values of m and n that make the differential equation for $f(\beta)$ depend only on the independent variable β, and hence show that $f'' + \left(\frac{1}{2\kappa D^2}\right)\beta f' = 0$ with $\beta = Dx/\sqrt{t}$.

4. Set $u = f(\beta)$, with $\beta = r/t^n$, and transform the unsteady radial heat equation $u_t = \kappa(u_{rr} + (1/r)u_r)$ to an equation in f. Find a value for n that reduces this to a differential equation for f depending only on β. Use this value of n to write down the form of β and the equation satisfied by f in order that $u = f(\beta)$ becomes a self-similar solution of the unsteady radial heat equation.

5. Explore the properties of the error function by showing that:

   (i) $\mathrm{erf}\,(x)$ is an odd function of x and plotting it for $-4 < x < 4$;

   (ii) $\int_a^b e^{-x^2} dx = \frac{1}{2}\sqrt{\pi}\,[\mathrm{erf}(b) - \mathrm{erf}(a)]$;

   (iii) in terms of the error function, the normal distribution in statistics $\Phi(x) = \frac{1}{\sqrt{2\pi}} \int_{-\infty}^{x} e^{-u^2/2} du$ can be written $\Phi(x) = \frac{1}{2}[1 + \mathrm{erf}(x/\sqrt{2})]$.

7.5 Fundamental Solution of the Heat Equation

The purpose of this section is to introduce the concept of the *fundamental solution* of the heat equation $u_t = ku_{xx}$, and to show how it leads to the solution of a general initial value problem for the equation for $-\infty < x < \infty$. To achieve this objective two preparatory steps are required. The first will be to use the results of Section 7.4 to solve a basic heat conduction problem in an infinitely long metal rod, and the second will be to remind the reader about an essential property of the Dirac delta function $\delta(x)$ prior to its use.

First we need to solve a problem different from that considered in Section 7.4, although it is closely related. This time, instead of considering a problem in a semi-infinite region, we will consider a problem involving an infinitely long metal rod with its sides thermally insulated and its diffusivity $\kappa = $ constant. The problem will be to find the temperature distribution in the rod as a function of x and t if the distance x is measured along the rod and at time $t = 0$ the part of the rod for $x < 0$ is at a uniform temperature $u = 0$, while the part of the rod for $x > 0$ is at a uniform temperature $u = 1$.

In physical terms, this can be thought of as finding the temperature distribution as a function of distance and time if at $t = 0$ two semi-infinite rods, one at zero temperature and the other at a unit temperature have their ends brought into perfect contact, after which a redistribution of heat occurs.

This problem is easily solved using the method of Section 7.4, because the analysis is identical up to Eq. (7.20), and as before

$$\frac{d^2 f}{d\beta^2} + f\frac{df}{d\beta} = 0,$$

although now $u = f(\beta)$.

The difference in this case lies in the boundary conditions, because the rod is infinite in length so the boundary conditions occur at $\pm\infty$, with

$$u(x, t) = \begin{cases} 0, & x \to -\infty \\ 1, & x \to \infty. \end{cases} \tag{7.29}$$

It follows from this that the boundary conditions for $u = f(\beta)$ will be

$$\lim_{\beta \to -\infty} f(\beta) = 0 \quad \text{and} \quad \lim_{\beta \to \infty} f(\beta) = 1. \tag{7.30}$$

Setting $g = df/d\beta$ as before, and integrating once, gives $g(\beta) = B\exp(-\beta^2/2)$, where B is an arbitrary constant. Integrating this result over the semi-infinite interval $-\infty < \beta \le x$ and using the condition $\lim_{\beta \to -\infty} f(\beta) = 0$ shows that

$$f(\beta) = B \int_{-\infty}^{x} \exp\left(-\frac{1}{2}s^2\right) ds.$$

Applying the boundary condition $\lim_{x \to \infty} f(x) = 1$ gives $1 = B \int_{-\infty}^{\infty} \exp(-\frac{1}{2}s^2)\, ds$, so using the standard integral $\int_{-\infty}^{\infty} \exp(-\frac{1}{2}s^2)\, ds = \sqrt{2\pi}$ we find that $B = 1/\sqrt{2\pi}$, and so

$$u(x, t) = f(\beta) = \frac{1}{\sqrt{2\pi}} \int_{-\infty}^{x} \exp\left(-\frac{1}{2}s^2\right) ds.$$

It is convenient to express this result in terms of the error function, and to do this we first rewrite it as

$$u(x, t) = \frac{1}{\sqrt{2\pi}} \left(\int_{-\infty}^{0} \exp\left(-\frac{1}{2}s^2 \right) ds + \int_{0}^{x} \exp\left(-\frac{1}{2}s^2 \right) ds \right).$$

Using the standard result $\int_{-\infty}^{0} \exp(-\frac{1}{2}s^2)\, ds = \sqrt{\pi/2}$ together with the definition of the error function in (7.24) gives the required solution

$$u(x, t) = \frac{1}{2}\left[1 + \mathrm{erf}\left(\frac{x}{2\sqrt{\kappa t}} \right) \right]. \tag{7.31}$$

We now refresh ideas about the Dirac delta function $\delta(x)$, although for the sake of simplicity we will only use an intuitive approach. It will be apparent from this that the delta function is not an ordinary function but a *generalized function*, meaning that it is an *operator* whose significance only becomes apparent when it occurs under an integral sign.

Consider the rectangular pulse function

$$\Delta(x, a) = \begin{cases} h, & a - 1/(2h) < x < a + 1/(2h) \\ 0, & \text{otherwise} \end{cases} \tag{7.32}$$

shown in Fig. 7.5, where later we will allow $h \to \infty$, causing the amplitude of the pulse to become very large and its width to become very small. It can be seen from this that, irrespective of the value of h, the integral of the rectangular pulse

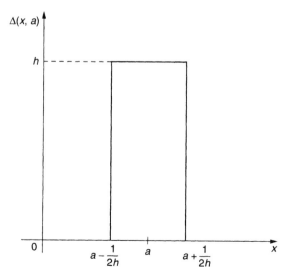

Figure 7.5   The pulse function $\Delta(x - a)$ used to define the delta function $\delta(x - a)$.

$\int_{\alpha}^{\beta} \Delta(x, a) \, dx = 1$ if its interval of definition $a - 1/(2h) < x < a + 1/(2h)$ lies in the interval $\alpha < x < \beta$, and that it is zero otherwise. For our intuitive definition of the Dirac **delta function** $\delta(x - a)$ located at the point $x = a$ we now set

$$\delta(x - a) = \lim_{h \to \infty} \Delta(x, a). \qquad (7.33)$$

To understand the significance of $\delta(x - a)$ we need to consider the integral

$$\int_{\alpha}^{\beta} f(x) \Delta(x - a) \, dx,$$

where $f(x)$ is an arbitrary continuous function over $\alpha < a < \beta$. It then follows from the first mean value theorem for integrals that

$$\int_{\alpha}^{\beta} f(x) \Delta(x - a) \, dx = \left[\frac{1}{2h} - \left(-\frac{1}{2h} \right) \right] \Delta(\xi) f(\xi) = \frac{1}{h} h f(\xi) = f(\xi),$$

where ξ is an unknown point such that $a - \frac{1}{2h} < \xi < a + \frac{1}{2h}$.

Proceeding to the limit $h \to \infty$ in definition (7.33) of the delta function causes $\Delta(x - a) \to \delta(x - a)$, and the unknown point ξ in the interval $a - \frac{1}{2h} < \xi < a + \frac{1}{2h}$ to be squeezed ever closer to a, so that in the limit $h \to \infty$ we see that $f(\xi) \to f(a)$. Thus we have derived the fundamental property of the delta function

$$\int_{\alpha}^{\beta} f(x) \delta(x - a) \, dx = f(a) \quad \text{for } \alpha < x < \beta. \qquad (7.34)$$

This shows the sievelike, or filtering, property of the delta function when it occurs under the integral sign, because from all the values of $f(x)$ in the interval of integration, $\delta(x - a)$ has selected the value $f(a)$ at the location of the delta function.

The fact that the delta function is not a function in the ordinary sense is clearly apparent from definition (7.33), because

$$\delta(x - a) = \begin{cases} 0 & \text{if } x \neq a \\ \infty & \text{if } x = a, \end{cases} \quad \text{and it is such that} \int_{-\infty}^{\infty} \delta(x - a) \, dx = 1. \quad (7.35)$$

Instead of using the limit of a sequence of ever-narrowing rectangular pulses of unit area when giving an intuitive definition of the delta function, any similar sequence of functions can be used, provided their integral is unity and their amplitude increases as their pulselike property narrows. However, in such cases establishing result (7.34) becomes harder. Typically, a sequence of

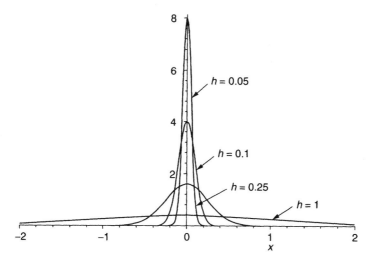

Figure 7.6 Some representative bell-shaped pulses.

bell-shaped pulses

$$\Delta(x - a) = \frac{1}{k\sqrt{2\pi}} \exp\left[-\frac{1}{2}\left(\frac{x - a}{k}\right)^2\right]$$

is used to define the delta function $\delta(x - a)$ as the limit

$$\delta(x - a) = \lim_{k \to 0} \frac{1}{k\sqrt{2\pi}} \exp\left[-\frac{1}{2}\left(\frac{x - a}{k}\right)^2\right]. \tag{7.36}$$

This is because the integral of $\Delta(x - a)$ is unity for all values of $k > 0$, and the bell-shaped pulses defined in this way become narrower as $k \to 0$. Some representative pulses are shown in Fig. 7.6 for $a = 0$ and different values of k.

We are now in a position to use solution (7.31) with the delta function to arrive at the fundamental solution of the heat equation $u_t = \kappa u_{xx}$ on the interval $-\infty < x < \infty$. It follows from solution (7.31) of the heat equation that if the initial condition is translated from $x = 0$ to $x = x_0$ with $u = 0$ for $x < x_0$ and $u = U$ for $x > x_0$, the solution will become

$$u(x, t) = \frac{1}{2}U\left[1 + \text{erf}\left(\frac{x - x_0}{2\sqrt{\kappa t}}\right)\right].$$

Similarly, if this same initial condition is translated to $x = x_1$ the solution will become

$$u(x, t) = \frac{1}{2}U\left[1 + \text{erf}\left(\frac{x - x_1}{2\sqrt{\kappa t}}\right)\right].$$

Now consider an initial condition such that $u(x, 0)$ is the rectangular pulse

$$u(x, 0) = U[H(x - x_0) - H(x - x_1)], \tag{7.37}$$

where $H(x - k)$ is the **Heaviside unit step function** defined as

$$H(x - k) = \begin{cases} 0, & x < k \\ 1, & x > k. \end{cases}$$

The linearity of the heat equation allows the solution corresponding to this rectangular pulse to be written

$$u(x, t) = \frac{1}{2} U \left[\text{erf} \left(\frac{x - x_0}{2\sqrt{\kappa t}} \right) - \text{erf} \left(\frac{x - x_1}{2\sqrt{\kappa t}} \right) \right]. \tag{7.38}$$

At time $t = 0$ let a fixed quantity of heat Q be released over the rectangular pulse $P(x) = H(x - x_0) - H(x - x_1)$. Then if the distribution of heat is uniform the corresponding temperature distribution will be

$$u(x, 0) = \frac{Q}{c\rho(x_1 - x_0)} [H(x - x_0) - H(x - x_1)],$$

so using (7.38) the solution of the associated Cauchy problem for the heat equation becomes

$$u(x, t) = \frac{Q}{c\rho} \frac{\left[\text{erf} \left(\frac{x - x_0}{2\sqrt{\kappa t}} \right) - \text{erf} \left(\frac{x - x_1}{2\sqrt{\kappa t}} \right) \right]}{x_1 - x_0}.$$

In the limit as $x_1 \to x_0$ this becomes

$$-\frac{Q}{c\rho} \frac{\partial}{\partial y} \text{erf} \left(\frac{x - y}{2\sqrt{\kappa t}} \right) = \frac{Q}{c\rho} \frac{1}{2\sqrt{\kappa t}} \exp \left(-\frac{(x - x_0)^2}{4\kappa t} \right).$$

Thus the solution of the heat equation corresponding to a delta function initial condition at $x = x_0$ is

$$G(x - x_0, t) = \frac{1}{2\sqrt{\pi \kappa t}} \exp \left(-\frac{(x - x_0)^2}{4\kappa t} \right). \tag{7.39}$$

The function $G(x - x_0, t)$ is called the **fundamental solution** of the heat equation

$$u_t = \kappa u_{xx}, \text{ subject to the initial condition } u(x, 0) = \delta(x - x_0), \tag{7.40}$$

for $-\infty < x < \infty$. Thus if an amount of heat $Q = c\rho$ is released at $x = a$ at time $t = 0$ in the infinitely long rod, the resulting temperature distribution will be $\frac{Q}{c\rho} G(x - a, t)$.

The fundamental solution is also called the **Green's function** for the heat equation. Inspection of (7.39) shows the fundamental solution to be a continuous function of x and t provided $t \neq 0$, and differentiation confirms that it satisfies the heat equation

$$G_t(x - x_0, t) = \kappa\, G_{xx}(x - x_0, t). \tag{7.41}$$

Note also that $G(x - x_0, 0) = \delta(x - x_0)$, where here the alternative definition of the delta function involving the limit of the sequence of bell-shaped pulses is used, as in (7.36).

An examination of the fundamental solution (7.39) shows that the release of heat at $x = x_0$ has an *immediate* effect on the temperature distribution over the entire real line. Thus the speed of propagation of heat as described by the heat equation is seen to be *infinite*. Clearly this is not a realistic feature of heat conduction since heat must travel at a finite speed, so this property of the solution reflects a fault in the assumptions used when deriving the heat equation. However, this failure of the heat conduction model is not important, because the decay of the exponential term in (7.39) is so rapid that the effect of heat release at a point $x = a$ becomes vanishingly small only a moderate distance from $x = a$.

Another feature of the fundamental solution is that it shows time cannot be reversed in solutions of the heat equation. That is, it is impossible to deduce from the solution at a time t_1 the solution at a previous time t_0, where $t_0 < t_1$. This is because when time is reversed the exponent in the exponential term in (7.39) becomes positive, causing the solution $u(x, t)$ to diverge (blow-up) as $|t| \to \infty$.

We are now in a position to use the fundamental solution to solve the following initial value problem on the entire real line:

$$u_t = \kappa u_{xx}, \text{ subject to the initial condition } u(x, 0) = U(x). \tag{7.42}$$

The function $U(x)$ can be considered to be a *heat source* with the property that at $t = 0$ it releases a quantity of heat $dQ = c\rho\, d\xi$ from an element of length $d\xi$ at $x = \xi$. Thus the temperature distribution along the line when $t > 0$ is given by

$$\frac{dQ}{c\rho} G(x - \xi, t) = U(\xi) G(x - \xi, t)\, d\xi.$$

Because of the linearity of the heat equation, it is to be expected that the temperature distribution caused by the initial condition $u(x, 0) = U(x)$ can be found by summing all such elements, which in the limit leads to the solution in the form of the integral

$$u(x, t) = \int_{-\infty}^{\infty} U(\xi) G(x - \xi, t)\, d\xi. \tag{7.43}$$

To show that (7.43) satisfies the initial condition in (7.42) we use the fact that $G(x - \xi, 0) = \delta(x - \xi)$ because by (7.34), when $t = 0$, Eq. (7.43) becomes

$$u(x, 0) = \int_{-\infty}^{\infty} U(\xi) G(x - \xi, 0)\, d\xi = \int_{-\infty}^{\infty} U(\xi)\delta(x - \xi)\, d\xi = U(x).$$

To show that (7.43) satisfies the heat equation we note that provided differentiation under the integral sign is permissible

$$u_t = \int_{-\infty}^{\infty} U(\xi) G_t(x - \xi, t)\, d\xi \quad \text{and} \quad u_{xx} = \int_{-\infty}^{\infty} U(\xi) G_{xx}(x - \xi, t)\, d\xi,$$

from which it then follows from (7.41) that

$$u_t - \kappa u_{xx} = \int_{-\infty}^{\infty} U(\xi)[G_t - \kappa G_{xx}]\, d\xi = 0.$$

To complete the demonstration that (7.43) gives the required solution of the initial value problem (7.42) it is necessary to establish the convergence of this last integral. This can be justified by making the reasonable assumption that $U(x)$ is a bounded function, although the proof is omitted and the details are left as an exercise for an interested reader.

Note from (7.38) that the heat equation has the effect of smoothing out immediately any discontinuities that may be present in the initial conditions. This means that discontinuities are not propagated, and this property also applies to a singularity like a delta function.

To conclude this discussion of the fundamental solution on the infinite interval $-\infty < x < \infty$ we draw attention to some of the main differences between solutions of the one-dimensional heat (diffusion) equation and those of wave equations, some of which have already been mentioned.

Wave Equation	Heat (Diffusion) Equation
(i) Wave propagation speed is finite.	Wave propagation speed is infinite.
(ii) Time reversal permissible as it reverses the direction of wave propagation.	Time reversal is not permitted as it leads to instability (blow-up of the solution).
(iii) Energy is conserved.	Energy is dissipated, causing the solution to decay.
(iv) Discontinuities in an initial condition persist and are transported along characteristics.	Singularities of all types are smoothed out immediately.
(v) Wave propagation is bidirectional because the equation is second order in time (waves propagate in both directions).	Heat propagation is unidirectional because the equation is first order in time (this is reflected in (ii)).

So far solutions of the heat equation using the fundamental solution have been confined to the infinite interval $-\infty < x < \infty$. By using a mathematical trick, a modified fundamental solution that allows initial value problems to be solved on the semi-infinite interval $0 \leq x < \infty$ can be defined. The idea is straightforward, and it uses either the symmetry or the skew symmetry property of solutions that occur when initial conditions are distributed in a symmetrical or skew-symmetrical manner along the x axis with respect to the origin.

Consider the effect on the solution of the heat equation if an initial condition involves delta functions of *opposite* signs located symmetrically about the origin, with a positive delta function at $x = x_0$ and a negative one at $x = -x_0$. The initial condition then becomes

$$u(x,0) = \delta(x - x_0) - \delta(x + x_0). \tag{7.44}$$

Using (7.43) the solution is seen to be given by

$$u(x,t) = \int_{-\infty}^{\infty} [\delta(\xi - x_0) - \delta(\xi + x_0)] G(x - \xi, t) \, d\xi$$

$$= G(x - x_0, t) - G(x + x_0, t). \tag{7.45}$$

Setting $x = 0$ and noting that $G(-x_0, t) = G(x_0, t)$ it follows directly that $u(0, t) = 0$.

Thus if consideration is restricted to the semi-infinite line $0 \leq x < \infty$, this shows that (7.45) provides the solution of the heat equation subject to the boundary condition $u(0, t) = 0$ and the initial condition $u(x, 0) = \delta(x - x_0)$. Hence the **fundamental solution** $\tilde{G}(x - x_0, t)$ for the semi-infinite interval $0 \leq x < \infty$ is found by solving

$$u_t = \kappa u_{xx} \tag{7.46}$$

subject to the boundary condition

$$u(0, t) = 0 \tag{7.47}$$

and the initial condition

$$u(x, 0) = \delta(x - x_0) - \delta(x + x_0). \tag{7.48}$$

Consequently the fundamental solution is just the *difference* of two fundamental solutions, one at $x = x_0$ and the other at $x = -x_0$, so

$$\tilde{G}(x - x_0, t) = \frac{1}{2\sqrt{\pi \kappa t}} \left[\exp\left(-\frac{(x - x_0)^2}{4\kappa t} \right) - \exp\left(-\frac{(x + x_0)^2}{4\kappa t} \right) \right]. \tag{7.49}$$

This result shows that the solution of an initial value problem for the heat equation on the semi-infinite interval $0 \leq x < \infty$ with the boundary condition

$u(0, t) = 0$ and the initial condition $u(x, 0) = U(x)$ is given by

$$u(x, t) = \int_0^\infty \tilde{G}(x - \xi, t)U(\xi)\, d\xi \quad \text{for } 0 \leq x < \infty. \tag{7.50}$$

A similar argument shows that if initial condition (7.44) is replaced by

$$u(x, 0) = \delta(x - x_0) + \delta(x + x_0), \tag{7.51}$$

the boundary condition at $x = 0$ becomes $u_x(0, t) = 0$. Thus the **fundamental solution** $\hat{G}(x - x_0, t)$ for the semi-infinite interval $0 \leq x < \infty$ subject to the boundary condition $u_x(0, t) = 0$ and the initial condition (7.51) is found by solving

$$u_t = \kappa u_{xx} \tag{7.52}$$

subject to the boundary condition

$$u_x(0, t) = 0 \tag{7.53}$$

and the initial condition

$$u(x, 0) = \delta(x - x_0) + \delta(x + x_0). \tag{7.54}$$

In this case the fundamental solution is just the *sum* of two fundamental solutions, one at $x = x_0$ and the other at $x = -x_0$, so

$$\hat{G}(x - x_0, t) = \frac{1}{2\sqrt{\pi \kappa t}}\left[\exp\left(-\frac{(x - x_0)^2}{4\kappa t}\right) + \exp\left(-\frac{(x + x_0)^2}{4\kappa t}\right)\right]. \tag{7.55}$$

Using this result shows that the solution of an initial value problem for the heat equation on a semi-infinite interval $0 \leq x < \infty$ with the boundary condition $u_x(0, t) = 0$ and the initial condition $u(x, 0) = U(x)$ is given by

$$u(x, t) = \int_0^\infty \hat{G}(x - \xi, t)U(\xi)\, d\xi \quad \text{for } 0 \leq x < \infty. \tag{7.56}$$

Geometrically, the meaning of the boundary condition $u(0, t) = 0$ used with $\tilde{G}(x - x_0, t)$ in (7.50) means that the initial condition $U(x)$ can be considered to have been extended to negative x as an *odd* function, so that $U(-x) = -U(x)$ in (7.43), although only the solution on $0 \leq x < \infty$ is considered. Similarly, the geometrical meaning of the boundary condition $u_x(0, t) = 0$ used with $\hat{G}(x - x_0, t)$ in (7.56) means that the initial condition $U(x)$ can be considered to have been extended to negative x as an *even* function, so that $U(-x) = U(x)$, although only the solution on $0 \leq x < \infty$ is considered.

In the derivation of these results, the delta function placed on the negative x axis to produce either the boundary condition $u(0, t) = 0$ or $u_x(0, t) = 0$ is called the **image** of the delta function on the positive x axis. The method of images has many applications, although its use will not be developed any further in this first account of the subject.

EXERCISES 7.5

1. Use differentiation of the fundamental solution (7.39) to verify that it satisfies the heat equation.

2. Use the fundamental solution of the heat equation to solve the IVP $u_t = \kappa u_{xx}$ subject to the initial condition $u(x,0) = \begin{cases} u_1, & x < 0 \\ u_0, & x \geq 0 \end{cases}$.

3. Use the fundamental solution of the heat equation to solve the IVP $u_t = \kappa u_{xx}$ subject to the initial condition $u(x,0) = \begin{cases} U, & 0 < x < 1 \\ 0, & \text{otherwise.} \end{cases}$

4. Use the fundamental solution of the heat equation to prove the uniqueness of the solution of an IVP.

5. Use the fundamental solution of the heat equation and computer algebra to solve the IVP

$$u_t = \kappa u_{xx} \text{ subject to the initial condition}$$

$$u(x,0) = \begin{cases} x, & 0 \leq x < 1 \\ 0, & \text{otherwise} \end{cases} \quad \text{for } -\infty < x < \infty.$$

Make a 3d computer plot of the solution and verify that the initial discontinuity is resolved immediately, and so does not propagate.

6. Use the fundamental solution of the heat equation and computer algebra to solve the IVP

$$u_t = \kappa u_{xx} \text{ subject to the initial condition}$$

$$u(x,0) = \begin{cases} 0, & x < 0 \\ U_1, & 0 \leq x \leq 1 \\ 0, & 1 < x < 3 \\ U_2, & 3 \leq x \leq 4 \\ 0, & x > 4 \end{cases} \quad \text{for } -\infty < x < \infty.$$

Make a 3d computer plot of the solution and verify that the initial discontinuity is resolved immediately, and so does not propagate.

7. Use computer algebra and the fundamental solution $\tilde{G}(x - x_0, t)$ for the heat equation on the semi-infinite interval $0 \leq x < \infty$ to solve the heat

equation subject to the boundary condition $U(x) = \begin{cases} 1, & 0 \le x \le 2 \\ 0, & x > 2 \end{cases}$. Make a 3d computer plot of the solution and verify that the initial discontinuity is resolved immediately, and so does not propagate.

7.6 Duhamel's Principle

In Chapter 6 the method of separation of variables was used to solve the homogeneous form of the heat equation subject to an initial condition and to boundary conditions that either were homogeneous or were of a special non-homogeneous time-dependent form. However, a more general situation arises when a solution where the boundary conditions are nonhomogeneous and a distributed heat source is present is required. It is the purpose of this section to show how such a solution can be found when constant nonhomogeneous Dirichlet or Neumann conditions occur at the ends of an interval of length L, and a distributed heat source $f(x, t)$ is present. In this case the heat equation with constant nonhomogeneous Dirichlet conditions takes the form

$$u_t = \kappa u_{xx} + f(x, t) \quad \text{for } 0 \le x \le L, \tag{7.57}$$

subject to the constant Dirichlet boundary conditions

$$u(0, t) = a \quad \text{and} \quad u(L, t) = b \quad \text{for } t \ge 0 \text{ (a and b constants),} \tag{7.58}$$

and the initial condition

$$u(x, 0) = g(x) \quad \text{for } 0 \le x \le L. \tag{7.59}$$

Because of the linearity of the heat equation, we will make use of the fact that this solution can be found by solving three simpler problems and adding the results. Before examining how such a composite solution is to be constructed, we first find *any* particular solution $u_1(x, t)$ of the very simple problem

$$u_{1_t} = \kappa u_{1_{xx}}, \tag{7.60}$$

subject only to to the *nonhomogeneous* boundary conditions

$$u_1(0, t) = a \quad \text{and} \quad u_1(L, t) = b. \tag{7.61}$$

As any particular solution will suffice, we will choose the simplest, which is independent of t, and so must satisfy the equation $u_{1xx} = 0$ subject to the boundary conditions $u_1(0) = a$ and $u_1(L) = b$, from which it follows at once that

$$u_1(x, t) = a + \left(\frac{b - a}{L}\right) x. \tag{7.62}$$

Let us now decompose the solution of IVBP (7.57) to (7.59) into a sum of three terms

$$u(x, t) = u_1(x, t) + u_2(x, t) + u_3(x, t), \tag{7.63}$$

where $u_1(x, t)$ has already been found.

Next we will require the function $u_2(x, t)$ to satisfy the *nonhomogeneous* IVBP

$$u_2(x, t) = \kappa u_{2_{xx}} + f(x, t), \tag{7.64}$$

subject to the *homogeneous* boundary conditions

$$u_2(0, t) = 0 \quad \text{and} \quad u_2(L, t) = 0 \tag{7.65}$$

and the *homogeneous* initial condition

$$u_2(x, 0) = 0. \tag{7.66}$$

Finally, we will require the function $u_3(x, t)$ to satisfy the IVBP for the *homogeneous* heat equation

$$u_{3_t} = \kappa u_{3_{xx}}, \tag{7.67}$$

subject to the *homogeneous* boundary conditions

$$u_3(0, t) = 0 \quad \text{and} \quad u_3(L, t) = 0 \tag{7.68}$$

and the initial condition $u_3(x, 0) = g(x) - u_1(x, t)$, which is equivalent to the initial condition

$$u_3(x, 0) = g(x) - a - \left(\frac{b - a}{L} \right) x. \tag{7.69}$$

The linearity of the heat equation allows the three functions $u_1(x, t)$, $u_2(x, t)$, and $u_3(x, t)$ to be added and the sum (7.63) to be a solution of Eq. (7.57), subject to boundary conditions that are the sum of the three respective sets of boundary conditions and an initial condition that is the sum of the three respective initial conditions. It is a simple matter to check that when these conditions are added they correspond to the boundary conditions and the initial condition for the nonhomogeneous heat equation (7.57). The uniqueness of a solution of the heat equation means that however a solution may be found, if it satisfies (7.57) and the auxiliary conditions (7.58) and (7.59), it must be the unique solution of the IVBP.

The solution of (7.67) for $u_3(x, t)$ follows from Exercise 1 of Section 6.3 as

$$u_3(x, t) = \sum_{n=1}^{\infty} A_n \sin \left(\frac{n\pi x}{L} \right) \exp \left(-\frac{n^2 \pi^2 \kappa t}{L} \right), \tag{7.70}$$

where

$$A_n = \frac{2}{L} \int_0^L \left[g(x) - a - \left(\frac{b-a}{L} \right) x \right] \sin\left(\frac{n\pi x}{L} \right) dx. \qquad (7.71)$$

To complete the determination of $u(x, t)$, the only task that remains is to find the solution $u_2(x, t)$, and to do this we will make use of an important result called **Duhamel's principle**. The principle can be proved in several different ways, one of which uses careful limiting arguments, while another requires the use of the Laplace transform. To go into the details of either of these methods would take us beyond the scope of this first account of the subject, so we will adopt a different approach. Instead of deriving Duhamel's principle, we will first state it, and then verify its correctness by direct calculation.

The problem to be considered involves solving the nonhomogeneous heat equation (7.64) subject to the homogeneous boundary conditions (7.65) and the homogeneous initial condition (7.66), where the heat source $f(x, t)$ is assumed to be a twice continuously differentiable function. Thus it is necessary to solve the IVBP

$$\frac{\partial u_2}{\partial t} = \kappa \frac{\partial^2 u_2}{\partial x^2} + f(x, t) \quad \text{for } 0 \leq x \leq L \text{ and } t \geq 0, \qquad (7.72)$$

subject to the homogeneous boundary conditions

$$u_2(0, t) = 0 \quad \text{and} \quad u_2(L, t) = 0 \qquad (7.73)$$

and the homogeneous initial condition

$$u_2(x, 0) = 0. \qquad (7.74)$$

The idea underlying Duhamel's principle is that $u_2(x, t)$ can be found in terms of the solution w of the much simpler problem

$$\frac{\partial w}{\partial t} = \kappa \frac{\partial^2 w}{\partial x^2} \quad \text{for } 0 \leq x \leq L \text{ and } t \geq 0, \qquad (7.75)$$

subject to the homogeneous Dirichlet boundary conditions

$$w(0, t : \tau) = 0 \quad \text{and} \quad w(L, t : \tau) = 0 \qquad (7.76)$$

and the nonhomogeneous initial condition

$$w(x, \tau : \tau) = f(x, \tau), \qquad (7.77)$$

where τ is a parameter whose role will become clear later.

Using either of the approaches mentioned previously, it can be shown that the solution for $u_2(x, t)$ is given in terms of w by the integral

$$u_2(x, t) = \int_0^t w(x, t : \tau)\, d\tau. \tag{7.78}$$

To understand why solving for w is simpler than solving for u_2, note first that w satisfies a *homogeneous* heat equation, and that now the heat source $f(x, \tau)$ appears in the *initial condition* (7.77) in terms of a parameter τ, but τ is independent of t so this heat source can be considered to be *fixed* with respect to x and t. The solution for $u(x, t)$ in (7.78) shows that $w(x, t : \tau)$ can be considered to be a distributed solution, the integral of which with respect to τ over the time interval $0 \le \tau \le t$ yields $u_2(x, t)$.

Duhamel's principle can be summarized as follows:

Duhamel's Principle

Step 1   Formulate the IVBP in (7.73) subject to the boundary conditions (7.74) and the initial condition (7.75).

Step 2   Find $w(x, t : \tau)$ from (7.76) subject to the boundary conditions (7.77) and the initial condition (7.78).

Step 3   Find the solution of the IVBP in Step 1 using (7.79)

To establish the correctness of solution (7.78) it is necessary to use a generalization of the first fundamental theorem of calculus. This is the Leibniz rule for the differentiation of a definite integral with respect to a parameter, when the parameter occurs both in the limits of the integral and in the integrand. In its most general form, when $f(\tau, t)$, $f_t(\tau, t)$, $\varphi(t)$, $\varphi'(t)$, $\psi(t)$, and $\psi'(t)$ are continuous functions, the Leibniz rule for differentiation of an integral with respect to a parameter t is

$$\frac{d}{dt} \int_{\psi(t)}^{\varphi(t)} f(\tau, t)\, d\tau = f(\varphi(t), t)\frac{d\varphi(t)}{dt} - f(\psi(t), t)\frac{d\psi(t)}{dt}$$

$$+ \int_{\psi(t)}^{\varphi(t)} \frac{\partial f(\tau, t)}{\partial t}\, d\tau. \tag{7.79}$$

This result has many different uses, one of which occurred in Section 2.4 when discussing shock solutions in nonlinear first-order equations, although its proof will be left as an exercise.

For what is to follow, (7.79) will only be needed in a much simpler form with $\psi(t) \equiv 0$ and $\varphi(t) = t$, causing it to reduce to

$$\frac{d}{dt} \int_0^t f(\tau, t)\, d\tau = f(t, t) + \int_0^t \frac{\partial f(\tau, t)}{\partial t}\, d\tau. \tag{7.80}$$

To verify that $u_2(x, t)$ given by (7.78) is the solution of (7.72) subject to conditions (7.73) and (7.74), note first from (7.78) that $u_2(x, t)$ satisfies the homogeneous initial condition $u_2(x, 0) = 0$, and that because $w(x, t)$ satisfies the homogeneous boundary conditions in (7.76), $u_2(x, t)$ must satisfy the homogeneous boundary conditions in (7.73). To complete the justification of the form of solution given in (7.78) we must now show that $u_2(x, t)$ satisfies (7.72). To do this we apply result (7.80) to (7.78) when we obtain

$$\frac{\partial u}{\partial t} = w(x, t : t) + \int_0^t \frac{\partial w(x, t : \tau)}{\partial t} \, d\tau.$$

Setting $\tau = t$ in (7.77) gives $w(x, t : t) = f(x, t)$, while from (7.75) $w_t = \kappa w_{xx}$, so the above result becomes

$$\frac{\partial u_2}{\partial t} = f(x, t) + \int_0^t \kappa \frac{\partial^2 w(x, t : \tau)}{\partial x^2} \, d\tau.$$

To complete our task we apply (7.80) to the integral in this last result, but this time with x as the parameter, when we find that

$$\frac{\partial u_2}{\partial t} = f(x, t) + \kappa \frac{\partial^2 u_2}{\partial t^2}.$$

This is the PDE in (7.72), so we have confirmed that (7.78) provides the solution of (7.72) subject to the boundary conditions (7.73) and the initial condition in (7.74).

An extension of Duhamel's principle to allow for time-dependent boundary conditions is possible, although it will not be considered here. There is, however, an extension for the nonhomogeneous heat equation that is almost immediate, and it corresponds to replacing the boundary conditions $u(0, t) = a$ and $u(L, t) = b$ by any one of the following pairs:

(i) $u(0, t) = a$ and $u_x(L, t) = b$;

(ii) $u_x(0, t) = a$ and $u(L, t) = b$;   (7.81)

(iii) $u_x(0, t) = a$ and $u_x(L, t) = b$.

When using any of the pairs of boundary conditions (i) to (iii) the following steps are involved:

Step 1   Replace the homogeneous Dirichlet boundary conditions for w in (7.76) with the corresponding homogeneous form of those chosen from (7.81).

Step 2   Determine a time-independent particular solution $u_1(x, t)$ subject to the nonhomogeneous boundary conditions chosen from (7.81).

Step 3   Find the solution from (7.78) using the form of w found in Step 1.

The proof that (7.78) gives the correct solution uses the same form of reasoning as before, so it will not be repeated. Duhamel's principle also applies to the wave equation, but it will not be discussed here, although the principle can be used to derive solution (4.40) of the nonhomogeneous one-dimensional wave equation obtained in a different way in Chapter 4.

To show how Duhamel's principle is used, and to make clear precisely how τ enters into the calculation, we solve the following simple problem.

Example 7.7. Use Duhamel's principle to solve the nonhomogeneous IVBP

$$u_t = \kappa u_{xx} + e^{-t} \sin \pi x, \quad 0 \leq x \leq 1, t \geq 0,$$

subject to the homogeneous Dirichlet boundary conditions

$$u(0, t) = u(1, t) = 0$$

and the homogeneous initial condition

$$u(x, 0) = 0, \quad 0 \leq x \leq 1.$$

Solution: By Duhamel's principle the solution of the problem is given by

$$u(x, t) = \int_0^t w(x, t : \tau) \, d\tau,$$

where $w(x, t : \tau)$ is a solution of the IVBP

$$w_t = \kappa w_{xx}, \quad 0 \leq x \leq 1, \ t \geq 0,$$

subject to the homogeneous boundary conditions

$$w(0, t : \tau) = w(1, t : \tau) = 0$$

and the nonhomogeneous initial condition

$$w(x, \tau : \tau) = e^{-\tau} \sin \pi x, \quad 0 \leq x \leq 1.$$

Separating variables in the equation for w by setting $w = X(x) T(t)$ and proceeding in the usual way leads to the result

$$\frac{T'}{\kappa T} = \frac{X''}{X} = -\lambda^2,$$

where λ^2 is a separation constant. Solving for X and T gives

$$X(x) = A_n(\tau) \cos \lambda x + B_n(\tau) \sin \lambda x \quad \text{and} \quad T(t) = \exp(-\lambda^2 \kappa t),$$

where the arbitrary multiplicative constant on the right of the expression for $T(t)$ has been absorbed into the arbitrary constants $A_n(\tau)$ and $B_n(\tau)$. If $X(x)$ is to satisfy the homogeneous Dirichlet boundary conditions it follows from $X(0) = 0$ that all the $A_n(\tau)$ are zero, and from $X(1) = 0$ that $\sin \lambda = 0$, so

$\lambda_n = n\pi$ for $n = 1, 2, \ldots$. Thus w must be of the form

$$w(x, t : \tau) = \sum_{n=1}^{\infty} B_n(\tau) \exp(-n^2 \pi^2 \kappa t) \sin n\pi x.$$

From the Duhamel principle, the function $w(x, \tau : \tau)$ must equal the nonhomogeneous term $f(x, t)$ with t replaced by τ, while in the function $\exp(-n^2 \pi^2 \kappa t)$ the variable t must also now be replaced by τ, showing that the equation determining the coefficients $B_n(\tau)$ is

$$e^{-\tau} \sin \pi x = \sum_{n=1}^{\infty} B_n(\tau) \exp(-n^2 \pi^2 \kappa \tau) \sin n\pi x.$$

Comparing coefficients of the sine functions to the left and right of this equation, or using the orthogonality of the sine functions over the interval $0 \leq x \leq L$, shows that

$$e^{-\tau} = B_1(\tau) \exp(-\pi^2 \kappa \tau), \quad \text{and} \quad B_n(\tau) = 0 \quad \text{for } n = 2, 3, \ldots.$$

Thus $B_1(\tau) = \exp[(\pi^2 \kappa - 1)\tau]$, so as all other $B_n(\tau)$ are zero, the series representation for $w(x, t : \tau)$ reduces to the single term

$$w(x, t : \tau) = \exp[(\pi^2 \kappa - 1)\tau] \exp(-\pi^2 \kappa t) \sin \pi x.$$

Hence from (7.79)

$$u(x, t) = \exp(-\pi^2 \kappa t) \sin \pi x \int_0^t \exp[(\pi^2 \kappa - 1)\tau] \, d\tau,$$

and so

$$u(x, t) = \frac{[e^{-t} - \exp(-\pi^2 \kappa t)]}{\pi^2 \kappa - 1} \sin \pi x.$$

This expression satisfies the nonhomogeneous heat equation $u_t = \kappa u_{xx} + e^{-t} \sin \pi x$, and also the homogeneous boundary conditions and the initial condition imposed on $u(x, t)$, so this confirms that $u(x, t)$ is the required solution. ∎

EXERCISES 7.6

1. Solve the IVBP

$$u_t = \kappa u_{xx} + e^{-2t} \sin \pi x \cos \pi x, \quad 0 \leq x \leq \pi, \ t \geq 0,$$

subject to the homogeneous Dirichlet boundary conditions

$$u(0, t) = u(1, t) = 0$$

and the homogeneous initial condition

$$u(x,0) = 0.$$

2. Solve the IVBP

$$u_t = \kappa u_{xx} + e^{-t} \sin 3\pi x, \quad 0 \le x \le \pi, \ t \ge 0,$$

subject to the homogeneous Dirichlet boundary conditions

$$u(0,t) = u(1,t) = 0$$

and the homogeneous initial condition

$$u(x,0) = 0.$$

3. Solve the IVBP

$$u_t = \kappa u_{xx} + e^{-t/2}(1 - \cos 2\pi x), \quad 0 \le x \le 1, \ t \ge 0,$$

subject to the homogeneous Dirichlet boundary conditions

$$u(0,t) = u(1,t) = 0$$

and the homogeneous initial condition

$$u(x,0) = 0.$$

4. Use direct calculation to confirm the validity of the extension of the Duhamel principle described in the text using any one of the pairs of boundary conditions in (7.81), and then Steps 1 through 3 following (7.81).

5. Solve the IVBP

$$u_t = \kappa u_{xx} + e^{-3t} \sin 2\pi x, \quad 0 \le x \le 1, t \ge 0,$$

subject to the nonhomogeneous Dirichlet boundary conditions

$$u(0,t) = 0 \quad \text{and} \quad u(1,t) = 1$$

and the nonhomogeneous initial condition

$$u(x,0) = \sin \frac{1}{2}\pi x.$$

Hyperbolic Systems, Riemann Invariants, Simple Waves, and Compound Riemann Problems

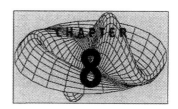

8.1 Properly Determined First-Order Systems of Equations

In previous chapters only solutions of PDEs involving a single dependent variable have been considered. The standard second-order hyperbolic, parabolic, and elliptic equations are all very important, since they occur in so many practical situations. However, it should be recognized that these linear PDEs involving a single dependent variable are very special cases of the PDEs that can arise with considering more general physical problems. When modeling different physical problems, it is frequently the case that more than one dependent variable is involved. If n dependent variables are involved, the most important case to be considered occurs when these are related by a *system* of n simultaneous *first-order* PDEs. This represents what is called a **properly determined system**, since in general the n first-order PDEs will give rise to unique solutions for the n dependent variables. It can happen, however, that n dependent variables are determined by m first-order PDEs with $n \neq m$. When this occurs the system is said to be **underdetermined** if $m < n$ and **overdetermined** if $m > n$, and although these cases are of interest we will only consider properly determined hyperbolic systems.

It may happen that all but one of the dependent variables can be eliminated from the system of PDEs, leading to a single higher-order PDE for the remaining dependent variable. In fact this occurred when deriving the standard second-order equations in Chapter 1. A typical example of how this can happen in more complicated situations was considered in Chapter 1, when small perturbations to the velocity vector and pressure were considered in the context of the two nonlinear first-order gas dynamic equations (1.69) and (1.70). This system is too complicated for it to be possible to eliminate either $\mathbf{v}$ or ρ from the pair of equations and to arrive at a higher-order PDE satisfied by only one of these dependent variables. In fact it was shown there that when sound waves are involved, the nonlinear gas dynamic equations

$$\mathbf{v}_t + \mathbf{v} \cdot \operatorname{grad} \mathbf{v} + (1/\rho)\operatorname{grad} p = \mathbf{0} \quad \text{(momentum equation)}$$

$$\rho_t + \operatorname{div}(\rho \mathbf{v}) = 0 \quad \text{(mass conservation equation)},$$

can be linearized. As a result, after the introduction of the velocity potential ϕ, which is related to the velocity vector $\mathbf{v}$ by $\mathbf{v} = \operatorname{grad} \phi$, the function ϕ was found to satisfy the linear wave equation $\phi_{tt} = c^2 \Delta \phi$, where c is the speed of sound in the gas. Thus, in this very special case, as a result of linearization two nonlinear first-order PDEs can be replaced by a single second-order PDE. Once the velocity potential has been determined, the velocity $\mathbf{v}$ can be found and through that the pressure or, equivalently, the density of the gas.

It is instructive to see how the one-dimensional wave equation $\phi_{tt} = c^2 \phi_{xx}$ can be represented as two first-order PDEs. To see this we define the two new dependent variables $u = \phi_t$ and $v = \phi_x$, as a result of which the wave equation becomes the first-order equation $u_t = c^2 v_x$. This equation is now underdetermined, because we have only one equation relating two dependent variables. To close the system and make it properly determined another equation relating u and v is needed. This is not difficult to find, because ϕ is a solution of a second-order equation so its second-order derivatives are continuous functions of the independent variables x and t. Consequently, from elementary calculus, the mixed second derivatives must be equal, so $\partial/\partial x(\phi_t) = \partial/\partial t(\phi_x)$. In terms of the new variables u and v this means that $u_x = v_t$, so the wave equation for ϕ can be replaced by the two first-order equations

$$u_t - c^2 v_x \quad \text{and} \quad v_t - u_x = 0. \tag{8.1}$$

To make further progress new notation must be introduced, and this is most easily accomplished by using matrix algebra. Let us consider the following rather general properly determined system of n first-order quasilinear PDEs

$$u_{1t} + a_{11}u_{1x} + a_{12}u_{2x} + \cdots + a_{1n}u_{nx} + b_1 = 0$$

$$u_{2t} + a_{21}u_{1x} + a_{22}u_{2x} + \cdots + a_{2n}u_{nx} + b_2 = 0$$

$$\cdot \quad \cdot \quad \cdot \quad \cdot \quad \cdot \quad \cdot \quad \cdot \quad \cdot \quad \cdot \quad \cdot \quad \cdot \quad \cdot \quad \cdot \quad \cdot \tag{8.2}$$

$$u_{nt} + a_{n1}u_{1x} + a_{n2}u_{2x} + \cdots + a_{nn}u_{nx} + b_n = 0.$$

In (8.2) x and t are the independent variables, and the n dependent variables are $u_1, u_2, \ldots, u_n$, with $u_i = u_i(x, t)$ for $i = 1, 2, \ldots, n$, and in general the a_{ij} and the b_i depend on x, t, and the dependent variables, so

$$a_{ij} = a_{ij}(x, t, u_1, u_2, \ldots, u_n) \quad \text{and} \quad b_i = b_i(x, t, u_1, u_2, \ldots, u_n). \tag{8.3}$$

The quasilinearity of system (8.2) arises from the fact that the equations are linear in the first derivatives of $u_1, u_2, \ldots, u_n$, but the coefficients of the derivatives depend on $u_1, u_2, \ldots, u_n$. Note that the system will be *semilinear* if only the $b_1, b_2, \ldots, b_n$ depend on $u_1, u_2, \ldots, u_n$, and *linear* when neither the a_{ij} nor the b_i are functions of the dependent variables.

Define the matrices **U**, **A**, and **B** by

$$
\mathbf{U} = \begin{bmatrix} u_1 \\ u_2 \\ u_3 \\ \vdots \\ u_n \end{bmatrix}, \qquad
\mathbf{U}_t = \begin{bmatrix} u_{1t} \\ u_{2t} \\ u_{3t} \\ \vdots \\ u_{nt} \end{bmatrix}, \qquad
\mathbf{U}_x = \begin{bmatrix} u_{1x} \\ u_{2x} \\ u_{3x} \\ \vdots \\ u_{nx} \end{bmatrix},
$$

$$
\mathbf{A} = \begin{bmatrix}
a_{11} & a_{12} & a_{13} & \cdots & a_{1n} \\
a_{21} & a_{22} & a_{23} & \cdots & a_{2n} \\
a_{31} & a_{32} & a_{33} & \cdots & a_{3n} \\
\vdots & \vdots & \vdots & \vdots & \vdots \\
a_{n1} & a_{n2} & a_{n3} & \cdots & a_{nn}
\end{bmatrix}, \qquad
\mathbf{B} = \begin{bmatrix} b_1 \\ b_2 \\ b_3 \\ \vdots \\ b_n \end{bmatrix}, \qquad (8.4)
$$

then the matrix form of system (8.2) becomes

$$ \mathbf{U}_t + \mathbf{A}\mathbf{U}_x + \mathbf{B} = \mathbf{0}. \qquad (8.5) $$

For example, in terms of this notation the linear system of equations in (8.1) becomes

$$ \mathbf{U}_t + \mathbf{A}\mathbf{U}_x = \mathbf{0}, \quad \text{with} \quad \mathbf{U} = \begin{bmatrix} u \\ v \end{bmatrix}, \quad \mathbf{A} = \begin{bmatrix} 0 & -c^2 \\ -1 & 0 \end{bmatrix} $$

where, of course, $\mathbf{0} = \begin{bmatrix} 0 \\ 0 \end{bmatrix}$. Another example, which will be important later, is provided by the gas dynamic equations (1.69) and (1.70) in one space dimension and time, which become

$$ \rho_t + v\rho_x + \rho v_x = 0 \quad \text{(conservation of mass)} $$
$$ v_t + vv_x + (c^2/\rho)\rho_x = 0 \quad \text{(conservation of momentum)}, \qquad (8.6) $$

where we have used the result grad $p = (dp/d\rho)\rho_x$, with $c^2 = dp/d\rho$. In matrix notation this becomes

$$ \mathbf{U}_t + \mathbf{A}\mathbf{U}_x = \mathbf{0}, \quad \text{with} \quad \mathbf{U} = \begin{bmatrix} \rho \\ v \end{bmatrix} \quad \text{and} \quad \mathbf{A} = \begin{bmatrix} v & \rho \\ c^2/\rho & v \end{bmatrix}. \qquad (8.7) $$

We saw earlier that *conservation laws* play an essential role in many physical situations, where they express the balance between the rate of outflow of a quantity from a volume V and the time rate of change of the amount of that quantity contained within V. For a scalar quantity q, such a conservation law will have the general divergence form

$$q_t + \text{div}\,\mathbf{h}(q),\tag{8.8}$$

where $\mathbf{h}(q)$ is some vector function of q, which may be either linear or non-linear. In one space dimension and time a **matrix conservation law** can be written

$$\mathbf{U}_t + (\mathbf{F}(\mathbf{U}))_x = \mathbf{0},\tag{8.9}$$

where $\mathbf{U}$ and $\mathbf{F}(\mathbf{U})$ are column vectors.

To illustrate matters, after some manipulation, including using the first equation in the second, the gas dynamic equations (8.6) become $\rho_t + (\rho v)_x = 0$ and $(\rho v)_t + (\rho v^2 + p)_x = 0$, so their matrix conservation form is

$$\mathbf{U}_t + \mathbf{F}_x = \mathbf{0}, \quad \text{with} \quad \mathbf{U} = \begin{bmatrix} \rho \\ \rho v \end{bmatrix} \quad \text{and} \quad \mathbf{F} = \begin{bmatrix} \rho v \\ \rho v^2 + p \end{bmatrix}.\tag{8.10}$$

In the gas dynamic case this makes explicit the nonlinear dependence of $\mathbf{F}$ on $\mathbf{U}$.

8.2 Hyperbolicity and Characteristic Curves

It is now necessary to define the meaning of hyperbolicity in the context of the system of Eqs. (8.2), and to do this we will need to work with the matrix form of the system given in (8.4):

$$\mathbf{U}_t + \mathbf{A}\mathbf{U}_x + \mathbf{B} = \mathbf{0}.\tag{8.11}$$

We will assume the dependent variables u_i in (8.11) are continuous, and that the elements a_{ij} and b_i are continuous functions of their arguments. Recalling how a change of variable was introduced when classifying second-order PDEs, we now change the independent variables x and t to a curvilinear coordinate ξ and a time variable t' by writing

$$\xi = \xi(x, t) \quad \text{and} \quad t' = t,\tag{8.12}$$

where for the moment ξ is an arbitrary function of x and t, or t'. There is, however, an important distinction between t and t' that will be used later. To understand this, observe that when considering partial derivatives of a function ϕ, the derivative ϕ_x is obtained from ϕ by differentiating it with respect to x

while holding t constant, while ϕ_ξ is obtained from ϕ by holding t' constant when differentiating it with respect to ξ. This means that t' varies along the curvilinear coordinate ξ, while $\partial\phi/\partial\xi$ is the differentiation of ϕ normal to the curve $\xi = $ constant.

Provided the Jacobian of the transformation (8.12) remains nonsingular, it follows from elementary calculus that the partial differential operators $\partial/\partial t$ and $\partial/\partial x$ will transform according to the laws

$$\frac{\partial}{\partial t} \equiv \frac{\partial\xi}{\partial t}\frac{\partial}{\partial\xi} + \frac{\partial t'}{\partial t}\frac{\partial}{\partial t'} \equiv \frac{\partial\xi}{\partial t}\frac{\partial}{\partial\xi} + \frac{\partial}{\partial t'}$$

$$\frac{\partial}{\partial x} \equiv \frac{\partial\xi}{\partial x}\frac{\partial}{\partial\xi} + \frac{\partial t'}{\partial x}\frac{\partial}{\partial t'} \equiv \frac{\partial\xi}{\partial x}\frac{\partial}{\partial\xi}.$$

Using these operators in (8.11) it becomes

$$\mathbf{U}_{t'} + \xi_t\mathbf{U}_\xi + \xi_x\mathbf{A}\mathbf{U}_\xi + \mathbf{B} = 0,$$

which can be written

$$\mathbf{U}_{t'} + (\xi_t\mathbf{I} + \xi_x\mathbf{A})\mathbf{U}_\xi + \mathbf{B} = 0, \tag{8.13}$$

where $\mathbf{I}$ is the $n \times n$ unit matrix.

If (8.13) is considered to be an algebraic equation relating $\mathbf{U}_{t'}$ and $\mathbf{U}_\xi$, we see immediately that $\mathbf{U}_\xi$ can only be determined in terms of the time derivative $\mathbf{U}_{t'}$ if the matrix $\xi_t\mathbf{I} + \xi_x\mathbf{A}$ is nonsingular. So far the function $\xi(x, t)$ has been arbitrary, but from this point onward we will take it to be a specific function $\xi(x, t) \equiv \varphi(x, t)$ with the property that the determinant of the matrix coefficient of $\mathbf{U}_\xi$ is

$$|\varphi_t\mathbf{I} + \varphi_x\mathbf{A}| = 0. \tag{8.14}$$

The significance of this result is that across curvilinear coordinate lines $\varphi = $ constant, although we have postulated that $\mathbf{U}$ is continuous, its derivative $\mathbf{U}_\xi$ may be indeterminate. In these circumstances it is possible for the derivative $\mathbf{U}_\xi$ to experience a *jump discontinuity* across the curvilinear coordinate line $\varphi = $ constant, in which case each derivative of u_i with respect to φ may also experience a jump discontinuity. As the variables u_i are assumed to be continuous, and a_{ij} and b_i are assumed to be continuous functions of their arguments, it follows that a_{ij} and b_i must be continuous across $\varphi = $ constant.

We now consider how matrix equation (8.13) changes as it crosses a curvilinear coordinate line $\varphi = k$ (constant). It is at this stage that the difference between derivatives with respect to x and t and φ and t' become significant. Whereas $\partial/\partial x$ involves differentiation with respect to x while holding t constant, $\partial/\partial\varphi$ indicates differentiation across the curve $\varphi = k$ while holding t' constant, where t' now changes along $\varphi = k$. Thus, while the derivative $\mathbf{U}_\varphi$ may be discontinuous across $\varphi = k$, the derivative $\mathbf{U}_{t'}$ will be continuous across $\varphi = k$.

Consequently, if we difference matrix equation (8.13) across the curvilinear coordinate line $\varphi = k$ at some point P on the line, it follows that the difference of $\mathbf{U}_{t'}$ must be zero, so as $\mathbf{A}$ and $\mathbf{B}$ are continuous we are left with the result

$$(\varphi_t \mathbf{I} + \varphi_x \mathbf{A})_P [[\mathbf{U}_\varphi]]_P = \mathbf{0}. \tag{8.15}$$

Here $[[\mathbf{U}_\varphi]]_P = (\mathbf{U}_{\varphi-})_P - (\mathbf{U}_{\varphi+})_P$ denotes the jump in $\mathbf{U}_\varphi$ across $\varphi = k$ at point P, while $\varphi+$ and $\varphi-$ indicate, respectively, that $\mathbf{U}_\varphi$ is to be evaluated immediately ahead of and immediately behind the curve $\varphi = k$. This is true at all points P on $\varphi =$ constant, so from now on the suffix P will be omitted.

Matrix equation (8.15) is a homogeneous system of algebraic equations relating the n elements of $[[\mathbf{U}_\varphi]]$, so if it is not to have the trivial solution $[[\mathbf{U}_\varphi]] = \mathbf{0}$, the determinant of its coefficients must vanish, leading to the condition

$$|\varphi_t \mathbf{I} + \varphi_x \mathbf{A}| = 0. \tag{8.16}$$

Along the lines $\varphi =$ constant we have, by differentiation,

$$\frac{\partial \varphi}{\partial t} + \frac{\partial \varphi}{\partial x} \frac{dx}{dt} = 0,$$

so the slope of these lines is

$$\frac{dx}{dt} = -\frac{\partial \varphi}{\partial t} \bigg/ \frac{\partial \varphi}{\partial x} \equiv \lambda \text{ (say).} \tag{8.17}$$

As φ_t and φ_x are scalars, and for the moment we will assume the curvilinear coordinate function φ is such that $\varphi_t \neq 0$, determinant (8.16) can be rewritten as $|\mathbf{A} + (\varphi_t/\varphi_x)\mathbf{I}| = 0$. After using (8.17) this determinant becomes

$$|\mathbf{A} - \lambda \mathbf{I}| = 0. \tag{8.18}$$

This shows the new curvilinear coordinate variable φ, which is a solution of (8.17), must be such that $\lambda = -(\varphi_t/\varphi_x)$ is an eigenvalue of the coefficient matrix $\mathbf{A}$. Using this result in (8.15) gives

$$(\mathbf{A} - \lambda \mathbf{I})[[\mathbf{U}_\varphi]] = \mathbf{0}. \tag{8.19}$$

This result is important, because it shows that for each eigenvalue λ, the column vector $[[\mathbf{U}_\varphi]]$ must be proportional to the corresponding eigenvector of $\mathbf{A}$. Consequently, for any eigenvalue λ, the ith element $[[\partial u_i/\partial \varphi]]$ of $[[\mathbf{U}_\varphi]]$ must be proportional to the ith element of the corresponding eigenvector.

As $\mathbf{A}$ is an $n \times n$ matrix it will have n eigenvalues $\lambda_1, \lambda_2, \ldots, \lambda_n$, which may be either real or complex values. We will be concerned only with the case that all of the λ values are real and distinct, when integration of (8.17) will give rise to n distinct families of real curves $C^{(1)}, C^{(2)}, \ldots, C^{(n)}$ defined by

$$C^{(i)}: dx/dt = \lambda_i, \quad \text{for } i = 1, 2, \ldots, n. \tag{8.20}$$

From (8.20) it is seen that when x is a space dimension (a length) and t is the time, as the λ values are real they will all have the dimensions of a *speed*. Thus any element of $\mathbf{U}$ moving along a curve $C^{(i)}$ will move with speed λ_i. Any one of the families of curves $C^{(i)}$ may be taken to be the curvilinear coordinate $\varphi = $ constant, and if the coordinate variable $\varphi(x, t)$ is required, it can be found from (8.17) once an initial condition has been given for φ. As φ serves only as a *coordinate variable*, the initial condition for φ can be any differentiable monotonic function of x, since this will ensure that initially (at least) the family of curves $\varphi = k$ (constant) will all be distinct, since each will correspond to a different value of k. The simplest initial condition with this property is $\varphi(x, 0) = x - x_0$, because this function is monotonic and differentiable, and the curve through x_0 corresponds to $\varphi = 0$, while φ is positive ahead of x_0 and negative behind it. Furthermore, with this choice for the parametrization of φ, the Jacobian of (8.12) will be nonvanishing initially, as is necessary if the change of variable is to give a one-to-one relationship between the (x, t) and (φ, t') coordinate systems. When the eigenvalues $\lambda_1, \lambda_2, \ldots, \lambda_n$ are all real and distinct, and there are n corresponding linearly independent **right eigenvectors** denoted by $\mathbf{r}^{(i)}$, the eigenvectors will satisfy the algebraic equation

$$\mathbf{A}\mathbf{r}^{(i)} = \lambda_i \mathbf{r}^{(i)} \quad \text{for } i = 1, 2, \ldots, n. \tag{8.21}$$

When this occurs system (8.11) is said to be **totally hyperbolic**. The corresponding families of curves $C^{(i)}$ are then called the n families of **characteristic curves** of the system. Later it will be convenient to make use of the **left eigenvectors** of $\mathbf{A}$ defined by the result

$$\mathbf{l}^{(i)}\mathbf{A} = \lambda_i \mathbf{l}^{(i)} \quad \text{for } i = 1, 2, \ldots, n, \tag{8.22}$$

which follows from (8.21) by forming its transpose.

If the eigenvalues are all complex the system will be **elliptic**, while if some are real and others are complex the system will be of **mixed hyperbolic and elliptic** type. The system will be **parabolic** if the eigenvalues are all real, but there are fewer than n linearly independent eigenvectors. These cases will not be considered here.

To check that this definition of hyperbolicity is in agreement with the definition used in Chapter 3, we consider two typical cases. Section 8.1 showed how the wave equation $\phi_{tt} = c^2 \phi_{xx}$ could be written in the form $\mathbf{U}_t + \mathbf{A}\mathbf{U}_x = 0$

by defining

$$\mathbf{U} = \begin{bmatrix} u \\ v \end{bmatrix} \quad \text{and} \quad \mathbf{A} = \begin{bmatrix} 0 & -c^2 \\ -1 & 0 \end{bmatrix}, \quad \text{with} \quad u = \phi_t \quad \text{and} \quad v = \phi_x.$$

In this case the eigenvalues of $|\mathbf{A} - \lambda \mathbf{I}| = 0$ are the roots λ_- and λ_+ of the characteristic equation $\lambda^2 - c^2 = 0$, showing that $\lambda_\pm = \pm c$. These eigenvalues are real and distinct, and it is easily checked that there are two linearly independent right eigenvectors (in this case they must exist), so the system is totally hyperbolic. This is, of course, equivalent to saying that the wave equation is totally hyperbolic, and since $\mathbf{A}$ does not depend on x and t the system, and hence the wave equation, is seen to be unconditionally hyperbolic. The two families of characteristic curves $C^{(1)}$ and $C^{(2)}$ follow by integrating

$$C^{(1)}: \frac{dx}{dt} = \lambda_- = -c \quad \text{and} \quad C^{(2)}: \frac{dx}{dt} = \lambda_+ = c,$$

leading to

$$C^{(1)}: x + ct = k_1 \quad \text{and} \quad C^{(2)}: x - ct = k_2,$$

with k_1 and k_2 being arbitrary integration constants. These results are in agreement with those found in Chapters 3 and 4 and, in particular, with those of Example 3.1.

As another example we consider the gas dynamic system in (8.7). The characteristic equation is given by the determinant

$$|\mathbf{A} - \lambda \mathbf{I}| = 0 \quad \text{or, equivalently, by} \quad \begin{vmatrix} v - \lambda & \rho \\ c^2/\rho & v - \lambda \end{vmatrix} = 0.$$

This has the eigenvalues $\lambda_1 = v - c$ and $\lambda_2 = v + c$, and the corresponding right eigenvectors

$$\mathbf{r}^{(1)} = \begin{bmatrix} 1 \\ -c/\rho \end{bmatrix} \quad \text{and} \quad \mathbf{r}^{(2)} = \begin{bmatrix} 1 \\ c/\rho \end{bmatrix}. \tag{8.23}$$

Thus this system is also totally hyperbolic, and its two families of characteristic curves are determined by integrating

$$C^{(1)}: dx/dt = v - c \quad \text{and} \quad C^{(2)}: dx/dt = v + c. \tag{8.24}$$

This last result shows that weak disturbances in a gas travel with speeds $\pm c$ relative to the gas speed v, where c is the speed of sound in the gas. We will see later that the situation will be entirely different when $\mathbf{U}$, instead of $\mathbf{U}_\varphi$, is discontinuous across some curve in the (x, t) plane, because then a shock wave will exist.

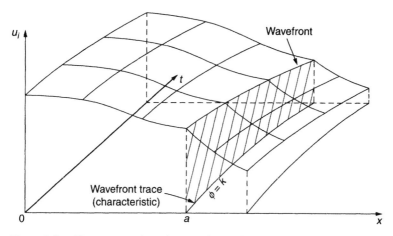

Figure 8.1   The propagation of a wavefront along a characteristic.

In summary, we have found that if matrix **A** of system (8.11) can be shown to have n real and distinct eigenvalues, to which there correspond n linearly independent eigenvectors, the system will be *totally hyperbolic*. It must, however, be remembered that in general $\mathbf{A} = \mathbf{A}(x, t, \mathbf{U})$, so it is possible for the classification of system (8.11) to depend on the solution at a point (x_0, t_0) in the (x, t) plane.

A characteristic curve $\varphi = k$, across which **U** is continuous though $\mathbf{U}_\varphi$ experiences a jump, will be called a **wavefront**. Thus a wavefront is an identifiable disturbance in the form of a cusp on the solution surface that propagates along the characteristic $\varphi = k$ as time increases. The characteristic along which such a wavefront propagates will be called the **wavefront trace**, since it is the projection of the wavefront onto the (x, t) plane. The relationship between the wavefront trace and the wavefront is illustrated in Fig. 8.1. This diagram shows the typical behavior of the ith element $u_i(x, t)$ of the solution vector **U** of system (8.11), corresponding to the arbitrary initial condition $\mathbf{U}(x, 0) = \mathbf{U}_0(x)$, with the ith element of the initial vector $\mathbf{U}_0$ denoted by $u_0^{(i)}$. A propagating disturbance with the property that **U** is continuous across the wavefront, but $\mathbf{U}_\varphi$ is discontinuous, will be called a **weak discontinuity**. This is to distinguish it from the case of a **shock** when **U** itself is discontinuous, which will be called a **strong discontinuity**.

The simplest example of a wavefront occurs when a disturbance described by a hyperbolic system of type (8.11) propagates into a region in which $\mathbf{U} = \mathbf{U}_0$, with $\mathbf{U}_0$ being a constant vector. In this case the surface to the right of the wavefront in Fig. 8.1 will be parallel to the (x, t) plane. As Eq. (8.11) must hold on both sides of the wavefront, and in general $\mathbf{B} = \mathbf{B}(\mathbf{U})$, the constant solution $\mathbf{U}_0$ ahead of the wavefront must satisfy the compatibility condition $\mathbf{B}(\mathbf{U}_0) = \mathbf{0}$.

In physical problems where a wavefront propagates into a region in which $U = U_0$ is constant, it is usual to say that the wave propagates into a **constant state**. This is because the elements of vector U describe the values of the physical quantities entering into the PDE as dependent variables at position x and time t, and when U_0 is constant the physical state characterized by U_0 is one in which all the variables are constant.

The fact that along a characteristic $\varphi = $ constant the weak discontinuity vector $[[U_\varphi]]$ is proportional to the corresponding right eigenvector can be used to provide qualitative information about the nature of a propagating disturbance. To see this, consider a weak discontinuity advancing into a constant state when, for example, the ith element of $[[\partial u_i/\partial \varphi]] = (\partial u_i/\partial \varphi)_- - (\partial u_i/\partial \varphi)_+ = (\partial u_i/\partial \varphi)_-$, because ahead of the wave $(\partial u_i/\partial \varphi)_+ = 0$ in the constant state.

Consequently, in the disturbed region immediately behind the advancing wavefront, the total derivative du_i of u_i reduces to

$$du_i = (\partial u_i/\partial \varphi)_- d\varphi.$$

Letting the ith element of the eigenvector corresponding to φ be $r_i^{(\varphi)}$, the fact that $[[U_\varphi]]$ is proportional to the right eigenvector $r^{(\varphi)}$ means that $du_i = \alpha r_i^{(\varphi)} d\varphi$, with α some constant of proportionality. It is convenient to choose α so the first element of $r^{(\varphi)}$ becomes unity, because then

$$du_j = r_j^{(\varphi)} du_1, \quad \text{for } j = 1, 2, \ldots, n. \tag{8.25}$$

To see how this works, let us examine the gas dynamic equations that gave rise to results (8.23) and (8.24). There, the first and second elements of the two-element column vector U are ρ and v, and from the right eigenvectors in (8.23), which are already normalized so their first elements are unity, we see that for $r^{(1)}$ the corresponding elements are 1 and $-c/\rho$, while for $r^{(2)}$ they are 1 and c/ρ. Thus if φ is the characteristic $C_0^{(1)}$ bounding the constant state in which $c = c_0$ and $v = v_0$, the change is

$$dv = -(c_0/\rho_0)d\rho, \quad \text{with} \quad C_0^{(1)}: x = x_0 + (v_0 - c_0)t,$$

while if φ is the characteristic $C_0^{(2)}$ bounding the constant state the change is

$$dv = (c_0/\rho_0)d\rho \quad \text{with} \quad C_0^{(2)}: x = x_1 + (v_0 + c_0)t,$$

with x_0 and x_1 being arbitrary integration constants.

These results show qualitatively how the speed v changes with the density ρ as either a $C_0^{(1)}$ characteristic or a $C_0^{(2)}$ characteristic bounding a constant state is crossed. This approach extends immediately to more general hyperbolic systems, and when little is known of their behavior it can provide useful insight into their qualitative properties.

Writing (8.25) in matrix form, and using the defining relationship $Ar = \lambda r$, shows that immediately adjacent to a constant state $U = U_0$, $(A_0 - \lambda_0 I)dU = 0$, where $A_0 = A(U_0)$ and $\lambda_0 = \lambda(U_0)$. Comparing this result with system (8.11) yields the following rule, which is sometimes useful when exploring changes that take place in dependent variables across a characteristic bounding a constant state U_0.

Changes in Elements of dU Across a Characteristic Bounding a Constant State U_0. To find the relationship between the elements du_i of dU for system (8.11) in the disturbed region immediately adjacent to a constant state region $U = U_0$ bounded by a characteristic corresponding to the eigenvalue λ, proceed as follows. Disregard the vector B, replace the undifferentiated variables by their constant state values, and make the replacements $\frac{\partial(\cdot)}{\partial t} \to -\lambda d(\cdot)$ and $\frac{\partial(\cdot)}{\partial x} \to d(\cdot)$.

When this rule is applied to the scalar equations (8.6) which formed the basis of the previous example, we obtain

$$-\lambda \, d\rho + v_0 \, d\rho + \rho_0 \, dv = 0 \quad \text{and} \quad -\lambda \, dv + v_0 \, du + \left(c_0^2/\rho_0\right) d\rho = 0.$$

Then, as $C_0^{(1)}$ corresponds to $\lambda_0^{(1)} = v_0 - c_0$, and $C_0^{(2)}$ to $\lambda_0^{(2)} = v_0 + c_0$, across $C_0^{(1)}$ we have $dv = -(c_0/\rho_0) \, d\rho$ while across $C_0^{(2)}$ we have $dv = (c_0/\rho_0) \, d\rho$.

EXERCISES 8.2

1. By introducing suitable new variables, write the Tricomi equation $xu_{tt} + u_{xx} = 0$ in the matrix form $U_t + AU_x = 0$, and hence find the eigenvalues and eigenvectors of A. Use the result to classify the type of equation involved in terms of the variable x.

2. The one-dimensional flow of water in the *shallow water approximation*, more properly called the *long wave approximation*, is described by the equations

$$u_t + uu_x + 2cc_x - H_x = 0 \quad \text{and} \quad 2c_t + 2uc_x + cu_x = 0.$$

Here the x axis is located in the equilibrium surface of the water, t is the time, u is the horizontal speed of flow in the x direction, c is the speed of propagation of a surface disturbance, $H(x) = gY(x)$ with g being the acceleration due to gravity, and $y + Y(x) = 0$ is the equation of the river or sea bed. Write this system in matrix form with U being a column vector with elements u and c, and show the system is unconditionally hyperbolic.

3. Write the equations in Exercise 2 in the conservation form

$$\frac{\partial F}{\partial t} + \frac{\partial G}{\partial x} = 0,$$

stating clearly the form of vectors F and G.

4. The one-dimensional unsteady equations of a polytropic gas with u being the gas speed in the x direction, ρ the gas density, $p = p(\rho, S)$ the gas pressure, S a thermodynamical quantity called the entropy, and $c^2 = \partial p / \partial \rho$ the square of the speed of sound in the gas are:

$$\rho_t + u\rho_x + \rho u_x = 0,$$
$$u_t + uu_x + (1/\rho)(\partial p/\partial \rho)\rho_x + (1/\rho)(\partial p/\partial S)S_x = 0,$$
$$S_t + uS_x = 0.$$

Write the system in the matrix form (8.11), and hence find the characteristic wave speeds determined by the eigenvalues of $\mathbf{A}$ and the corresponding right eigenvectors. Write down, but do not attempt to integrate, the equations determining the families of characteristic curves. Relate $d\rho$, du, and dS across a characteristic adjacent to a constant state ρ_0, u_0, and S_0.

5. Steady two-dimensional irrotational isentropic gas flow is governed by the equations

$$(c^2 - u^2)u_x - uv(u_y + v_x) + (c^2 - v^2)v_y = 0 \quad \text{and} \quad v_x - u_y = 0,$$

where c is the speed of sound in the gas and u and v are the components of the gas velocity vector $\mathbf{q} = u\mathbf{i} + v\mathbf{j}$, with $\mathbf{i}$ and $\mathbf{j}$ being unit vectors in the x and y directions. By defining the **Mach number** M of the flow as $M = (u^2 + v^2)^{1/2}/c$, show the flow is only totally hyperbolic when it is supersonic $(M > 1)$. Comment on flows for which $M = 1$ and $0 < M < 1$.

8.3 Riemann Invariants

Many important practical problems are described by totally hyperbolic systems involving two quasilinear first-order partial differential equations with independent variables x and y, one of which may or may not be the time, and two dependent variables, u_1 and u_2. From among this class of problems are to be found systems with the specially simple form

$$\frac{\partial u_1}{\partial x} + a_{11}\frac{\partial u_1}{\partial y} + a_{12}\frac{\partial u_2}{\partial y} = 0,$$

$$\frac{\partial u_2}{\partial x} + a_{21}\frac{\partial u_1}{\partial y} + a_{22}\frac{\partial u_2}{\partial y} = 0,$$

(8.26)

which contains no terms involving the undifferentiated dependent variables u_1 and u_2, and where the coefficients a_{ij} may be *explicit* functions of u_1 and u_2, but not of x and y, which only occur *implicitly* through their occurrence in $u_1(x, y)$ and $u_2(x, y)$.

Hyperbolic systems of this type are said to be **reducible**, because by inter-changing the roles of dependent and independent variables it is possible to convert them to an equivalent linear system. However, this reduction, involving what is called the **hodograph transformation**, comes at a price, because it is difficult to determine the way the boundary transforms. Although this approach is interesting, it will not be pursued here, and instead we will introduce the concept of *Riemann invariants*. These invariant quantities provide valuable insight into the nature of the solution of such systems, and when the necessary integrations can be performed they can also lead to explicit solutions.

In what is to follow we will consider reducible systems of the type shown in (8.26) where, because we will be concerned with time-dependent wave propagation, the independent variables will be taken to be x (a space dimension) and the time t,

$$\frac{\partial u_1}{\partial t} + a_{11}\frac{\partial u_1}{\partial x} + a_{12}\frac{\partial u_2}{\partial x} = 0,$$

$$\frac{\partial u_2}{\partial t} + a_{21}\frac{\partial u_1}{\partial x} + a_{22}\frac{\partial u_2}{\partial x} = 0. \tag{8.27}$$

This system is quasilinear when, as is usually the case, $a_{ij} = a_{ij}(u_1, u_2)$, although if the coefficients a_{ij} are absolute constants the system reduces to a homogeneous system of first-order PDEs for u_1 and u_2.

If we define the matrices

$$\mathbf{A} = \begin{bmatrix} a_{11} & a_{12} \\ a_{21} & a_{22} \end{bmatrix} \quad \text{and} \quad \mathbf{U} = \begin{bmatrix} u_1 \\ u_2 \end{bmatrix}, \tag{8.28}$$

system (8.27) becomes

$$\mathbf{U}_t + \mathbf{A}\mathbf{U}_x = \mathbf{0}, \tag{8.29}$$

and the fact that it is assumed to be totally hyperbolic means that the two eigenvalues of the characteristic determinant

$$|\mathbf{A} - \lambda \mathbf{I}| = 0 \tag{8.30}$$

will have the two distinct eigenvalues $\lambda^{(i)}$, for $i = 1, 2$, and two corresponding *left* eigenvectors $\mathbf{l}^{(i)} = 1, 2$, satisfying the algebraic equation

$$\mathbf{l}^{(i)}\mathbf{A} = \lambda^{(i)}\mathbf{l}^{(i)}, \quad i = 1, 2. \tag{8.31}$$

We now make use of the left eigenvectors of $\mathbf{A}$ by multiplying (8.29) from the left by $\mathbf{l}^{(i)}$ (premultiplying it by $\mathbf{l}^{(i)}$), and using (8.31) to replace $\mathbf{l}^{(i)}\mathbf{A}$ by $\lambda^{(i)}\mathbf{l}^{(i)}$ obtain

$$\mathbf{l}^{(i)}(\mathbf{U}_t + \lambda^{(i)}\mathbf{U}_x) = 0, \quad \text{for } i = 1, 2. \tag{8.32}$$

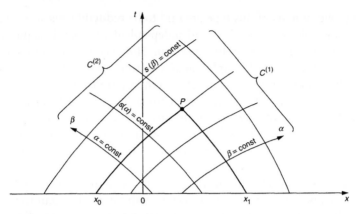

Figure 8.2 $C^{(1)}$ and $C^{(2)}$ families of characteristics as functions of α and β.

Examination of the bracketed term in (8.32) shows it is the *directional derivative* of **U** with respect to time t along the family of characteristic curves $C^{(i)}$ defined by

$$C^{(i)}: dx/dt = \lambda^{(i)}, \quad \text{for } i = 1, 2. \tag{8.33}$$

If we now denote differentiation with respect to time along the $C^{(1)}$ family of characteristics by $d/d\alpha$ and along the $C^{(2)}$ family of characteristics by $d/d\beta$, results (8.32) become *along the $C^{(1)}$ family of characteristics*

$$\mathbf{l}^{(1)} \frac{d\mathbf{U}}{d\alpha} = 0, \tag{8.34}$$

and *along the $C^{(2)}$ family of characteristics*

$$\mathbf{l}^{(2)} \frac{d\mathbf{U}}{d\beta} = 0. \tag{8.35}$$

This situation is illustrated in Fig. 8.2 where the family of $C^{(1)}$ characteristics moves to the right as α increases, while the family of $C^{(2)}$ characteristics moves to the left as β increases.

The left eigenvectors of **A** are two element row vectors with $\mathbf{l}^{(i)} = [l_1^{(i)}, l_2^{(i)}]$, with $i = 1, 2$, so (8.34) and (8.35) can be written as

$$l_1^{(1)} \frac{du_1}{d\alpha} + l_2^{(i)} \frac{du_2}{d\alpha} = 0 \quad \text{along } C^{(1)} \text{ characteristics}, \tag{8.36}$$

and

$$l_1^{(2)} \frac{du_1}{d\beta} + l_2^{(2)} \frac{du_2}{d\beta} = 0 \quad \text{along } C^{(2)} \text{ characteristics}. \tag{8.37}$$

The coefficients a_{ij} of system (8.27) are, by supposition, dependent only on u_1 and u_2, so its eigenvectors $\mathbf{l}^{(i)}$ can also only be functions of u_1 and u_2. In general the two ordinary differential equations (8.36) and (8.37) cannot be integrated as they stand, because in the first equation $l_1^{(1)}$ may be a function of both u_1 and u_2, instead of being only a function of u_1, and $l_2^{(1)}$ may also be a function of u_1 and u_2 instead of being only a function of u_2, with similar results for the second equation. However, it is known from the study of first-order ordinary differential equations that integrating factors $\mu_1(u_1, u_2)$ and $\mu_2(u_1, u_2)$ always exist, with the property that the equations can be integrated after the first equation has been multiplied by μ_1 and the second by μ_2. Unfortunately, the theory that ensures the existence of such integrating factors does not show how they can be determined in particular cases. However, in most applications this is unimportant because the factors are usually so obvious that they can be found by inspection. This will be illustrated later in an example.

So, assuming the integrating factors are known, it follows from (8.36) and (8.37) that after multiplication by μ_1 and μ_2, respectively, integration gives *along $C^{(1)}$ characteristics*

$$\int \mu_1 l_1^{(1)} du_1 + \int \mu_1 l_2^{(1)} du_2 = r(\beta), \qquad (8.38)$$

and *along $C^{(2)}$ characteristics*

$$\int \mu_2 l_1^{(2)} du_1 + \int \mu_2 l_2^{(2)} du_2 = s(\alpha), \qquad (8.39)$$

where r and s are called **Riemann invariants** of the system and, until initial values are specified for u_1 and u_2, they are arbitrary functions of their respective arguments β and α. The characteristics $C^{(1)}$ and $C^{(2)}$, along which these respective expressions are constant, follow by integration of the equations in (8.33). Note that to find the characteristics it is necessary to know the solution, but in general this is not known. However, in the next section, we will see that in certain important and useful special cases the families of characteristics can be determined, enabling the system of equations to be solved analytically.

To understand the importance of *Riemann invariants*, and how they can lead to a solution, it is necessary to reexamine Fig. 8.2. Consider the $C^{(1)}$ characteristic originating from the initial line at the point $(x_0, 0)$, the $C^{(2)}$ characteristic originating from the initial line at the point $(x_1, 0)$, and their intersection at point P located at the point (x_P, t_P) in the (x, t) plane. We see from (8.38) that $r(\beta)$ is a function of u_1 and u_2, so let it be denoted by $R(u_1, u_2)$ and, similarly, let $s(\alpha)$, which is some other function of u_1 and u_2, be denoted by $S(u_1, u_2)$.

Then, as β only varies along $C^{(2)}$ characteristics, and so is constant along $C^{(1)}$ characteristics, it follows that the combination of dependent variables u_1 and u_2 forming the function $R(u_1, u_2)$ must be **invariant** (constant) along the $C^{(1)}$ characteristic through P. Its value will be

$$R(u_1, u_2) = R(\bar{u}_1(x_0), \bar{u}_2(x_0)), \tag{8.40}$$

where $u_1(x, 0) = \bar{u}_1(x_0)$ and $u_2(x, 0) = \bar{u}_2(x_0)$ are the *initial* values of u_1 and u_2 on the x axis. A corresponding argument shows that along the $C^{(2)}$ characteristic through P, along which only α varies,

$$S(u_1, u_2) = S(\bar{u}_1(x_1), \bar{u}_2(x_1)). \tag{8.41}$$

The expressions on the left of (8.40) and (8.41) are known functions of u_1 and u_2, and each is equal to the constant value on the right. Thus Eqs. (8.40) and (8.41) form two implicit expressions involving u_1 and u_2, so as the values of u_1 and u_2 at P must satisfy both (8.40) and (8.41), if these can be solved for u_1 and u_2 these will be the values of the solution of the system at the point P. Having found the solution at this point, it can be found at other points in the (x, t) plane by changing the values of x_0 and x_1, so in principle the solution can be found. The following examples illustrate the arguments involved in two very different cases, one involving a linear system and the other a quasi-linear system.

Example 8.1. Use Riemann invariants to solve the system of linear equations

$$\frac{\partial u_1}{\partial t} + \frac{\partial u_2}{\partial x} = 0 \quad \text{and} \quad \frac{\partial u_2}{\partial t} + 4\frac{\partial u_1}{\partial x} = 0,$$

subject to the initial conditions

$$u_1(x, 0) = \sinh 2x, \quad u_2(x, 0) = \cosh 2x.$$

Solution: In matrix form the system becomes $\mathbf{U}_t + \mathbf{A}\mathbf{U}_x = \mathbf{0}$, where

$$\mathbf{U} = \begin{bmatrix} u_1 \\ u_2 \end{bmatrix} \quad \text{and} \quad \mathbf{A} = \begin{bmatrix} 0 & 1 \\ 4 & 0 \end{bmatrix}.$$

The eigenvalues of $\mathbf{A}$ are $\lambda_1 = 2$ and $\lambda_2 = -2$, and the corresponding left eigenvectors, which satisfy the homogeneous system of algebraic equations $\mathbf{l}^{(i)}\mathbf{A} = \lambda_i\mathbf{l}^{(i)}$ with $i = 1, 2$, are $\mathbf{l}^{(1)} = [2, 1]$ and $\mathbf{l}^{(2)} = [2, -1]$, so $l_1^{(1)} = 2, l_2^{(1)} = 1, l_1^{(2)} = 2$, and $l_2^{(2)} = -1$.

Thus the two characteristic curves $C^{(1)}$ and $C^{(2)}$ are given by

$$C^{(1)}: dx/dt = 2, \quad \text{with the solution } x = x_0 + 2t, \quad \text{so } x_0 = x - 2t$$

and

$$C^{(2)}: dx/dt = -2, \quad \text{with the solution } x = x_1 - 2t, \quad \text{so } x_1 = x + 2t.$$

Equations (8.38) and (8.39) determining the Riemann invariants become

$$2\frac{du_1}{d\alpha} + \frac{du_2}{d\alpha} = 0 \quad \text{along } C^{(1)} \text{ characteristics}$$

and

$$2\frac{du_1}{d\beta} - \frac{du_1}{d\beta} = 0 \quad \text{along } C^{(2)} \text{ characteristics.}$$

No integrating factors are required in this case and direct integration gives

$$2u_1 + u_2 = r(\beta) \quad \text{along } C^{(1)} \text{ characteristics}$$

and

$$2u_1 - u_2 = s(\alpha) \quad \text{along } C^{(2)} \text{ characteristics.}$$

Using the initial conditions we find that at $(x_0, 0)$

$$r(\beta) = \text{constant} = 2u_1(x_0, 0) + u_2(x_0, 0) = 2\sinh 2x_0 + \cosh 2x_0,$$

while at $(x_1, 0)$

$$s(\alpha) = \text{constant} = 2u_1(x_1, 0) - u_2(x_1, 0) = 2\sinh 2x_1 - \cosh 2x_1.$$

In terms of the Riemann invariants

$$u_1(x, t) = \frac{1}{4}(r(\beta) + s(\alpha)) \quad \text{and} \quad u_2(x, t) = \frac{1}{2}(r(\beta) - s(\alpha)),$$

so using the expressions for $r(\beta)$ and $s(\alpha)$ gives

$$u_1(x, t) = \frac{1}{2}[\sinh 2(x - 2t) + \sinh 2(x + 2t)]$$
$$+ \frac{1}{4}[\cosh 2(x - 2t) - \cosh 2(x + 2t)]$$

$$u_2(x, t) = [\sinh 2(x - 2t) - \sinh 2(x + 2t)]$$
$$+ \frac{1}{2}[\cosh 2(x - 2t) + \cosh 2(x + 2t)].$$

Note that when solving this system we have been solving the wave equation, because eliminating $u_2(x, t)$ between the original PDEs gives $(u_1)_{tt} = 4(u_1)_{xx}$, showing that u_1 satisfies the wave equation, with a similar result holding for $u_2(x, t)$. ∎

Example 8.2. Find the Riemann invariants belonging to the gas dynamic equations in (8.6), which in matrix form become $U_t + AU_x = 0$, where

$$U = \begin{bmatrix} \rho \\ v \end{bmatrix} \quad \text{and} \quad A = \begin{bmatrix} v & \rho \\ c^2/\rho & v \end{bmatrix},$$

and using the gas law $p = k\rho^\gamma$, find ρ and v in terms of the Riemann invariants.

Solution: It was shown in Section 8.2 that $\lambda_1 = v - c$ and $\lambda_2 = v + c$, so the families of characteristic curves $C^{(1)}$ and $C^{(2)}$ are determined by

$$C^{(1)}: dx/dt = v - c \quad \text{and} \quad C^{(2)}: dx/dt = v + c.$$

A simple calculation shows that the corresponding left eigenvectors are

$$1^{(1)} = [1, -\rho/c] \quad \text{and} \quad 1^{(2)} = [1, \rho/c].$$

Thus Eqs. (8.38) and (8.39) determining the Riemann invariants become

$$\frac{d\rho}{d\alpha} - \frac{\rho}{c}\frac{dv}{d\alpha} = 0 \quad \text{along } C^{(1)} \text{ characteristics}$$

and

$$\frac{d\rho}{d\beta} + \frac{\rho}{c}\frac{dv}{d\beta} = 0 \quad \text{along } C^{(2)} \text{ characteristics}.$$

These equations cannot be integrated as they stand because of the multiplier ρ/c present in their second terms, where $c = c(\rho)$. However, multiplication by the integrating factor c/ρ resolves the problem, because the equations then become

$$\frac{c}{\rho}\frac{d\rho}{d\alpha} - \frac{dv}{d\alpha} = 0 \quad \text{along } C^{(1)} \text{ characteristics}$$

and

$$\frac{c}{\rho}\frac{d\rho}{d\beta} + \frac{dv}{d\beta} = 0 \quad \text{along } C^{(2)} \text{ characteristics}.$$

Thus, after integration, these become

$$\int \frac{c}{\rho}d\rho - v = r(\beta) \quad \text{along } C^{(1)} \text{ characteristics}$$

and

$$\int \frac{c}{\rho}d\rho + v = s(\alpha) \quad \text{along } C^{(2)} \text{ characteristics}.$$

Using the gas law $p = k\rho^{\gamma}$ shows that $c^2 = dp/d\rho = k\gamma\rho^{\gamma-1}$, and so

$$\frac{2c}{\gamma - 1} - v = r(\beta) \quad \text{and} \quad \frac{2c}{\gamma - 1} + v = s(\alpha).$$

So, expressing v and ρ in terms of the Riemann invariants we find that

$$c = \frac{1}{4}(\gamma - 1)(s + r) \quad \text{and} \quad v = \frac{1}{2}(s - r).$$

■

EXERCISES 8.3

1. Use Riemann invariants to solve the linear system of equations $(u_1)_t + (u_2)_x = 0$ and $(u_2)_t + (u_1)_x = 0$ subject to the initial conditions $u_1(x, 0) = e^x$ and $u_2(x, 0) = e^{-x}$.

2. Consider the nonhomogeneous system of quasi-linear equations

$$u_t + 2uu_x + 4vv_x + m = 0 \quad \text{and} \quad v_t + vu_x + 2uv_x + n = 0,$$

where m and n are constants. Modify the form of argument used to obtain the Riemann invariants for the homogeneous form of this system of equations to show that in this special case they take the form

$$u - 2v + (m - 2n)t = \text{constant},$$
$$\text{along the characteristics } C^{(1)}: dx/dt = 2(u - v)$$

and

$$u + 2v + (m + 2n)t = \text{constant},$$
$$\text{along the characteristics } C^{(2)}: dx/dt = 2(u + v).$$

This system of equations is related to the system of long wave equations in shallow water when waves advance up a shelving beach with a constant slope.

3. Use Riemann invariants to solve the linear system of equations $(u_1)_t + (u_2)_x = 0$ and $(u_2)_t + (u_1)_x = 0$ subject to the initial conditions $u_1(x, 0) = 1$ and $u_2(x, 0) = \sin x$.

4. Find the Riemann invariants of the quasi-linear system

$$u_t + (1/k^2)v_x = 0 \quad \text{and} \quad v_t + (1/u^2)u_x = 0 \quad (k = \text{constant}),$$

and express u and v in terms of the invariants.

8.4 Simple Waves

An important and useful class of problems arises when one of the Riemann invariants of Section 8.3 is an absolute constant. When this happens, the solutions are called **simple wave solutions** or just **simple waves**. Thus a simple wave occurs when either $r(\beta) = r_0$ is an absolute constant or $s(\alpha) = s_0$ is an absolute constant.

Suppose, for example, that $s(\alpha) \equiv s_0$, then Eqs. (8.38) and (8.39) become

$$f_{11}(u_1) + f_{12}(u_2) = r(\beta) \quad \text{along the } C^{(1)} \text{ characteristics} \qquad (8.42)$$

and

$$f_{21}(u_1) + f_{22}(u_2) = s_0 \quad \text{along the } C^{(2)} \text{ characteristics,} \qquad (8.43)$$

where

$$f_{ij}(u_i) = \int \mu_i l_j^{(i)} du_j. \qquad (8.44)$$

We see from these equations that everywhere along a $C^{(1)}$ characteristic specified by $\beta = \beta_0 = $ constant, say, u_1 and u_2 must also be constant, because they are the solutions of the nonlinear system of simultaneous equations

$$f_{11}(u_1) + f_{12}(u_2) = r(\beta_0) \quad \text{and} \quad f_{21}(u_1) + f_{22}(u_2) = s_0.$$

The constant values of u_1 and u_2, say $u_1 = \tilde{u}_1(\xi)$ and $u_2 = \tilde{u}_2(\xi)$, are determined by the values of the initial data $u_1(x, 0) = \tilde{u}_1(x)$ and $u_2(x, 0) = \tilde{u}_2(x)$ at the point $(\xi, 0)$ on the initial line $t = 0$ from which the $C^{(1)}$ characteristic originates. Thus any function of u_1 and u_2 will be constant along this characteristic and, in particular, the eigenvalue $\lambda_1(u_1, u_2)$ will become $\lambda_1(\tilde{u}_1(\xi), \tilde{u}_2(\xi)) = \Lambda_1(\xi)$, say. The $C^{(1)}$ characteristic through the point $(\xi, 0)$ is given by

$$C^{(1)} : dx/dt = \Lambda_1(\xi),$$

so as $\Lambda_1(\xi) = $ constant the $C^{(1)}$ characteristic must be the straight line

$$x = \xi + t\Lambda_1(\xi). \qquad (8.45)$$

The values of β_0 and ξ were arbitrary, so by allowing ξ to move along its permitted interval on the initial line result (8.45) will generate a straight line family of $C^{(1)}$ characteristics, while a corresponding result applies if $r(\beta) \equiv r_0$ is an absolute constant.

The behavior of a family of straight line characteristics belongs to one of two different categories. In the first category members of the family of characteristics originating from the interval $x_1 \le x \le x_2$ on the initial line diverge from one another as t increases, as shown in Fig. 8.3a, while in the second category they converge and form an envelope, as shown in Fig. 8.3b. In the first case a

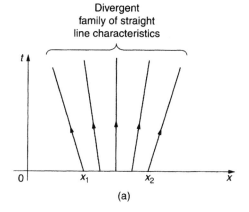

(a)

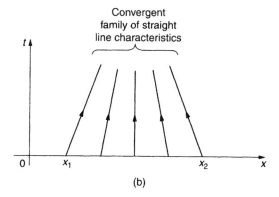

(b)

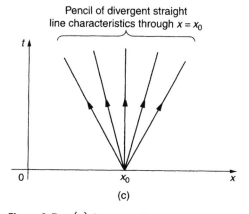

(c)

Figure 8.3   (a) An expansion wave.
(b) A compression wave. (c) A centered
simple wave.

simple wave solution is defined throughout the wedge-shaped region bounded by characteristics through the points $(x_1, 0)$ and $(x_2, 0)$, while in the second case the solution is only defined and unique until the time $t = t_c$ when the cusp first forms in the envelope. By analogy with gas dynamics, the simple wave illustrated in Fig. 8.3a is called an **expansion** or **rarefaction wave**, while the simple wave illustrated in Fig. 8.3b is called a **compression wave**, and a **shock wave** forms when the cusp forms in the envelope of characteristics. In a gas the pressure *decreases* across an advancing expansion wave, whereas in a compression wavecompression wave the pressure across an advancing wave *increases*.

The simple wave property characterized by the dependent variables u_1 and u_2 being constant along a straight line characteristic means that a boundary of a simple wave region can occur adjacent to a constant state region. This property is particularly valuable when piecing together a complicated solution where, for example, one constant state region needs to be joined to a different constant state solution.

On occasion, the straight line characteristics belonging to an expansion wave all originate from a single point on the initial line, as illustrated in Fig. 8.3c. Such solutions are called **centered simple waves**, with their center at the point $(x_0, 0)$.

To illustrate the way a simple wave solution is used to form the transition between two different constant states we will consider the so-called **piston problem** in gas dynamics. The physical model for this problem is a long tube filled with gas with density ρ_1 in which the sound speed is c_1 at rest at one end, sealed at some point along the tube by a piston that can be withdrawn along the length of the tube, as shown in Fig. 8.4. The equations governing the gas flow are the gas dynamic equations already given in matrix form in Example 8.2 of Section 8.3, where the characteristics were seen to be

$$C^{(1)}: dx/dt = v - c \text{ corresponding to the eigenvalue } \lambda_1 = v - c \text{ of } \mathbf{A}$$

and

$$C^{(2)}: dx/dt = v + c \text{ corresponding to the eigenvalue } \lambda_2 = v + c \text{ of } \mathbf{A}.$$

Suppose that at time $t = 0$ the piston is withdrawn from rest in a smoothly accelerated manner until at time t_1 the withdrawal speed is V, and thereafter the piston withdrawal speed V remains constant. We will assume that V is less

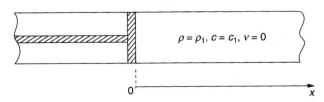

Figure 8.4   Gas at rest in a long tube sealed at $x = 0$ by a piston.

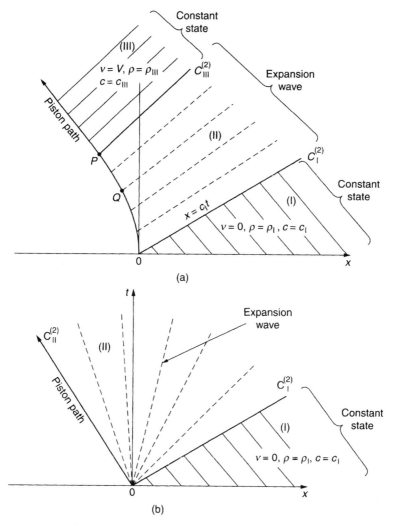

Figure 8.5 (a) Smooth piston withdrawal. (b) Impulsive piston withdrawal.

than the speed of sound in the gas at the piston face, so the expanding gas will be able to remain in contact with the piston. The pattern of the characteristics is illustrated in Fig. 8.5a, where they are seen to behave differently in three separate regions. In region (I) the gas is at rest with $v = 0, \rho = \rho_{\mathrm{I}}, c = c_{\mathrm{I}}$, so all of the characteristics in this region are straight lines parallel to the characteristic $C_{\mathrm{I}}^{(2)}$. In region (II) the characteristics diverge and form an expansion simple wave, while in region (III) the gas is again in a constant state where this time $v = V$, $\rho = \rho_{\mathrm{III}}$, and $c = c_{\mathrm{III}}$, so once again all of the characteristics are parallel straight lines, although now they are parallel to the characteristic $C_{\mathrm{III}}^{(2)}$ through the point P.

The gas density and speed is constant along each straight line characteristic in region (II), although different along different characteristics. As the density and speed are related by a Riemann invariant, the piston speed at any point Q along the piston path joining the origin to the point P will determine the gas density on the characteristic through Q.

A related problem arises if, when the piston is at rest, it is withdrawn impulsively with speed V less than the speed of sound at the piston face. This situation is illustrated in Fig. 8.5b, where it is seen there are now only two regions involved. Region (I) is again the equilibrium region in which $v = 0, \rho = \rho_I$, and $c = c_I$, so in this region all characteristics are parallel to the characteristic $C_I^{(2)}$ through the origin-bounding region (I). In region (II) all characteristics radiate out from the origin, showing that in this region the gas flow is described by a centered simple expansion wave bounded at the left by the characteristic $C_{II}^{(2)}$ corresponding to the piston path. There is a singularity at the origin, because characteristics associated with *different* Riemann invariants all pass through the point O, but this singularity is immediately resolved by the centered simple wave away from the origin.

To find the nature of the flow in Region (II) of Fig. 8.5b note that the characteristics through the origin all have the equation $x/t = \xi$, with ξ being a parameter, and that the $C^{(2)}$ characteristics are determined by $dx/dt = v + c$, so as $v = 0$ and $c = c_I$ on $C_I^{(2)}$, it must have the equation $x = c_I t$, or equivalently $\xi = c_I$, while the equation of $C_{II}^{(2)}$ must be $x = -Vt$, where the negative sign is necessary because the piston is being *withdrawn*.

All $C^{(1)}$ characteristics must enter the constant state region (I) where $v = v_0 = 0$ and $c = c_I$, so from Example 8.2 this means that in this region the Riemann invariant $r(\beta) = 2c_I/(\gamma - 1)$ is an absolute constant. In region (II) the Riemann invariant is $2c/(\gamma - 1) - v$, so equating this to $r(\beta)$ gives $2c/(\gamma - 1) - v = 2c_I/(\gamma - 1)$, showing that $v = 2(c - c_I)/(\gamma - 1)$.

However, $dx/dt = \xi$ on a $C^{(2)}$ characteristic, which is itself defined by $dx/dt = v + c$, so after eliminating v between the previous expression and $\xi = v + c$, the sound speed c in region (II) is found to be given parametrically in terms of ξ by

$$c = [(\gamma - 1)\xi + 2c_I]/(\gamma + 1).$$

Solving for v in terms of ξ using $\xi = v + c$ and $v = 2(c - c_I)/(\gamma - 1)$ gives

$$v = 2(\xi - c_I)/(\gamma + 1).$$

The expressions for the sound speed c and gas speed v in region (II) in terms of x and t are found by replacing ξ by x/t when we obtain

$$v = 2(x/t - c_I)/(\gamma + 1) \quad \text{and} \quad c = [(\gamma - 1)(x/t) + 2c_I]/(\gamma + 1).$$

These equations express v and c in terms of the quotient x/t, which lies between the value of ξ on $C_I^{(2)}$ and that on $C_{II}^{(2)}$, so that $-V \le x/t \le c_I$.

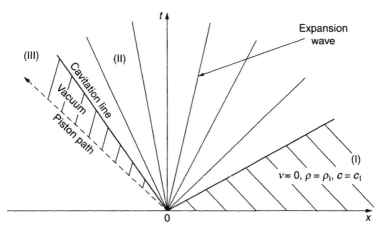

Figure 8.6 The development of a vacuum region between the withdrawn piston and the advancing gas front.

As $p = k\rho^\gamma$ and $c^2 = dp/d\rho = \gamma k\rho^{\gamma-1}$, it follows that $\rho = 0$ when $c = 0$, and from the expression for c this is seen to happen when the withdrawal speed equals $v_c = -2c_1/(\gamma - 1)$. This is called the **cavitation speed** of the piston, because if the piston is withdrawn any faster than v_c the expanding gas will be unable to keep up with the piston face, causing a vacuum region to develop between the expanding gas front and the piston face, as shown in Fig. 8.6.

To complete the examination of the piston problem we need to determine what happens if the piston is pushed into the gas compressing it. From $v = 2(x/t - c_1)/(\gamma + 1)$ we have $v_t = -2x/[t^2(1 + \gamma)]$ and $v_x = 2/[t(\gamma + 1)]$, so $v_t + (x/t)v_x = 0$, but rearranging $c = [(\gamma - 1)(x/t) + 2c_1]/(\gamma + 1)$ shows that $x/t = \frac{1}{2}[c_1 + (\gamma + 1)v]$, so the gas speed v is determined by the equation

$$v_t + [c_1 + (\gamma + 1)v/2]v_x = 0.$$

From the Riemann invariant the sound speed $c = \frac{1}{2}[2c_1 + (\gamma - 1)v]/(\gamma + 1)$ so c and hence the gas density ρ are determined by the solution v of a first-order quasi-linear equation. If the gas density is required this follows by noting that as $p = k\rho^\gamma$, and $c^2 = dp/d\rho = \gamma k\rho^{\gamma-1}$, $(c/c_1)^{1/2} = (\rho/\rho_1)^{(\gamma-1)/2}$, so from the expression for c the gas density ρ is determined in terms of v by

$$\rho = \rho_1[1 + (\gamma - 1)v/(2c_1)]^{2/(\gamma-1)}.$$

Equations of the type determining v were examined in detail in Chapter 2, where it was shown that when the characteristic curves of such a PDE converge, as happens when the piston is accelerated *smoothly* from rest *into* the gas, the solution will eventually become nonunique, causing the development of a shock. More will be said about this in the next section, although for the moment it will suffice to note that a shock wave is formed immediately if the piston is pushed into the gas in an impulsive manner.

EXERCISES 8.4

1. Using the equation $\frac{\partial v}{\partial t} + (c_1 + \frac{1}{2}(\gamma - 1)v)\frac{\partial v}{\partial x} = 0$ derived in the text, consider the case where at time t the piston position is given by $x = \sigma(t)$, where $\sigma(t)$ is a known function of t such that $\sigma(0) = 0$ and $\sigma'(0) = 0$, corresponding to the piston being accelerated *smoothly* from rest at time $t = 0$. Determine the family of straight line characteristics associated with the problem using the time $t = \tau$ as a parameter, and by finding the envelope of the characteristics show that a cusp will only form if $\sigma''(0) > 0$, corresponding to the piston being pushed *into* the gas, compressing it. Show also that in this case the cusp will form at a time t_P and position x_P given by

$$t_P = \frac{2c_1}{(\gamma + 1)\sigma''(0)} \quad \text{and} \quad x_P = c_1 t_P.$$

(Hint: You will find it useful to use the result if $f(x, t, \alpha) = 0$ describes a family of curves with parameter α; in the case that the family of curves generates an envelope it is found by eliminating α between the equations $f(x, t, \alpha) = 0$ and $\partial/\partial\alpha[f(x, t, \alpha)] = 0$.)

2. The one-dimensional long wave approximation for water waves in a channel of constant depth h is described by the equations

$$v_t + vv_x + 2cc_x = 0 \quad \text{and} \quad 2c_t + 2vc_x + cv_x = 0,$$

where v is the horizontal water speed, and $c = \sqrt{(gh)}$ is the surface wave propagation speed with g being the acceleration due to gravity. Find the Riemann invariants for this system and express v and c in terms of them. Express the equations for the $C^{(1)}$ and $C^{(2)}$ characteristics in terms of $r(\beta), s(\alpha), \alpha$, and β, and by using the equality of the mixed derivatives $\partial^2 x/\partial s \partial r = \partial^2 x/\partial r \partial s$, derive a linear second-order equation for the time t with r and s as independent variables.

3. The current $i(x, t)$ and voltage $v(x, t)$ in a transmission line obey the equations

$$Li_t + v_x + Ri = 0 \quad \text{and} \quad Cv_t + i_x + Gv = 0.$$

Write these equations in the matrix form $\mathbf{U}_t + \mathbf{AU}_x = \mathbf{0}$, and determine their Riemann invariants in terms of i, v and $c = (LC)^{-1/2}$ for the case of a lossless line in which $R = G = 0$. Show that when the line is distortionless, so $R/L = G/C$, the Riemann invariants can be integrated to give

$$i = \frac{1}{2cL} \exp\left(-\frac{2R}{L}x\right)[f(x - ct) + g(x + ct)]$$

and

$$v = \frac{1}{2} \exp\left(-\frac{2R}{L}x\right)[f(x - ct) - g(x + ct)],$$

where f and g are arbitrary differentiable functions of their arguments.

4. Consider the smooth piston withdrawal problem examined in Section 8.4, but suppose initially the piston moves to the left with constant speed v_I, and the gas to its right follows the piston in a constant state with speed v_I and constant sound speed c_I. If at time $t = 0$ the piston is accelerated smoothly to the left from speed v_I, determine how the equations

$$v_t + [c_I + (\gamma + 1)v/2]v_x = 0 \quad \text{and} \quad \rho = \rho_I[1 + (\gamma - 1)v/(2c_I)]^{2/(\gamma-1)}$$

derived in the text must be modified. Is there a simple way of solving this problem?

8.5 Shocks and the Riemann Problem

The concept of a conservation law is central to the study of shock wave solutions of hyperbolic systems. This is because when the solution of a hyperbolic system becomes discontinuous, it turns out that the propagation speed of the discontinuity, and the values of the discontinuous-dependent variables on each side of the discontinuity, can only be related if the system can be written in conservation form. The discussion of first-order quasi-linear equations showed the solution of an initial value problem becomes nonunique when characteristics intersect, and that such intersection indicates the development of a discontinuous solution.

A fundamental feature of the discontinuous solutions considered here is that they will be called **strong discontinuities**, or **shock waves**, when they are produced by the intersection of characteristics, but not otherwise. This definition may seem obvious in the light of the results in Chapter 2, but it will be seen later that the nonlinearity of hyperbolic systems permits discontinuous solutions that are not unique to arise, so some selection principle is necessary to identify physically meaningful shock waves from the set of mathematically possible but nonphysical discontinuous solutions.

A general quasi-linear hyperbolic system in three space dimensions and time is a hyperbolic system that can be written in the divergence form

$$\mathbf{U}_t + \text{div } \mathbf{F}(\mathbf{U}) = 0, \tag{8.46}$$

where $\mathbf{U} = \mathbf{U}(\mathbf{r}, t)$ is an n-element column vector with scalar elements depending on the position vector $\mathbf{r}$ and the time t, and $\mathbf{F}(\mathbf{U})$ is an n-element column vector with vector elements depending on $\mathbf{U}(\mathbf{r}, t)$. When only one space dimension is involved, (8.46) simplifies to

$$\frac{\partial U}{\partial t} + \frac{\partial F(U)}{\partial x} = 0, \tag{8.47}$$

where $U(x, t)$ is an n-element column vector, and $F(U)$ is an n-element column vector with scalar elements that are functions of $U(x, t)$. In general, the element $f_1, f_2, \ldots, f_n$ of $F(U)$ will depend on the elements $u_1, u_2, \ldots, u_n$, so when the second term of (8.47) is expanded, it becomes

$$\frac{\partial F(U)}{\partial x} = AU_x, \quad \text{where} \quad A = \begin{bmatrix} \partial f_1/\partial u_1 & \cdot & \cdot & \partial f_1/\partial u_n \\ \cdot & \cdot & \cdot & \cdot \\ \cdot & \cdot & \cdot & \cdot \\ \partial f_n/\partial u_1 & \cdot & \cdot & \partial f_n/\partial u_n \end{bmatrix}. \tag{8.48}$$

Thus a quasi-linear hyperbolic system in one space dimension and time $U_t + AU_x = 0$ will be equivalent to a system of conservation laws only if it can be written in the form (8.47).

In physical problems, the derivation of the governing equations often leads to a system of equations in conservation form. However, if a hyperbolic system of equations arises directly in the form $U_t + AU_x = 0$, there is no simple way by which to determine whether it is a conservation system, other than by showing directly it can be written in the divergence form (8.47).

To illustrate this we consider the general gas dynamic equations in terms of the gas velocity $\mathbf{v}$, the gas density ρ, and the gas pressure $p = p(\rho)$,

$$\rho_t + \text{div}(\rho \mathbf{v}) = 0 \tag{8.49}$$

$$\mathbf{v}_t + \mathbf{v} \cdot \text{grad}\, \mathbf{v} + (1/\rho)\text{grad}\, p = 0, \tag{8.50}$$

which are known to be hyperbolic, and restrict them to one space dimension. The first equation is already in the required one-dimensional divergence form, because it reduces to

$$\rho_t + (\rho v)_x = 0.$$

The one-dimensional form of the second equation is not in divergence form, but if it is multiplied by ρ and added to v times the first equation it becomes the conservation equation

$$(\rho v)_t + (\rho v^2 + p)_x = 0.$$

Using this last result allows the system to be written in the one-dimensional matrix conservation form

$$U_t + \{F(U)\}_x = 0, \quad \text{with} \quad U = \begin{bmatrix} \rho \\ \rho v \end{bmatrix} \quad \text{and} \quad F(U) = \begin{bmatrix} \rho v \\ \rho v^2 + p \end{bmatrix}. \tag{8.51}$$

To determine the equation governing discontinuous solutions of general systems we need only consider the typical scalar conservation law

$$u_t + (F(u))_x = 0, \tag{8.52}$$

which was examined in Section 2.4. It was shown there that if u experiences a jump discontinuity, it must propagate with the speed U determined by

$$U[[u]] = [[F(u)]], \tag{8.53}$$

where $[[\alpha]] = \alpha_L - \alpha_R$ with α_L is the value of α to the immediate left of the jump discontinuity and α_R is the value to the immediate right.

As each of the n equations in (8.47) are of the form (8.52), it follows that the **jump condition** for system (8.47) is

$$U[[\mathbf{U}]] = [[\mathbf{F}(\mathbf{U})]]. \tag{8.54}$$

An equation of this type, called the **Rankine–Hugoniot** jump condition, arose in the study of gas dynamics before general discontinuous solutions were considered, so (8.54) is sometimes called the **generalized Rankine–Hugoniot** jump condition.

Matrix equation (8.54) relates the n values $u_{1L}, u_{2L}, \ldots, u_{nL}$ of the elements of vector $\mathbf{U}$ to the left of the discontinuity and the n values $u_{1R}, u_{2R}, \ldots, u_{nR}$ to the right to the speed U with which the discontinuity propagates. Thus if, for example, $n + 1$ of these quantities are known, the remaining n can, in principle, be found from (8.54).

However, it is important to recognize that when $\mathbf{F}(\mathbf{U})$ is a nonlinear function of the elements of $\mathbf{U}$, the solution of the n scalar equations for $2n + 1$ quantities, when $n + 1$ of them are known, will *not* necessarily have a unique solution. Thus when more than one discontinuous solution is possible, they must be checked to discover which is a physical solution and which, although mathematical solutions, have no physical significance. As already mentioned, a discontinuous solution will *only* be called a **shock** when it is produced by the intersection of characteristics.

With the aid of this definition of a shock we are now in a position to complete our examination of the **Riemann problem** introduced in Chapter 2 (see Fig. 2.22). The approach to be used is illustrated by the following simple example.

Example 8.3. Given the equation $u_t + uu_x = 0$, solve the Riemann problems using the initial conditions

$$\text{(a) } u(x, 0) = \begin{cases} 1, & x < 0 \\ 1/2, & x > 0 \end{cases} \quad \text{and} \quad \text{(b) } u(x, 0) = \begin{cases} 1/2, & x < 0 \\ 1, & x > 0. \end{cases}$$

Solution: The equation is not in conservation form, but it is easily seen that it becomes a conservation equation when written $u_t + (\frac{1}{2}u^2)_x = 0$. In the notation of (8.52), $u = u$ and $F(u) = \frac{1}{2}u^2$, so the jump condition becomes $U(u_L - u_R) = \frac{1}{2}(u_L^2 - u_R^2)$, or $U = \frac{1}{2}(u_L + u_R)$, showing the propagation speed of the discontinuous solution is the average of the values of u to the left and right of the discontinuity. To complete the analysis it is necessary to examine the behavior of the characteristics to be sure this discontinuous solution is a shock wave.

In case (a) the straight line characteristics issuing out from points on the initial line to the left of the origin have the slope $dx/dt = 1$, while those issuing out from points on the initial line to the right of the origin have the slope $dx/dt = \frac{1}{2}$, so the two sets of characteristics converge. Remember that when the characteristics are drawn in the (x, t) plane, x is the abscissa and t the ordinate, so in this plane the characteristics to the left of the origin have slope $dt/dx = 1$, while those to the right have slope $dt/dx = 2$. Thus initial conditions (a) produce a genuine shock that propagates with speed $U = \frac{1}{2}(1 + \frac{1}{2}) = \frac{3}{4}$ along the chain-dashed line in Fig. 8.7a.

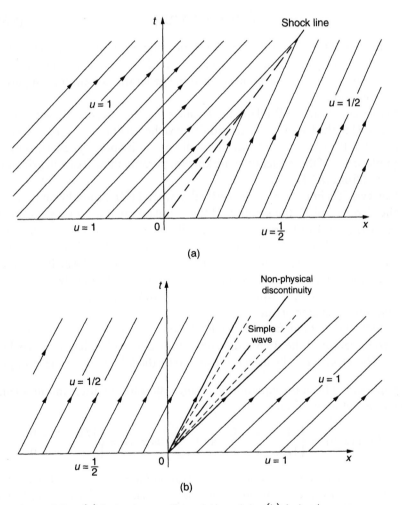

(a)

(b)

Figure 8.7   (a) A shock wave through the origin. (b) A simple wave region and a nonphysical discontinuity.

In case (b) the characteristics to the left of the origin have slope $dx/dt = \frac{1}{2}$ while those to the right have slope $dx/dt = 1$, so in this case the characteristics diverge and a simple wave lies in the wedge-shaped region in Fig. 8.7b, allowing the solution to make a smooth transition between the two constant states. The jump condition again predicts that a discontinuous solution will propagate with speed $U = \frac{3}{4}$, but this line, shown as a dashed line in Fig. 8.7b, is seen to lie in the simple wave region where *no* characteristics intersect, so this is a nonphysical discontinuity, and so must be rejected. A more careful analysis of this nonphysical discontinuity shows it to be unstable, so even if it started to develop, it would immediately decay into the stable simple wave solution. ■

EXERCISES 8.5

1. An alternative form of the equations for the long wave approximation for water waves is

$$h_t + uh_x + hu_x = 0 \quad \text{and} \quad u_t + uu_x + gh_x = 0,$$

where u is the horizontal speed of the water, g is the acceleration due to gravity, and h is the depth of the water. A **tidal bore** in shallow water is the propagation of a step change in height of the water as shown in Fig. 8.8, where the water advances to the left with speed s. Tidal bores are a common phenomenon in some estuaries.

   Express the equations in conservation form and hence find the conditions for the steady motion of a bore using a suffix 0 to indicate conditions to the left of the bore and a suffix 1 to indicate conditions to the right, where the water height to the left is h_0 and to the right is h_1. Find the speed s of the bore and the speed u_1 of the water behind the bore.

2. Obtain in implicit form the solution of $u_t + u^2 u_x = 0$ given that

$$u(x,0) = \begin{cases} -1, & x < -1 \\ x, & -1 \le x \le 3 \\ 3, & x > 3. \end{cases}$$

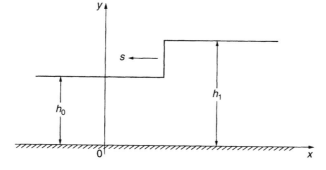

Figure 8.8   A tidal bore advancing to the left with speed s.

Solve the following Riemann problems for the equation:

$$\text{(a) } u(x,0) = \begin{cases} 2, & x < 0 \\ 3, & x > 0 \end{cases} \quad \text{and} \quad \text{(b) } u(x,0) = \begin{cases} 3, & x < 0 \\ 2, & x > 0. \end{cases}$$

In Exercises 3 through 5 solve the compound Riemann problem and explain why the solution found using the method of Section 8.5 will only exist for a finite time t_c, and find t_c.

3. $u_t + uu_x = 0$ given that $u(x,0) = \begin{cases} 1, & x < 0 \\ 2, & 0 < x < 1 \\ 1, & x > 1. \end{cases}$

4. $u_t + u^{1/2}u_x = 0$ given that $u(x,0) = \begin{cases} 4, & x < -1 \\ 1, & -1 \le x \le 3 \\ 9, & x > 3. \end{cases}$

5. $u_t + u^2 u_x = 0$ given that $u(x,0) = \begin{cases} 3, & x < 1 \\ 2, & 1 < x < 2 \\ 1, & x > 2. \end{cases}$

Answers to Odd-Numbered Exercises

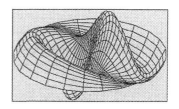

Chapter 1

Solutions 1.1

1. By the chain rule $u_x = u_r r_x + u_\theta \theta_x$, $u_y = u_r r_y + u_\theta \theta_y$, $u_z = u_z$. However, $r = (x^2 + y^2 + z^2)^{1/2}$ and $\theta = \arctan(y/x)$, so $r_x = x/r$, $\theta_x = -y/r^2$, $r_y = y/r$, $\theta_y = x/r^2$, giving $u_x = x(u_r/r) - y(u_\theta/r^2)$, $u_y = y(u_r/r) + x(u_\theta/r^2)$, and $u_z = u_z$. Thus we have

$$\frac{\partial}{\partial x} \equiv \frac{x}{r}\frac{\partial}{\partial r} - \frac{y}{r^2}\frac{\partial}{\partial \theta} \quad \text{and} \quad \frac{\partial}{\partial y} \equiv \frac{y}{r}\frac{\partial}{\partial r} + \frac{x}{r^2}\frac{\partial}{\partial \theta}.$$

Using this result to differentiate u_x with respect to x and u_y with respect to y, and differentiating u_z with respect to z, the result follows after using the equality of mixed derivatives $u_{r\theta} = u_{\theta r}$ and forming the sum $u_{xx} + u_{yy} + u_{zz}$.

3. $\frac{\partial}{\partial r}(r^2 \frac{\partial u}{\partial r}) + \frac{1}{\sin \theta}\frac{\partial}{\partial \theta}(\sin \theta \frac{\partial u}{\partial \theta}) = 0$. As the azimuthal angle ϕ does not appear in the Laplacian this implies that the solution is independent of ϕ and so has rotational symmetry about the z axis. Consequently, for any fixed angle θ, the solution depends only on r.

5. Start by setting $u = \alpha v$ and $t = \beta \tau$ where α and β are scale constants to be determined. If u becomes v under this change of variable, the KdV equation is transformed into $(\frac{\alpha}{\beta})v_\tau + \alpha^2 v v_x + \mu \alpha v_{xxx} = 0$. The required reduction then follows by setting $\alpha = \mu$ and $\beta = \frac{1}{\mu}$.

Solutions 1.2

1. $u_t = -cf'(x - ct)$ and $u_x = f'(x - ct)$, where a prime indicates differentiation of f with respect to its argument, that is, $f'(\zeta) = df/d\zeta$. So $u_t + cu_x = 0$.

3. From Exercise 1 the PDE has a solution of the form $u(x, t) = f(x - 2t)$. Now $u_t = -2f'(x - 2t)$, but $u_t(x, 0) = x$, so setting $t = 0$ shows that $f'(x) = -x/2$. Integrating this with respect to t (regarding x as a constant) gives $f(x) = -x^2/4 + h(x)$, where $h(x)$ is an *arbitrary* differentiable function of x. Replacing x by $x - 2t$ in $f(x)$ shows that the solution is given by $u(x, t) = -\frac{1}{4}(x - 2t)^2 + h(x - 2t)$, where h is an *arbitrary* function of $x - 2t$.

5. $u_t = -cf'(x - ct) + cg'(x + ct)$ and $u_{tt} = c^2 f''(x - ct) + c^2 g''(x + ct)$, while $u_x = f'(x - ct) + g'(x + ct)$ and $u_{xx} = f''(x - ct) + g''(x + ct)$, so $u_{tt} = c^2 u_{xx}$.

7. All that is necessary is to replace the expression $T_u = T_0 u_x$ when it first appears by $T_u = T(x)u_x$, and thereafter to keep $T(x)$ under the integral signs.

9. The condition that $S(x)$ must be slowly varying is necessary to ensure that what is essentially a three-dimensional problem can be approximated by a one-dimensional one. If $S(x)$ changes rapidly, the solution at $x = x_1$, say, will depend on y and z, and so cannot be approximated by the solution in a plane area.

15. $u(r) = A + B \ln r$.

Solutions 1.3

1. Show by differentiation that the given function satisfies the PDE and the second-order partial derivatives are continuous.

3. Show by differentiation that the given function satisfies the PDE and the second-order partial derivatives are continuous.

5. $L[w] = L[u] + L[v] = f$.

Chapter 2

Solutions 2.1

1. The equation of characteristics through $(\xi, 0)$ is $y = \frac{5}{2}(\xi - x)$. The integral of the compatibility condition gives $u + k(\xi) = 2x$. The Cauchy condition shows $k(\xi) = \xi$, so the solution is $u(x, y) = x - \frac{2}{5}y$.

3. The equation of the characteristics is $y/x^2 = \xi$, so they form a family of curves through the origin. The integral of the compatibility condition gives $uk(\xi) = x^3$. The Cauchy condition shows $k(\xi) = 1/\cos\xi$, so the solution is $u(x, y) = x^3 \cos(y/x^2)$.

5. The equation of characteristics is $2x + y = \xi$. Integration of the compatibility curve gives $\ln(u - 1) = x + \ln k(\xi)$, so $u = 1 + k(\xi)e^x$. On the Cauchy data line $x = ky$, so $2x + y = \xi$ becomes $y = \frac{\xi}{(1+2k)}$ and as $x = ky$ we have $x = \frac{k\xi}{(1+2k)}$. Using the result $u = 2y$ on the Cauchy data line gives $\frac{2\xi}{1+2k} = 1 + k(\xi) \exp \frac{k\xi}{(1+2k)}$. Finally, solving for $k(\xi)$ and using it in the expression for u with $\xi = 2x + y$ gives

$$u(x, y) = 1 + \frac{4x + 2y - 2k - 1}{1 + 2k} \exp\left(\frac{x - ky}{2k + 1}\right).$$

The solution is valid for $k \neq \frac{-1}{2}$, and this corresponds to the Cauchy data line coinciding with a characteristic.

7. The equation of the characteristics is $y = 2x + \xi$. Integration of the compatibility condition gives $(1+u)k(\xi) = e^x$. As $u = \sin x$ on $y = 3x+1$ this gives $(1 + \sin x)k(x + 1) = e^x$, so replacing x by $x - 1$ we find that $k(x) = \frac{\exp(x-1)}{1+\sin(x-1)}$. Finally, replacing x by ξ followed by use of the result $\xi = y - 2x$ gives $u(x, y) = \exp(3x - y + 1)[1 + \sin(y - 2x - 1)] - 1$.

9. Use the same method as in Example 2.5, taking into account the fact that the Cauchy condition is $u(0, y) = \tanh y$, so now it is imposed on the y axis. The solution is

$$u(x, y) = \frac{\tanh(y - x)}{1 - x\tanh(y - x)}.$$

There is again a critical curve determined by the vanishing of the denominator, and it is obtained from the curve in Example 2.5 by interchanging the x and y axes.

11. $\frac{dy}{dx} = 2$ and $\frac{du}{dx} = u$ so the characteristic through $(\xi, 0)$ is $y = 2x - 2\xi$ and along this characteristic $u(x, y) = f(\xi)e^x$, so $u(x, y) = f(x - \frac{y}{2})e^x$. As $u(x, 0) = 3x$ for $x \geq 0$, we have $3x = f(x)e^x$, so $f(x) = 3xe^{-x}$. Thus $f(\xi) = 3(x - \frac{y}{2}) \exp(-x + \frac{y}{2})$, and so $u(x, y) = 3(x - \frac{y}{2})e^{y/2}$ in the region in the first quadrant with $x \geq 0$ and below the characteristic $y = 2x$. Similarly, the characteristic through the point $(0, \eta)$ is $y - \eta = 2x$, and $u(x, y) = g(\eta)e^x$. As $u(0, y) = \sin y$ for $y \geq 0$ we see that $\sin y = g(y)$, so $g(\eta) = \sin(y - 2x)$, from which it follows that $u(x, y) = e^x \sin(y - 2x)$ in the first quadrant with $y \geq 0$ and above the characteristic $y = 2x$.

13. (a) $\frac{dy}{dx} = c$, so the characteristic through the point (x_0, y_0) is $y = y_0 + c(x - x_0)$. However, the initial line Γ has the equation $y = x - 1$, so the

characteristic through the point $(\xi, \xi - 1)$ is $y = cx + (1 - c)\xi$, so that $\xi = \frac{y - cx + 1}{1 - c}$. Along this characteristic $\frac{du}{dx} = 0$, so $u(x, y) = u(\xi)$, but the initial condition is $u = x$ on Γ, so $u(\xi) = \xi$, whence $u(x, y) = \frac{y - cx + 1}{1 - c}$.

(b) As $u = x$ on Γ, and the general solution is $u(x, y) = f(y - cx)$ it follows that $x = f((1 - c)x - 1)$. Thus to find the function $f(x)$ we set $X = (1 - c)x - 1$, from which it follows that $x = \frac{X + 1}{1 - c}$, and so $f(y - cx) = \frac{y - cx + 1}{1 - c}$, and so $u(x, y) = \frac{y - cx + 1}{1 - c}$. This is valid provided $c \neq 1$. Here c is the slope of the parallel characteristics, whereas the slope of the initial line is 1, so when $c = 1$ a characteristic coincides with the initial line.

15. $dy/dx = 2x$, so the characteristic through the point $(\xi, \xi^2/2)$ has the equation $y = x^2 - \xi^2/2$, showing that $\xi^2 = 2(x^2 - y)$. We have $du/dx = 2xu$, and so on this characteristic we have $\ln u - \ln k(\xi) = x^2 - \xi^2$, so $u(x, y) = k(\xi^2) \exp(x^2 - \xi^2)$. However, $u = \xi^2$ when $x = \xi$, and so $k(\xi) = \xi^2$. Thus $u(x, y) = \xi^2 \exp(x^2 - \xi^2)$, and after substituting for ξ^2 we find that $u(x, y) = 2(x^2 - y) \exp(2y - x^2)$.

Solutions 2.2

1. Use the form of reasoning that gave result (2.25) to obtain the required implicit solution $u = g(x - tf(u))$. Partial differentiation of this result with respect to x gives

$$\frac{\partial u}{\partial x} = \frac{g'(x - tf(u))}{1 + tg'(x - tf(u)) f'(u)}.$$

This last result shows that $u_x(x, t)$ is defined in the upper half of the (x, t) plane provided $1 + tg'(x - tf(u)) f'(u) \neq 0$ for all $t > 0$.

3. The compatibility equation is $\frac{du}{dt} = e^t$, so along the characteristic through the point $(\xi, 0)$ we have $u(x, t) = e^t - 1 - \xi$. Using this in $\frac{du}{dt} = e^t$ to find the equation of this characteristic gives $x = -e^t + 1 + t + \xi(1 + t)$. Elimination of ξ between these two equations gives the explicit solution $u(x, t) = e^t - \frac{e^t + x}{1 + t}$. This solution is valid for all $t > 0$.

5. The compatibility condition is $du/dt = t$, so using the Cauchy condition and integrating this result show that along the characteristic through the point $(\xi, 0)$ the solution is $u(x, t) = 2\xi + \frac{1}{2}t^2$. When this result is used in the characteristic equation and the result is integrated, the equation of this characteristic is found to be $x = (1 - 2t)\xi + t^3/6$. Elimination of ξ between these two equations shows the solution to be

$$u(x, t) = \frac{\frac{1}{2}t^2 - \frac{2}{3}t^3 - 2x}{1 - 2t}.$$

The denominator vanishes when $t = \frac{1}{2}$, so the solution is valid for $0 < t < \frac{1}{2}$.

7. First we solve for v, which is easily seen to be $v(x, t) = x + ct$. Substituting into the equation for u then shows that $u_t + 2uu_x = ct$, so $dx/dt = 2u$ and $du/dt = ct$. Integration of this last equation from the point $(\xi, 0)$ gives $u - k(\xi) = ct^2/2$, but $u(x, 0) = x$ when $t = 0$, so $k(\xi) = \xi$ and hence $u = \xi + ct^2/2$. Using this expression for u in $dx/dt = 2u$, and integrating along the characteristic through $(\xi, 0)$, shows that $x - ct^3/3 = (1 + 2t)\xi$. Eliminating ξ now shows that $u(x, t) = \frac{1}{6}(6x + 3ct^2 + 4ct^3)/(1 + 2t)$, and we already have $v(x, t) = x + ct$. There is no restriction on t in the half-plane $t > 0$.

Solutions 2.3

1. $u(x, t) = \begin{cases} -1, & t - x > 1 \\ x/(1 - t), & 0 < t < 1 \\ 1, & t - x < 1. \end{cases}$

3. The PDE is semi-linear so characteristics through all points $(\xi, 0)$ have the equation $x = t + \xi$. Integration of $\frac{du}{dt} = u^2$ gives $\frac{-1}{u} = f(\xi) + t$, so the solution corresponding to the initial condition in the interval $-a < x < a$ is $u(x, t) = \frac{x - t}{1 + t^2 - xt}$. The solution corresponding to the initial condition in the interval $-\infty < x < -a$ is $u(x, t) = \frac{-a}{1 + at}$ and the solution corresponding to the initial condition in the interval $a < x < \infty$ is $u(x, t) = \frac{a}{1 - at}$, which only exists for $0 < t < \frac{1}{a}$. Consequently, piecing together these three results gives the solution which only exists for $0 < t < \frac{1}{a}$.

5. The implicit solution corresponding to the initial condition in the interval $-1 \le x \le 4$ is $x = u + 3u^3t$. The solution corresponding to the initial condition in the interval $-\infty < x < -1$ is $u(x, t) = -1$ and the solution corresponding to the initial condition in the interval $4 < x < \infty$ is $u(x, t) = 4$.

7. $u(x, t) = \begin{cases} 2, & -\infty < \frac{x}{t} < 4 \\ \left(\frac{x}{t}\right)^{1/2}, & 4 \le \frac{x}{t} \le 9 \\ 3, & 9 < \frac{x}{t} < \infty. \end{cases}$

9. The solution uses the fact that the fan of characteristics radiating out from the point $(-1, 0)$ has the equation $\frac{x+1}{t} = \alpha$, while the fan of characteristics radiating out from the point $(1, 0)$ has the equation $\frac{x-1}{t} = \beta$, where α and β are parameters. The solution corresponding to the initial

condition in the interval $-\infty < x < -1$ is $u(x, t) = 1$. The solution in the transition region centered on $(-1, 0)$ is $u(x, t) = \frac{x+1}{t}$ for $1 \le \frac{x+1}{t} \le 2$. The solution corresponding to the initial condition in the interval $-1 < x < 1$ is $u(x, t) = 2$. The solution in the transition region centered on $(1, 0)$ is $u(x, t) = \frac{x-1}{t}$ for $2 \le \frac{x-1}{t} \le 3$. The solution corresponding to the initial condition in the interval $1 < x < \infty$ is $u(x, t) = 3$.

Solutions 2.4

1. (a) Here $u_L < u_R$ so no shock will form. There is a centered simple wave located at the origin where the solution is $u(x, t) = (\frac{x}{t})^{1/2}$ for $4 \le \frac{x}{t} \le 9$. The solution is $u(x, t) = 2$ for $-\infty < \frac{x}{t} \le 4$ and $u(x, t) = 3$ for $9 \le \frac{x}{t} < \infty$. An attempt to fit a shock in this case will produce a nonphysical solution because the discontinuity line lies inside the centered simple wave region so it does not correspond to a line along which characteristics impinge from either side.

   (b) Here $u_L > u_R$ so a shock will form from the discontinuity at the origin. The conservation form of the PDE is $u_t + (\frac{1}{3}u^3)_x = 0$, so the jump condition is $U[[u]] = [[\frac{1}{3}u^3]]$. Substituting for u_L and u_R shows the shock speed to be $U = \frac{19}{3}$. The shock will originate from the origin, so the equation of the shock line is $x = \frac{19}{3}t$.

3. Care is required with this Problem, because an equation of the form $u_t + f(u)u_x = 0$ with piecewise constant initial data will have a shock form when $f(u)$ is a *decreasing* function, so it will be necessary that $f(u_L) > f(u_R)$. However, here $f(u) = -2u$, so the previous condition must be reversed if a shock is to occur. Examination of the Cauchy condition shows there to be a shock wave originating from the origin, and a centered simple wave located at the point $(1, 0)$. Expressing the PDE in conservation form it becomes $u_t + (-u^2)_x = 0$, so the jump condition at the origin is $U[[u]] = [[-u^2]]$. Setting $u_L = \frac{1}{2}$ and $u_R = 1$ shows that the shock speed to be $U = -\frac{3}{2}$, so the equation of the shock line through the origin is $t = -\frac{2}{3}x$. Using the method of Example 2.9, the centered simple wave located at $(1, 0)$ is found to have the solution $u(x, t) = \frac{1-x}{2t}$ for $-1 \le \frac{1-x}{2t} \le -\frac{1}{2}$. The solution is $u(x, t) = \frac{1}{2}$ to the left of the shock, $u(x, t) = 1$ after the shock and before the centered simple wave, and again $u(x, t) = \frac{1}{2}$ after the simple wave. The shock line is $t = -\frac{2}{3}x$, and the left boundary of the centered simple wave region is $t = \frac{1}{2}(1 - x)$, and these intersect when $x = -3$ and $t = 2$, so the solution before this interaction takes place is only valid for $0 < t < 2$. The solution is illustrated in Fig. 2.4.1.

5. Using the method of Section 2.1, the solution of the PDE corresponding to the initial condition in the interval $-\infty < x < 1$ is seen to be

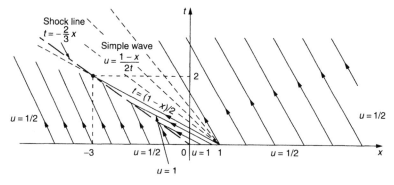

Figure 2.4.1

$u(x, t) = 2(x - t^2) \exp(-t^2)$. Similarly, the solution corresponding to the initial condition in the interval $1 < x < \infty$ is seen to be $u(x, t) = (x - t^2) \exp(-t^2)$. The characteristic through the point $(1, 0)$ on the initial line where the Cauchy data is discontinuous is $x = 1 + t^2$. The initial discontinuity will propagate along this line, so the solution on the immediate left is $u_L(t) = 2 \exp(-t^2)$ and the solution on the immediate right is $u_R(t) = \exp(-t^2)$. Thus $[[u]] = u_L - u_R = \exp(-t^2)$, and this can be seen to tend to 0 as time increases.

Chapter 3

Solutions 3.1

3. Parabolic: $u_{\eta\eta} = -2u_\xi - 3u_\eta, \quad \xi = x + y, \quad \eta = x,$ or

$$u_{\eta\eta} = u_\eta - 2u_\xi, \quad \xi = x + y, \quad \eta = y.$$

5. Parabolic: $u_{\eta\eta} = -3(\xi + 3\eta)u_\xi, \quad \xi = y - 3x, \quad \eta = x,$ or

$$u_{\eta\eta} = -\frac{1}{3}\eta(u_\xi + u_\eta), \quad \xi = y - 3x, \quad \eta = y.$$

7. Elliptic: $u_{\xi\xi} + u_{\eta\eta} = \sin \eta\, u_\xi - u, \quad \xi = \sin x + y, \quad \eta = x.$

9. Hyperbolic: $d = xy > 0 \quad$ for $x > 0, y > 0.$

$$u_{\xi\eta} = \frac{1}{3}\left(\frac{\eta}{\xi^2 - \eta^2}\right)u_\xi - \frac{1}{3}\left(\frac{\xi}{\xi^2 - \eta^2}\right)u_\eta,$$

$$\xi = \frac{2}{3}(x^{3/2} + y^{3/2}), \quad \eta = \frac{2}{3}(y^{3/2} - x^{3/2}).$$

11. Elliptic: $u_{\xi\xi} + u_{\eta\eta} = -\dfrac{1}{3}\left(\dfrac{u_\xi}{\xi} - \dfrac{u_\eta}{\eta}\right),$ $\xi = \dfrac{2}{3}y^{3/2},$ $\eta = -\dfrac{2}{3}x^{3/2}.$

13. Elliptic: $u_{\xi\xi} + u_{\eta\eta} = \dfrac{1}{2\eta(\xi + \eta)}\{\xi u_\xi - (\xi + 2\eta)u_\eta\},$

$$\xi = \frac{1}{2}(y^2 - x^2), \quad \eta = \frac{1}{2}x^2.$$

15. Hyperbolic: $u_{\xi\eta} = \left[\dfrac{\xi - \eta}{2(\xi - \eta)^2 - 12}\right](u_\xi - u_\eta),$

$$\xi = y - x - \cos x, \quad \eta = y - x + \cos x.$$

Solutions 3.2

1. Elliptic: A possible reduction is $Q = 2(x + z)^2 + (y + z)^2 + 2(x - 2y)^2.$
   Eigenvalues are 0.07577, 4.70607, 11.21816.

3. Hyperbolic: A possible reduction is $Q = (x+2y)^2 + 3(y-2z)^2 - (x+z)^2.$
   Eigenvalues are -0.57775, 3.0, 15.57775.

Chapter 4

Solutions 4.1

1. $Ax + By + Cz - ct = k$ with k being a constant is the equation of
   a plane in (x, y, z) space that moves as t increases. Hence for any fixed
   value of k the function ϕ will be constant, showing that this is a plane wave
   that moves with time. A unit normal $\hat{\mathbf{n}} = (\frac{A}{m})\mathbf{i} + (\frac{B}{m})\mathbf{j} + (\frac{C}{m})\mathbf{k}$, where
   $m^2 = A^2 + B^2 + C^2.$ If the wavelength is λ, the wave speed is $\frac{\lambda c}{2\pi}$ with
   $\lambda = \frac{2\pi}{m}.$

3. $\phi = a \sin px \sin cpt$ for any $a, p \neq 0.$

5. It is sufficient to consider a harmonic wave of the form $\phi = a \cos(x - ct + k)$ with a and k being arbitrary. Then the sum

$$h(x, t, k) = a\left[\cos(x \pm ct + k) + \cos\left(x \pm ct + k + \frac{2\pi}{3}\right)\right.$$
$$\left. + \cos\left(x \pm ct + k + \frac{4\pi}{3}\right)\right]$$

represents three waves of equal amplitude, moving in either direction,
with their respective phases differing by $\frac{2\pi}{3}.$ The use of trigonometric identities to express $h(x, t, k)$ as the sum of terms $\cos(x \pm ct)$ and

$\sin(x \pm ct)$ shows $h(x, t, k) \equiv 0$, confirming that the sum is identically zero.

Solutions 4.2

1. The result follows by using Eq. (4.40). In case (a), when $c \neq \frac{\omega}{k}$,

$$\phi(x, t) = -\frac{1}{(\omega^2 - k^2 c^2)} \sin(kx - \omega t) + \frac{(kc - \omega)}{2kc(\omega^2 - k^2 c^2)} \sin[k(x + ct)]$$
$$+ \frac{(kc + \omega)}{2kc(\omega^2 - k^2 c^2)} \sin[k(x - ct)].$$

In case (b) when $c = \frac{\omega}{k}$,

$$\phi(x, t) = \frac{1}{4} \sin(x - t) - \frac{1}{4} \sin(x + t) + \frac{1}{2} t \cos(x - t).$$

In this section the 3d computer plots were made using MAPLE V, Release 5.1 software. To generate the first term in the D'Alembert solution, a unit rectangular pulse of width $2a$ moving to the right with speed c was produced by using two Heaviside unit step functions as

Heaviside(x − ct + a) − Heaviside(x − ct − a).

A function $f(x)$ moving to the right with speed c is then the function **f(x − ct)**. Finally, a 3d plot of the function

$$h(x) = \begin{cases} 0, & x < -a \\ f(x), & -a < x < a \\ 0, & x > a \end{cases}$$

moving to the right with speed c was made by plotting

eq1: = (Heaviside(x − ct + a) − Heaviside(x − ct − a))* h(x − ct).

A plot of the same function, but this time moving to the left with speed c, is obtained by reversing the sign of c. The integral term in the D'Alembert solution is obtained in similar fashion by integrating a function $k(s)$, which vanishes outside the interval $-a \le s \le a$, between the limits $x - ct \le s \le x + ct$ as follows:

eq2: = int((Heaviside(s + a) − Heaviside(s − a))
***k(s), s = x − ct .. x + ct).**

This approach is easily adapted for use with different symbolic algebra software.

3. See Fig. 4.2.1.

5. See Fig. 4.2.2.

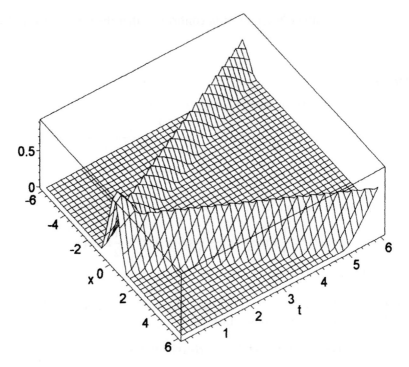

Figure 4.2.1

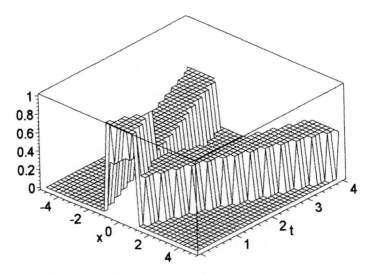

Figure 4.2.2

7. The first result follows by substituting $\phi = u \exp \frac{-pt}{2}$ in the equation $c^2\phi_{xx} = \phi_{tt} + p\phi_t$. The second result follows by observing that when p^2 can be neglected, the modified telegraph equation reduces to the wave equation in u. The third result is a direct consequence of the second result.

Solutions 4.3

1. Let the points of intersection of two pairs of parallel characteristics be at $A(x_A, t_A), B(x_B, t_B), C(x_C, t_C)$, and $D(x_D, t_D)$ shown in Fig. 4.5. Using the fact that the two families of characteristics are each families of parallel straight lines with $dx/dt = \pm c$, it is easily established that

$$x_B - ct_B = x_C - ct_C, x_A - ct_A = x_D - ct_D, x_A + t_A$$
$$= x_B + ct_B, x_D + ct_D = x_C + ct_C.$$

Thus we have

$$\phi(A) + \phi(C) = f(x_A - ct_A) + g(x_A + ct_A)$$
$$+ f(x_C - ct_C) + g(x_C + ct_C),$$
$$\phi(B) + \phi(D) = f(x_B - ct_B) + g(x_B + ct_B)$$
$$+ f(x_D - ct_D) + g(x_D + ct_D).$$

An application of the previous results then shows, as required, that

$$\phi(A) + \phi(C) = \phi(B) + \phi(D).$$

3. The approach used here is to ask the question "What conditions must be imposed on the initial conditions $h(x)$ and $k(x)$ in the ordinary D'Alembert solution in order that $\phi_x(0, t) = 0$?" Once the conditions have been determined, the functions $h_1(x)$ and $k_1(x)$ will be taken to coincide with $h(x)$ and $k(x)$ for $x > 0$, and by restricting the D'Alembert solution to the first quadrant the required solution will be obtained.

As $\phi(x, t) = f(x - ct) + g(x + ct)$ we have $\phi_x(x, t) = f'(x - ct) + g'(x + ct)$, so if $\phi_x(0, t) = 0$ it follows that $f'(-ct) = -g'(ct)$. Replacing ct by x gives $f'(-x) = -g'(x)$. Differentiation of (4.29) with respect to x for a function $h(x)$ defined for all x gives $f'(x) = \frac{1}{2}h'(x) - \frac{1}{2c}[k(x) - k(a)]$ and, similarly, differentiating (4.30) with respect to x gives $g'(x) = \frac{1}{2}h'(x) + \frac{1}{2c}[k(x) - k(a)]$. However, $f'(-x) = -g'(x)$, so $\frac{1}{2}h'(-x) - \frac{1}{2c}[k(-x) - k(a)] = -\frac{1}{2}h'(x) - \frac{1}{2c}[k(x) - k(a)]$.

The functions $h(x)$ and $k(x)$ are arbitrarily assigned initial conditions, so this last result can only be true if $h'(-x) = -h'(x)$ and $k(-x) = k(x)$, which are the conditions that $h(x)$ and $k(x)$ are *even* functions. So the solution in the first quadrant is obtained by extending $h_1(x)$ and $k_1(x)$ as even functions, and then only using the resulting D'Alembert solution in the first quadrant.

Solution 4.4

1.

$$|\phi(P, t) - \psi(P, t)| \le \frac{1}{4\pi c^2} \left| \frac{\partial}{\partial t} \left[\frac{1}{t} \iint_{S_p^{ct}} \varepsilon \, dS \right] \right| + \frac{1}{4\pi c^2 t} \left| \iint_{S_p^{ct}} \varepsilon \, dS \right|$$

$$= \frac{1}{4\pi c^2 t^2} \varepsilon 4\pi c^2 t^2 + \frac{1}{4\pi c^2 t} \varepsilon 4\pi c^2 t^2 = (1 + t)\varepsilon.$$

Thus, for any fixed $t = T$, the expression $|\phi(P, t) - \psi(P, t)|$ can be made arbitrarily small by taking $\varepsilon > 0$ sufficiently small. Hence the solution depends continuously on its initial conditions. If one space variable is kept constant, the result remains true, so the continuous dependence of the solution on the initial conditions is also true for the two-dimensional wave equation.

Solution 4.6

1. Suppose, if possible, that two solutions ϕ and ψ exist such that $\phi_{tt} = c^2\phi_{xx}$ and $\psi_{tt} = c^2\psi_{xx}$, where $\phi(x,0) = \psi(x,0) = h(x)$ and $\phi_t(x,0) = \psi_t(x,0) = k(x)$. Setting $\Phi(x,t) = \phi(x,t) - \psi(x,t)$, it follows from linearity that $\Phi(x,t)$ will satisfy the homogeneous initial conditions $\Phi(x,0) = \Phi_t'(x,0) \equiv 0$. Then from the D'Alembert solution we have that $\Phi(x,t) = 0 + \frac{1}{2c} \int_{x-ct}^{x+ct} 0 \, ds \equiv 0$. Thus $\phi(x,t) \equiv \psi(x,t)$, and uniqueness with respect to Cauchy conditions has been established.

Chapter 5

Solutions 5.1

3.

$$b_n = \frac{1}{L} \int_{-L}^{L} f(x) \sin \frac{n\pi x}{L} \, dx$$

$$= \frac{1}{L} \int_{-L}^{0} f(x) \sin \frac{n\pi x}{L} \, dx + \frac{1}{L} \int_{0}^{L} f(x) \sin \frac{n\pi x}{L} \, dx$$

$$= -\frac{1}{L} \int_{0}^{L} f(-u) \sin \frac{n\pi u}{L} \, du + \frac{1}{L} \int_{0}^{L} f(x) \sin \frac{n\pi x}{L} \, dx$$

$$= \frac{2}{L} \int_{0}^{L} f(x) \sin \frac{n\pi x}{L} \, dx,$$

because $\sin(\frac{n\pi x}{L})$ is an odd function, u is a dummy variable and so may be replaced by x, and $f(x)$ is an odd function so $f(-x) = -f(x)$. The fact that $a_n = 0$, for $n = 0, 1, \ldots$, follows in similar fashion after using the fact that $\cos(\frac{n\pi x}{L})$ is an even function while $f(x)$ is an odd function.

13. Replace the sine and cosines in (5.2) by their exponential forms $\exp(\frac{in\pi x}{L})$ and $\exp(\frac{-n\pi x}{L})$ and group terms with corresponding values of n.

15. $c_0 = \frac{\pi}{4}, c_1 = 0, c_2 = \frac{5}{2}(3 - \pi), c_3 = 0, c_4 = \frac{1}{8}(153\pi - 480)$, so
$f(x) \approx \frac{\pi}{4} P_0(x) + \frac{5}{2}(3 - \pi) P_2(x) + \frac{1}{8}(153\pi - 480) P_4(x)$.
Note that the odd coefficients vanish, leaving only the even coefficients. This is to be expected, because $f(x) = \frac{1}{1+x^2}$ is an *even* function.

Solutions 5.2

1. Proceed as described in the problem.

3. $f(x) = \frac{\pi}{2} - \frac{4}{\pi} \sum_{n=1}^{\infty} \frac{\cos(2n-1)x}{(2n-1)^2}$. The function $f(x)$ is continuous at $x = 0$, where $f(x) = 0$, so setting $x = 0$ in the Fourier series gives
$0 = \frac{\pi}{2} - \frac{4}{\pi} \sum_{n=1}^{\infty} \frac{1}{(2n-1)^2}$, so $\pi^2 = 8 \sum_{n=1}^{\infty} \frac{1}{(2n-1)^2}$.

5. $e^{ax} = \frac{2\sinh a\pi}{\pi}[\frac{1}{2a} + \sum_{n=1}^{\infty} (-1)^n(\frac{a\cos nx - n\sin nx}{a^2+n^2})]$, $-\pi < x < \pi$. e^{ax} is continuous at $x = 0$ where it equals 1, so $1 = \frac{2\sinh a\pi}{\pi}[\frac{1}{2a} + \sum_{n=1}^{\infty} \frac{(-1)^n a}{a^2+n^2}]$, so $\operatorname{csch} a\pi = \frac{2}{\pi}[\frac{1}{2a} + \sum_{n=1}^{\infty} \frac{a}{a^2+n^2}]$.

7. $f(x) = \frac{\pi^2}{12} + \sum_{n=1}^{\infty} \frac{\cos nx}{n^2}$. The function is continuous at $x = 0$ where it equals $\frac{\pi^2}{4}$, so setting $x = 0$ in the Fourier series and using this result gives $\frac{1}{6}\pi^2 = \sum_{n=1}^{\infty} \frac{1}{n^2}$.

9. The function is continuous for $-\pi \le x \le \pi$ with $f(-\pi) = f(\pi)$. $f'(x)$ is continuous except at $x = 0$, so differentiation of the Fourier series for $f(x)$ will yield the Fourier series for $f'(x)$ except at $x = 0, \pm\pi$. $|\sin x| = \frac{2}{\pi} - \frac{4}{\pi} \sum_{n=1}^{\infty} \frac{\cos 2nx}{(4n^2-1)}$, so differentiation gives $f'(x) = \frac{8}{\pi} \sum_{n=1}^{\infty} \frac{n\sin 2nx}{(4n^2-1)}$ for $x \ne 0, \pm\pi$.

11. The calculations involved are routine. $\pi^4 = 90 \sum_{n=1}^{\infty} \frac{1}{n^4}$.

Solutions 5.3

For Fourier–Legendre expansions where $f(x) = \sum_{n=0}^{\infty} a_n P_n(x)$ with $-1 \le x \le 1$, $a_n = (\frac{2n+1}{2}) \int_{-1}^{1} f(x) P_n(x)\, dx$. For all but the simplest functions the coefficients a_n must be evaluated numerically.

1. $a_0 = 0.3183$, $a_1 = 0.6079$, $a_2 = 0.3088$, $a_3 = -0.1124$, $a_4 = -0.1822$, $a_5 = 0.0046$, $a_6 = 0.0875$, $a_7 = -0.0001$, $a_8 = -0.0544$, $a_9 = 0$.

3. $a_0 = 1$, $a_1 = 0$, $a_2 = -0.625$, $a_3 = 0$, $a_4 = 0.1875$, $a_5 = 0$, $a_6 = -0.1015$, $a_7 = 0$.

For Fourier–Bessel expansions, where $f(x) = \sum_{r=0}^{\infty} a_r J_n(j_{n,r} x/R)$ with $0 \leq x/R \leq 1$, $a_r = (2/[J_{n+1}(j_{n,r})]^2) \int_0^R x f(x) J_n(j_{n,r} x/R) dx$. For all but the simplest functions the coefficients a_n must be evaluated numerically.

5. $a_1 = 1.2960$, $a_2 = -0.9499$, $a_3 = 0.7873$, $a_4 = -0.6874$, $a_5 = 0.6181$, $a_6 = -0.5662$, $a_7 = 0.5256$, $a_8 = -0.4927$.

Chapter 6

Solutions 6.1

1. Reason as in Example 6.1 to show that $\lambda_n = \frac{n\pi}{a}$ for $n = 0, 1, \ldots$. Use the boundary conditions to show the solution corresponding to $n = 0$ is $X_0 = c_0$ and that the solutions corresponding to $n > 1$ are $X_n = A_n \cos(\frac{n\pi x}{a})$. Next show that $Y_0 = c_0(b - y)$ and the solutions $Y_n = D_n[\sinh(\frac{n\pi y}{a}) - \tanh(\frac{n\pi b}{a})\cosh(\frac{n\pi y}{a})]$ for $n > 0$. Thus the general form of the solution is

$$u(x, y) = c_0(b - y) + \sum_{n=1}^{\infty} D_n\left[\sinh\left(\frac{n\pi y}{a}\right)\right.$$
$$\left. - \tanh\left(\frac{n\pi b}{a}\right)\cosh\left(\frac{n\pi y}{a}\right)\right]\cos\left(\frac{n\pi x}{a}\right).$$

Next set $y = 0$ and use the boundary condition $u(x, 0) = \frac{1}{2}\cos^2\frac{\pi x}{a}$. Fourier series need not be used when determining the constants c_0 and D_n if we note that $\frac{1}{2}\cos^2\frac{\pi x}{a} = \frac{1}{4} + \frac{1}{4}\cos\frac{2\pi x}{a}$. Setting $y = 0$ in $u(x, y)$ and using this last condition gives

$$\frac{1}{4} + \frac{1}{4}\cos\frac{2\pi x}{a} = c_0 b - \sum_{n=1}^{\infty} D_n \tanh\frac{n\pi b}{a}\cos\frac{n\pi x}{a}.$$

For this to become an identity the coefficients of corresponding terms must be equal, so

$$\frac{1}{4} = c_0 b, \quad \frac{1}{4} = -D_2 \tanh\frac{2\pi b}{a} \quad \text{and} \quad D_n = 0 \quad \text{for } n = 1, 3, 4, \ldots.$$

Thus $u(x, y) = \frac{1}{4b}(b - y) + \frac{1}{4}(\cosh\frac{2\pi y}{b} - \coth\frac{2\pi b}{a}\sinh\frac{2\pi y}{a})\cos\frac{2\pi x}{a}$. This is a finite term (closed form) solution, because it satisfies the PDE and the boundary conditions.

In the last part of the problem the two solutions can be added because:

(i) the PDE is linear and homogeneous and the regions are identical;
(ii) the first problem has homogeneous conditions on three sides of the rectangle with the exception of $y = 0$, whereas like the first problem the second problem has homogeneous conditions on the sides $x = 0$

and $x = a$, the same nonhomogeneous condition on $y = 0$, and a nonhomogeneous condition on $y = b$. These conditions are compatible and define a composite problem, so the two solutions can be added.

3. $\lambda_n = (2n-1)^2\pi^2/(4a^2)$ for $n = 1, 2, \ldots, X_n(x) = \sin[(2n-1)\pi x/(2a)]$.

5. $\lambda_n = (n\pi/a)^2$ for $n = 1, 2, \ldots, X_n(x) = A_n \cos(n\pi x/a) + B_n \sin(n\pi x/a)$.

7. The result follows after a routine change of variable.

9. The result follows after a routine change of variable and the introduction of separation constants.

Solutions 6.3

1. $X_n(x) = \sin \lambda_n x$, $\lambda_n = n\pi/L$, $n = 1, 2, \ldots$, and $T_n(t) = A_n \exp(-\lambda_n^2 kt)$.
   Thus $u(x, t) = \sum_{n=1}^{\infty} A_n \sin \frac{n\pi x}{L} \exp(-\frac{n^2\pi^2 kt}{L^2})$, but as $u(x, 0) = f(x)$ we have $f(x) = \sum_{n=1}^{\infty} A_n \sin \frac{n\pi x}{L}$, and so $A_n = \frac{2}{L} \int_0^L f(x) \sin \frac{n\pi x}{L} dx$.

3. $X_n(x) = \sin \lambda_n x$, $\lambda_n = \frac{n\pi}{a}$, $n = 1, 2, \ldots$, and $Y_n(y) = A_n \cosh \lambda_n y + B_n \sinh \lambda_n y$. Thus $u(x, y) = \sum_{n=1}^{\infty} \sin \frac{n\pi x}{a} (A_n \cosh \frac{n\pi y}{a} + B_n \sinh \frac{n\pi y}{a})$.
   Condition $u(x, b) = 0$ gives $0 = \sum_{n=1}^{\infty} \sin \frac{n\pi x}{a}(A_n \cosh \frac{n\pi b}{a} + B_n \sinh \frac{n\pi b}{a})$, or $B_n = -\coth \frac{n\pi b}{a}$, while condition $u(x, 0) = f(x)$ gives $f(x) = \sum_{n=1}^{\infty} A_n \sin \frac{n\pi x}{a}$, and so $A_n = \frac{2}{a} \int_0^a f(x) \sin \frac{n\pi x}{a} dx/\sinh(n\pi b/a)$. Combining results and simplifying using the hyperbolic identity $\sinh(A - B) = \sinh A \cosh B - \cosh A \sinh B$ shows that $u(x, y) = \sum_{n=1}^{\infty} A_n \sin \frac{n\pi x}{a} \sinh \frac{n\pi(b-y)}{a}$.

5. Use the result of Exercise 1 with $u(x, 0) = x^2(\pi - x)$ to obtain
$$u(x, t) = 4 \sum_{n=1}^{\infty} \frac{[2(-1)^{n+1} - 1]}{n^3} \sin nx \exp(-n^2 kt).$$

7. (a) $B_1(t) = t$, $B_n(t) = 0$, $n \geq 2$. $C_1(t) = t - \sin t$, $C_n(t) = 0$, $n \geq 2$. $v(x, t) = (t - \sin t) \sin x$, $U(x, t) = 0$, so $u(x, t) = U(x, t) + v(x, t) = (t - \sin t) \sin x$.
   (b) $B_n(t) = \frac{2}{\pi} \int_0^\pi x \sin x \sin nx \, dx$ so $B_1(t) = \frac{\pi}{2}$, $B_n = -\frac{4(1+(-1)^n)}{\pi n^2(n^2-1)^2}$.
   $C_n'' + n^2 C_n = B_n$, so $C_1(t) = \frac{\pi}{2}(1 - \cos t)$, $C_n = \frac{4(1+(-1)^n)}{\pi n^2(n^2-1)^2}(\cos nt - 1)$.
   Thus $u(x, t) = \frac{1}{2}\pi(1 - \cos t) \sin x + \frac{4}{\pi} \sum_{n=2}^{\infty} \frac{(1+(-1)^n)}{n^2(n^2-1)^2}(\cos nt - 1) \sin nx$.

9. The result follows from the result of Exercise 2 by setting
$$f(x) = \begin{cases} hx/a, & 0 \leq x \leq a \\ h(L - x)/(L - a), & a \leq x \leq L \end{cases}$$

and $g(x) = 0$, from which it follows that

$$u(x, t) = \frac{2hL^2}{\pi^2 a(L - a)} \sum_{n=1}^{\infty} \frac{1}{n^2} \sin \frac{n\pi a}{L} \sin \frac{n\pi x}{L} \cos \frac{n\pi act}{L}.$$

11. $u(x, y, t) = T(x) V(x, y)$, $\frac{T''}{c^2 T} = \frac{V_{xx} + V_{yy}}{V} = -k^2$, so $T'' + k^2 c^2 T = 0$ and $V_{xx} + V_{yy} + k^2 V = 0$. Now set $V(x, y) = X(x) Y(y)$, when we find that $\frac{Y''}{Y} + k^2 = -\frac{X''}{X}$, leading to the results $X'' + k_1^2 X = 0$ and $Y'' + k_2^2 Y = 0$, where k_1^2 and k_2^2 are separation constants with $k^2 = k_1^2 + k_2^2$. Thus

$$X(x) = A \cos k_1 x + B \sin k_1 x, \qquad Y(y) = C \cos k_2 y + D \sin k_2 y.$$

The homogeneous boundary conditions require that $X(x) = \sin k_1 x$ and $Y(y) = \sin k_2 y$ where $\sin k_1 \alpha = 0$ and $\sin k_2 \beta = 0$. Thus $k_{1,m} = \frac{m\pi}{\alpha}$ and $k_{2,n} = \frac{n\pi}{\beta}$ with $m, n = 1, 2, \ldots$. Hence $(k_{m,n})^2 = \pi^2(\frac{m^2}{\alpha^2} + \frac{n^2}{\beta^2})$, showing that the eigenfunctions are

$$V_{m,n}(x, y) = \sin \frac{m\pi x}{\alpha} \sin \frac{n\pi y}{\beta}.$$

The general form of the solution is thus

$$u(x, y, t) = \sum_{m,m=1}^{\infty} [A_{mn} \cos(k_{m,n} ct) + B_{mn} \sin(k_{m,n} ct)] \sin \frac{m\pi x}{\alpha} \sin \frac{n\pi y}{\beta}.$$

From the initial conditions it then follows that

$$A_{mn} = \frac{4}{\alpha\beta} \int_0^{\alpha} \int_0^{\beta} f(x, y) \sin \frac{m\pi x}{\alpha} \sin \frac{n\pi y}{\beta} \, dx \, dy \quad \text{and}$$

$$B_{mn} = \frac{4}{ck_{mn}\alpha\beta} \int_0^{\alpha} \int_0^{\beta} g(x, y) \sin \frac{m\pi x}{\alpha} \sin \frac{n\pi y}{\beta} \, dx \, dy.$$

Substituting $f(x, y) = 2 \sin(\frac{3\pi x}{\alpha}) \sin(\frac{\pi y}{\beta})$ and $g(x, y) = 0$, the solution is seen to be $u(x, y, t) = 2 \sin \frac{3\pi x}{\alpha} \sin \frac{\pi y}{\beta} \cos(k_{3,1} ct)$ with $(k_{3,1})^2 = \pi^2(\frac{9}{\alpha^2} + \frac{1}{\beta^2})$. The solution is exact and simple because the initial condition $u(x, y, 0)$ is proportional to an eigenfunction while $g(x, y) = 0$, so the series solution reduces to a single term.

13. $u(r, t) = R(r) T(t)$ and after the introduction of a separation constant λ^2, $\frac{T''}{c^2 T} = R'' + \frac{1}{r} R' = -\lambda^2$, so $T'' + \lambda^2 c^2 T = 0$, $r R'' + R' - \lambda^2 r R = 0$. Thus, $T(t) = C \cos \lambda ct + D \sin \lambda ct$, and $R(r) = A J_0(\lambda r) + B Y_0(\lambda r)$. $R(r)$ must remain finite at the origin so $B = 0$, showing that $R(r) = J_0(\lambda r)$. The boundary condition $u(R, t) = 0$ means that $J_0(\lambda R) = 0$, so $\lambda_n = \frac{j_{0,n}}{R}$, with $j_{0,n}$ being the nth zero of $J_0(r)$. Thus the general form of

the solution is

$$u(r, t) = \sum_{n=1}^{\infty} J_0\left(\frac{j_{0,n}r}{R}\right)\left(C_n \cos \frac{j_{0,n}ct}{R} + D_n \sin \frac{j_{0,n}ct}{R}\right).$$

Initially the membrane is in equilibrium so $u(r, 0) = 0$, so hence $0 = \sum_{n=1}^{\infty} C_n J_0\left(\frac{j_{0,n}r}{R}\right)$, but this is only possible if $C_n = 0$ for $n = 1, 2, \ldots$, so

$$u(r, t) = \sum_{n=1}^{\infty} D_n J_0\left(\frac{j_{0,n}r}{R}\right) \sin \frac{j_{0,n}ct}{R} \quad \text{and}$$

$$u_t(r, t) = \sum_{n=1}^{\infty} \frac{j_{0,n}c}{R} D_n J_0\left(\frac{j_{0,n}r}{R}\right) \cos \frac{j_{0,n}ct}{R}.$$

The initial speed is $u_t(r, 0) = h(1 - \frac{r^2}{R^2})$, so

$$h\left(1 - \frac{r^2}{R^2}\right) = \sum_{n=1}^{\infty} D_n \frac{j_{0,n}c}{R} J_0\left(\frac{j_{0,n}r}{R}\right).$$

Using the orthogonality properties of Bessel functions we find that

$$D_n = \frac{hR}{j_{0,n}c} \frac{\int_0^R r\left(1 - \frac{r^2}{R^2}\right) J_0\left(\frac{j_{0,n}r}{R}\right) dr}{\int_0^R r\left[J_0\left(\frac{j_{0,n}r}{R}\right)\right]^2 dr}.$$

As $\int_0^R r\left(1 - \frac{r^2}{R^2}\right) J_0\left(\frac{j_{0,n}r}{R}\right) dr = \frac{4R^2 J_1(j_{0,n})}{(j_{0,n})^3}$, $D_n = \frac{8hR^3}{c(j_{0,n})^4 J_1(j_{0,1})}$, and so

$$u(r, t) = \frac{8hR^3}{c} \sum_{n=1}^{\infty} \frac{J_0\left(\frac{j_{0,n}r}{R}\right)}{(j_{0,n})^4 J_1(j_{0,1})} \sin \frac{j_{0,n}ct}{R}.$$

15. Reason as in Exercise 14 to show that

$$u(r, \theta) = \frac{4U}{\pi} \sum_{n=0}^{\infty} \frac{1}{(2n+1)} \left(\frac{r}{\rho}\right)^{2n+1} \sin(2n+1)\theta.$$

Note that this solution could have been deduced directly from the result of Exercise 14 because that solution is an odd function of θ, so if the solution is extended to $\pi < \theta < 2\pi$ it will automatically assign the value $-U$ to the lower boundary of the circle.

17. The solution outside the sphere is of the form $u(r, \phi, \theta) = \sum_{n=0}^{\infty} \frac{B_n}{r^{n+1}} P_n(\xi)$, so from the boundary condition on $r = 1$, using the fact that $\sin\theta = (1 - \xi^2)^{1/2}$, it follows that $U(1 - \xi^2)^{1/2} = \sum_{n=0}^{\infty} B_n P_n(\xi)$. Using the orthogonality property of Legendre polynomials we have $U \int_{-1}^{1} (1 - \xi^2)^{1/2} P_n(\xi) d\xi = \frac{2}{2n+1} B_n$, so if $P_n(\xi)$ is an odd function of n and as $(1 - \xi^2)^{1/2}$ is an even function, we have $0 = B_1 = B_3 = \ldots$. Integration

gives $B_0 = \frac{\pi U}{4}$, $B_2 = \frac{-5\pi U}{32}$, $B_4 = -\frac{9\pi U}{256}$, ..., so the required solution is

$$u(r,\phi,\theta) = \frac{\pi U}{4}\left(\frac{1}{r}P_0(\xi) - \frac{5}{8}\frac{1}{r^3}P_2(\xi) - \frac{9}{64}\frac{1}{r^5}P_4(\xi) - \cdots\right).$$

19. The solution between the two spheres is of the form

$$u(r,\theta,\phi) = \sum_{n=1}^{\infty}\left(\frac{A_n r^n + B_n}{r^{n+1}}\right)P_n(\xi) \quad \text{with} \quad \xi = \cos\theta.$$

Matching the boundary condition on $r = \frac{1}{2}$ gives,

$$U = \sum_{n=0}^{\infty}\left[A_n\frac{1}{2}^n + B_n(2)^{n+1}\right].$$

Using the orthogonality property of Legendre polynomials gives $U\int_{-1}^{1}$ $P_n(\xi)d\xi = (\frac{2}{2n+1})[A_n\frac{1}{2}^n + B_n(2)^{n+1}]$, and so $A_0 + 2B_0 = U, \frac{1}{2}A_1 + 4B_1 = 0, \frac{1}{4}A_2 + 8B_2 = 0, \frac{1}{8}A_3 + 16B_3 = 0, \ldots$. Repeating this process for the boundary condition on $r = 1$ gives $A_n + B_n = (\frac{2n+1}{2})U$ $[\int_{-1}^{0}P_n(\xi)d\xi + \int_{0}^{1}(1-\xi^2)^{1/2}P_n(\xi)d\xi]$. After using numerical integration to evaluate $A_n + B_n$ it follows that $A_0 = 0.7845U$, $B_0 = 0.1073U$, $A_1 = -0.2857U$, $B_1 = 0.0357U$, $A_2 = -0.2533U$, $B_2 = 0.0079U$, $A_3 = -0.1365U$, $B_3 = 0.0011U, \ldots$. Thus the solution is

$$\frac{u(r,\phi,\xi)}{U} = \left(0.7854 + \frac{0.1073}{r}\right)P_0(\xi) + \left(-0.2857r + \frac{0.0357}{r^2}\right)P_1(\xi)$$
$$+ \left(-0.2533r^2 + \frac{0.0079}{r^3}\right)P_2(\xi) + \cdots$$

21. Setting $u(r, z) = R(r)Z(z)$ and introducing the separation constant λ^2 lead to $rR'' + R' - \lambda^2 R = 0$ and $Z'' + \lambda^2 Z = 0$ and so $R(r) = AJ_0(\lambda r) + BY_0(\lambda r)$ while $Z(z) = C\cos\lambda z + D\sin\lambda z$. For the solution to be finite when $r = 0$ we must set $B = 0$ so that $R(r) = AJ_0(\lambda r)$. The boundary condition $(u_r + hu)|_{r=\rho} = 0$ shows that the eigenvalue λ_n must be a root of the equation $-\lambda J_1(\lambda\rho) + hJ_0(\lambda\rho) = 0$. So if λ_n is the nth positive root of this equation, the solution will be of the form

$$u(r, z) = \sum_{n=1}^{\infty} J_0(\lambda_n r)[C_n\cos\lambda_n z + D_n\sin\lambda_n z].$$

To satisfy the boundary condition $u(r, 0) = 0$ we must have $0 = \sum_{n=1}^{\infty} C_n J_0(\lambda_n r)$ for $0 \leq r \leq \rho$, which is only possible if $C_n = 0$ for $n = 1, 2, \ldots$, and so $u(r, z) = \sum_{n=1}^{\infty} D_n J_0(\lambda_n r)\sin\lambda_n z$. The other boundary condition requires that $u(r, H) = U$, so $U = \sum_{n=1}^{\infty} D_n J_0(\lambda_n r)\sin\lambda_n H$.

Using the orthogonality property of the Bessel function $J_0(r)$ gives

$$U \int_0^\rho r J_0(\lambda_n r)\, dr = D_n \sin \lambda_n H \int_0^\rho r [J_0(\lambda_n r)]^2\, dr,$$

so using the result $\int_0^\rho r[J_0(\lambda_n r)]^2\, dr = \frac{\rho^2}{2}[\lambda_n^2(J_1(\lambda_n \rho))^2 + (J_0(\lambda_n \rho))^2]$ with the boundary condition $-\lambda J_1(\lambda \rho) + h J_0(\lambda \rho) = 0$ shows that

$$D_n = \frac{2U}{\lambda_n \rho(1 + h^2)} \frac{1}{[J_0(\lambda_n \rho)]^2 \sin \lambda_n H}.$$

Hence the solution is

$$u(r, z) = \frac{2U}{\rho(1 + h^2)} \sum_{n=1}^{\infty} \frac{J_0(\lambda_n r) \sin \lambda_n z}{\lambda_n [J_0(\lambda_n \rho)]^2 \sin \lambda_n H}.$$

23. Set $u(x, t) = X(x)T(t)$. After introducing a separation constant λ^2, the variables can be separated and the PDE written as $\frac{T'}{(\kappa T)} + \frac{\alpha}{\kappa} = \frac{X''}{X} = -\lambda^2$. Thus $X(x) = A\cos(\lambda x) + B\sin(\lambda x)$ and $T(t) = Ce^{-\alpha t}\exp(-\kappa\lambda^2 t)$. The first boundary condition $u_x(0, t) = 0$ shows that $B = 0$ while the second boundary condition $u_x(L, t) = 0$ shows that $\lambda = 0$ or $\sin(\lambda L) = 0$, so $\lambda_n = n\pi$ for $n = 0, 1, 2, \dots$. Thus as $X_n(x) = A_n \cos \frac{n\pi x}{L}$ and $T_n(t) = e^{-\alpha t} \exp[\frac{-\kappa n^2\pi^2 t}{L^2}]$, the solution is of the form

$$u(x, t) = e^{-\alpha t} \sum_{n=1}^{\infty} A_n \cos\left(\frac{n\pi x}{L}\right) \exp\left[\frac{-\kappa n^2\pi^2 t}{L^2}\right].$$

Using the initial condition $u(x, 0) = hx(L - x)$ this becomes

$$hx(L - x) = \sum_{n=1}^{\infty} A_n \cos\left(\frac{\lambda_n x}{L}\right),$$

showing that A_n is the nth term in the half-range Fourier cosine series expansion of $hx(L - x)$ over the interval $0 \le x \le L$. As $A_0 = \frac{hL^2}{6}$ and $A_n = -2hL^2(\frac{1+(-1)^n}{n^2\pi^2})$ for $n > 1$, the solution becomes

$$u(x, t) = hL^2 e^{-\alpha t}\left[\frac{1}{6} - 2\sum_{n=1}^{\infty}\left(\frac{1 + (-1)^n}{n^2\pi^2}\right) \cos\frac{n\pi x}{L} \exp\left(-\frac{\kappa n^2\pi^2 t}{L^2}\right)\right].$$

This corresponds to a rod of length L with its ends at $x = 0$ and $x = L$ thermally insulated, with heat being removed along the rod at a rate equal to αu, with the initial temperature distribution given by $u(x, 0) = hx(L - x)$.

25. The temperature distribution will be of the form

$$T(r,\theta) = \sum_{n=0}^{\infty} \left(a_n r^n + \frac{b_n}{r^{n+1}} \right) P_n(\cos\theta).$$

To satisfy the boundary conditions we must have $\sum_{n=0}^{\infty}(a_n R_1^n + \frac{b_n}{R_1^{n+1}}) P_n(\cos\theta) = T_0$ on the inner boundary, and $\sum_{n=0}^{\infty}(a_n R_2^n + \frac{b_n}{R_2^{n+1}}) P_n(\cos\theta) = T_0|\cos\theta|$ on the outer boundary. Expanding $|\cos\theta|$ in a Legendre series and setting $x = \cos\theta$ gives $|x| = \sum_{n=0}^{\infty} c_n P_n(x)$ for $-1 \leq x \leq 1$, with $c_n = (\frac{2n+1}{2})\int_{-1}^{1} |x|\, dx$, from which it follows that $c_0 = \frac{1}{2}, c_1 = 0, c_2 = \frac{5}{8}, c_3 = 0, c_4 = \frac{-3}{16}, \ldots$ Returning to the boundary conditions we have

$$a_0 + \frac{b_0}{R_1} = T_0, \quad a_n R_1^n + \frac{b_n}{R_1^{n+1}} = 0, \quad \text{for } n = 1, 2, \ldots \quad (\text{at } r = R_1)$$

and

$$a_0 + \frac{b_0}{R_2} = \frac{T_0}{2} \qquad a_2 R_2^2 + \frac{b_2}{R_2^3} = \frac{5T_0}{8}$$

$$a_1 R_2 + \frac{b_1}{R_2^2} = 0, \qquad a_3 R_2^3 + \frac{b_3}{R_2^4} = 0 \qquad (\text{at } r = R_2),$$

$$a_4 R_2^4 + b_4/R_2^5 = -3T_0/16, \ldots.$$

Thus

$$a_0 = \frac{(R_2 - 2R_1)T_0}{2(R_2 - R_1)}, \qquad b_0 = \frac{R_1 R_2 T_0}{2(R_2 - R_1)}$$

$$a_2 = -\frac{5T_0}{8(R_1^2 - R_2^2)}, \qquad b_2 = \frac{5R_1^2 R_2^3 T_0}{8(R_1^2 - R_2^2)}$$

$$a_4 = \frac{3T_0}{16(R_1^4 - R_2^4)}, \qquad b_4 = -\frac{3R_1^4 R_2^5 T_0}{16(R_1^4 - R_2^4)},$$

$$\cdots$$

The temperature distribution now follows by substituting a_n and b_n into

$$T(r,\theta) = \sum_{n=0}^{\infty} \left(a_n r^n + \frac{b_n}{r^{n+1}} \right) P_n(\cos\theta).$$

27. Using the data of Example 6.12 it follows that the $\bar{X}_i(x)$ are unchanged, but the coefficients R_i become $R_1 = 1.6490$, $R_2 = -0.1783$, $R_3 = 0.9745$, $R_4 = 0.0270$, $R_5 = -0.2931$. Plot $\frac{V(x,t)}{T_0}$ for $t = 0, 15, 50$, and 75 s using a five-term approximation. The Gibbs phenomena at $x = 0$ and $x = 10$ are different because different materials are involved, but

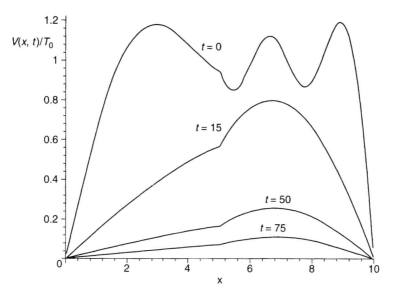

Figure 6.3.1

by $t = 15$ s they are no longer apparent and subsequent temperature distributions are similar to those in Fig. 6.8a, although, as would be expected, the respective magnitudes are greater.

Chapter 7

Solutions 7.2

1. The calculation is straightforward and is performed as indicated in the problem.

3. If $f(\psi) = k$, then for $0 \leq r \leq r_0$,

$$u(r,\theta) = \frac{1}{2\pi} \int_0^{2\pi} \frac{k}{r_0^2 - 2rr_0 \cos(\psi - \theta) + r^2}\, d\psi$$

$$= \frac{k}{2\pi} \int_0^{2\pi} P(r_0, r, \psi - \theta)\, d\psi = \frac{2k\pi}{2\pi} = k.$$

5. Use the method of Example 7.1. Using trigonometric identities shows that $u(1,\theta) = \sin^2 \theta + 2\cos^3 \theta = \frac{1}{2} - \frac{1}{2}\cos 2\theta + \frac{3}{2}\cos\theta + \frac{1}{2}\cos 3\theta$, but on $r = 1$ this must equal $u(1,\theta) = a_0 + \sum_{n=1}^{\infty} (a_n \cos n\theta + b_n \sin n\theta)$. So equating coefficients of corresponding terms gives $a_0 = \frac{1}{2}$, $a_1 = \frac{3}{2}$, $a_2 = -\frac{1}{2}$, $a_3 = \frac{1}{2}$, $a_n = 0$, $n > 3$, and $b_n = 0$, $n = 1, 2, \ldots$. Thus the required solution is

$$u(r,\theta) = \frac{1}{2} + \frac{3}{2}r\cos\theta - \frac{1}{2}r^2 \cos 2\theta + \frac{1}{2}r^3 \cos 3\theta \quad \text{for } 0 \leq r \leq 1.$$

7. The proof of uniqueness follows immediately by supposing, if possible, that two different harmonic solutions u and v both satisfy the same boundary condition in a region D. The linearity of the equation means that $w = u - v$ is also a solution, but this time satisfying the condition $w = 0$ on the boundary Γ of D. The maximum/minimum theorem for harmonic functions then asserts that the maximum and the minimum of w must be zero throughout D, and so $u \equiv v$ in D, thereby establishing the uniqueness of the solution.

9. From the Poisson integral formula for the half-plane

$$u(x, y) = \frac{y}{\pi} \int_{-2}^{2} \frac{ds}{(x - s)^2 + y^2}$$

$$= \frac{1}{\pi} \left\{ \text{Arctan} \left[\frac{(x + 2)}{y} \right] - \text{Arctan} \left[\frac{(x - 2)}{y} \right] \right\}.$$

11. From the Poisson integral formula for the half-plane

$$u(x, y) = \frac{y}{\pi} \int_{-1}^{0} \frac{1 + s}{(x - s)^2 + y^2} \, ds + \frac{y}{\pi} \int_{0}^{1} \frac{1 - s}{(x - s)^2 + y^2} \, ds$$

$$= -\frac{y}{\pi} \int_{-1}^{0} \frac{(x - s)}{(x - s)^2 + y^2} \, ds + \frac{(1 + x)}{\pi} \int_{-1}^{0} \frac{y}{(x - s)^2 + y^2} \, ds$$

$$+ \frac{y}{\pi} \int_{0}^{1} \frac{(x - s)}{(x - s)^2 + y^2} \, ds + \frac{(1 - x)}{\pi} \int_{0}^{1} \frac{y}{(x - s)^2 + y^2} \, ds$$

$$= -\frac{y}{\pi} \int_{-1}^{0} \frac{(x - s)}{(x - s)^2 + y^2} \, ds + \frac{y}{\pi} \int_{0}^{1} \frac{(x - s)}{(x - s)^2 + y^2} \, ds$$

$$+ \frac{2x}{\pi} \text{Arctan} \left(\frac{y}{x} \right) - \left(\frac{1 + x}{\pi} \right) \text{Arctan} \left\{ \frac{y}{x + 1} \right\}$$

$$+ \left(\frac{1 - x}{\pi} \right) \text{Arctan} \left\{ \frac{y}{x - 1} \right\}.$$

Thus, after evaluating the first two integrals, the solution becomes

$$u(x, y) = \frac{y}{2\pi} \ln \left[\frac{(x^2 + y^2)^2}{\{(x + 1)^2 + y^2\}\{(x - 1)^2 + y^2\}} \right] + \frac{2x}{\pi} \text{Arctan} \left(\frac{y}{x} \right)$$

$$- \left(\frac{1 + x}{\pi} \right) \text{Arctan} \left\{ \frac{y}{x + 1} \right\} + \left(\frac{1 - x}{\pi} \right) \text{Arctan} \left\{ \frac{y}{x - 1} \right\}.$$

Solutions 7.3

1. Set $\tau = \kappa t$, leave x unchanged, and write $u(x, t) = v(x, \tau)$. Then $u_t = \kappa v_\tau$ and $u_{xx} = v_{xx}$, causing the equation $u_t = \kappa u_{xx}$ to simplify to $v_\tau = v_{xx}$.

3. It is simplest to substitute U into the equation $U_t = U_{\xi\xi}$ to show this leads to the equation $u_t = u_{xx} - b(t)u_x$. Partial differentiation of $U(\xi, t) = u(\xi + \int_0^t b(\tau)d\tau, t)$ with respect to t gives $U_t = u_x \frac{\partial}{\partial t}(\xi + \int_0^t b(\tau)d\tau) + u_t$, but from the fundamental theorem of calculus $\frac{\partial}{\partial t}(\xi + \int_0^t b(\tau)d\tau) = b(t)$, so $U_t = b(t)u_x + u_t$. Partial differentiation of $U(\xi, t)$ with respect to x gives $U_\xi = \frac{\partial u}{\partial x}\frac{\partial x}{\partial \xi} = u_x$, because $x = \xi + \int_0^t b(\tau)d\tau$ and, similarly, $U_{\xi\xi} = u_{xx}$. Substitution into $U_t = U_{\xi\xi}$ transforms it into $u_t = u_{xx} - b(t)u_x$, thereby establishing the validity of the transformation.

5. The calculation is routine, because $K_t = K_{xx} = \frac{1}{8}\frac{(x^2 - 2t)}{t^{5/2}\sqrt{\pi}}\exp[\frac{-x^2}{4t}]$. The redefined function K is the fundamental solution of $K_t = \kappa K_{xx}$, which is the heat equation with diffusivity κ.

Solutions 7.4

1. Setting $u = \frac{df}{dx}$, the equation for u has the solution $u(x) = B\exp(\frac{-kx^2}{2})$, and integrating over $0 \le s \le x$ gives $f(x) = f(0) + B\int_0^x \exp(-\frac{1}{2}ks^2)ds$. Now $\int_0^x \exp(-\frac{1}{2}s^2)ds = \sqrt{\frac{\pi}{2}}\mathrm{erf}(x/\sqrt{2})$, so $f(x) = f(0) + B\sqrt{\frac{\pi}{2}}\mathrm{erf}(x\sqrt{\frac{k}{2}})$, and after redefining constants $f(x) = A + B\,\mathrm{erf}(x\sqrt{\frac{k}{2}})$.

3. Set $\frac{u - U_0}{U_1 - U_0} = f(\beta)$ with $\beta = \frac{Dt^m}{x^n}$, then substitution into the heat equation $u_t = \kappa u_{xx}$ followed by simplification gives

$$\frac{m}{t}f'(\beta) = \frac{\kappa n(n+1)}{x^2}f'(\beta) + \frac{\kappa n^2 Dt^m}{x^{n+2}}f''(\beta).$$

Setting $n = -1$, substituting for x in terms of β, and setting $m = -\frac{1}{2}$ make t vanish and $\beta = \frac{Dx}{\sqrt{t}}$. The equation for $f(\beta)$ then becomes the same as in (7.19) apart from the symbol D in place of C.

Solutions 7.5

3.
$$u(x, t) = \frac{1}{2}\left[\mathrm{erf}\left(\frac{x}{2\sqrt{\kappa t}}\right) - \mathrm{erf}\left(\frac{x - 1}{2\sqrt{\kappa t}}\right)\right].$$

5.
$$u(x, t) = \frac{1}{2\sqrt{\pi}}\left[2\sqrt{\kappa t}\left\{\exp\left(\frac{-x^2}{4\kappa t}\right) - \exp\left(\frac{-(x-1)^2}{4\kappa t}\right)\right\}\right.$$
$$\left. + x\sqrt{\pi}\left\{\mathrm{erf}\left(\frac{x}{2\sqrt{\kappa t}}\right) - \mathrm{erf}\left(\frac{x-1}{2\sqrt{\kappa t}}\right)\right\}\right]$$

When making a 3d plot of $u(x, t)$ choose a value of $\kappa > 0$, choose a finite space interval for x about the origin, and as the solution is not valid when $t = 0$ use a finite time interval starting from some small positive value of t.

7.

$$u(x, t) = \frac{1}{2}\left[\text{erf}\left(\frac{x+2}{2\sqrt{\kappa t}}\right) - \text{erf}\left(\frac{x-2}{2\sqrt{\kappa t}}\right)\right].$$

When making a 3d plot of $u(x, t)$ choose a value of $\kappa > 0$, choose a finite space interval of the form $0 \le x \le L$, and as the solution is not valid when $t = 0$ use a finite time interval starting from some small positive value of t.

Solutions 7.6

1. The reasoning proceeds as in the text until the equation determining $B_n(\tau)$, and after using the identity $\sin \pi x \cos \pi x = \frac{1}{2} \sin 2\pi x$ this becomes $\frac{1}{2} e^{-2x} \sin 2x = \sum_{n=1}^{\infty} B_n \exp(-n^2\kappa\tau) \sin nx$. Comparing terms in $\sin nx$ shows that $B_2(\tau) = \frac{1}{2} \exp[(4\kappa - 2)\tau]$, and all other $B_n(\tau)$ are zero. Thus $w(x, t : \tau) = \frac{1}{2} \exp[(4\kappa - 2)\tau - 4\kappa t] \sin 2x$, and so $u(x, t) = \frac{1}{4}\left(\frac{e^{-2t} - e^{-4\kappa t}}{2\kappa - 1}\right) \sin 2x$.

3. The reasoning proceeds as in the text until the equation determining $B_n(\tau)$, which now becomes $e^{-\tau/2}(1 - \cos 2\pi x) = \sum_{n=1}^{\infty} B_n(\tau) \exp(-n^2\pi^2\kappa\tau) \sin n\pi x$. The sine series expansion of $1 - \cos 2\pi x$ over $0 \le x \le 1$ is

$$1 - \cos 2\pi x = \begin{cases} 0, & n = 2 \\ \frac{8((-1)^n - 1)}{n\pi(n^2 - 4)}, & n \neq 2. \end{cases}$$

Thus $B_2(\tau) = 0$ and $B_n(\tau) = \frac{8((-1)^n - 1)}{n\pi(n^2 - 1)} \exp[(n^2\pi^2\kappa - \frac{1}{2})\tau], n \neq 2$. Hence $w(x, t : \tau) = \frac{16}{3\pi} \exp[(\pi^2\kappa - \frac{1}{2})\tau - \pi^2\kappa t] \sin \pi x + \frac{8}{\pi} \sum_{n=3}^{\infty} \frac{((-1)^n - 1)}{n(n^2 - 1)} \exp[(n^2\pi^2\kappa - \frac{1}{2})\tau] \exp(-n^2\pi^2\kappa t) \sin n\pi x$.

By (7.79) the solution $u(x, t)$ follows from this expression after substituting the results

$$\int_0^t \exp\left[\left(\pi^2\kappa - \frac{1}{2}\right)\tau\right] d\tau = 2\frac{\exp\left[\left(\pi^2\kappa - \frac{1}{2}\right)t\right] - 1}{2\pi^2\kappa - 1},$$

and

$$\int_0^t \exp\left[\left(n^2\pi^2\kappa - \frac{1}{2}\right)\tau\right] d\tau = 2\frac{\exp\left[\left(n^2\pi^2\kappa - \frac{1}{2}\right)t\right] - 1}{2n^2\pi^2\kappa - 1}.$$

5. It follows immediately from (7.69) that $u_1(x, t) = x$. Reasoning as in the text shows that

$$u_2(x, t) = \frac{\left(e^{-3t} - e^{-4\pi^2\kappa t}\right)}{4\pi^2\kappa - 3} \sin 2\pi x.$$

Now $u_3(x, t)$ must satisfy the equation $u_{3t} = \kappa u_{3xx}$ with $u_3(0, t) = u_3(1, t) = 0$ and the initial condition $u_3(x, 0) = \sin\frac{1}{2}x - x$. Solving this IVBP gives

$$u_3(x, t) = \frac{2}{3\pi} e^{-\pi^2\kappa t} \sin \pi x + \frac{2}{\pi} \sum_{n=2}^{\infty} \frac{(-1)^{n+1} e^{-n^2\pi^2\kappa t}}{n(4n^2 - 1)} \sin n\pi x.$$

The required solution is then $u(x, t) = u_1(x, t) + u_2(x, t) + u_3(x, t)$.

Chapter 8

Solutions 8.2

1. Set $u_t = v$, $u_x = w$, then the Tricomi equation becomes $xv_t + w_x = 0$, or $v_t + (\frac{1}{x})w_x = 0$. A further equation is required to close the system, and this is provided by the equality of mixed derivatives $v_x = w_t$. The Tricomi equation then becomes the system

$$\mathbf{U}_t + \mathbf{A}\mathbf{U}_x = 0 \quad \text{with } \mathbf{U} = \begin{bmatrix} v \\ w \end{bmatrix}, \quad \mathbf{A} = \begin{bmatrix} 0 & \frac{1}{x} \\ -1 & 0 \end{bmatrix}.$$

This has eigenvalues given by $\lambda^2 = \frac{-1}{x}$ for $x \neq 0$. The eigenvalues are real and distinct when $x < 0$, corresponding to a hyperbolic system; they are imaginary when $x > 0$, corresponding to an elliptic system; and the system is degenerately parabolic when $x = 0$.

3. The first equation is in conservation form when written in the form $u_t + (\frac{1}{2}u^2 + c^2 - H)_x = 0$. The second equation can be converted to a conservation equation after multiplication by c when it becomes $(c^2)_t + (uc^2)_x = 0$. Thus its conservation form is $\mathbf{F}_t + \mathbf{G}_x = 0$ with

$$\mathbf{F} = \begin{bmatrix} u \\ c \end{bmatrix} \quad \text{and} \quad \mathbf{G} = \begin{bmatrix} \frac{1}{2}u^2 + c^2 - H \\ uc^2 \end{bmatrix}.$$

5. Write the system as

$$\begin{bmatrix} (c^2 - u^2) & -uv \\ 0 & 1 \end{bmatrix} \begin{bmatrix} u \\ v \end{bmatrix}_x + \begin{bmatrix} -uv & (c^2 - v^2) \\ -1 & 0 \end{bmatrix} \begin{bmatrix} u \\ v \end{bmatrix}_y = \begin{bmatrix} 0 \\ 0 \end{bmatrix}.$$

Premultiplying the system by the inverse of the first matrix and solving for the eigenvalues give

$$\lambda^{(1)} = \frac{-uv + c\sqrt{u^2 + v^2 - c^2}}{c^2 - u^2} \quad \text{and} \quad \lambda^{(2)} = \frac{-uv - c\sqrt{u^2 + v^2 - c^2}}{c^2 - u^2}.$$

Defining the **Mach number** of the flow by $M = (u^2 + v^2)^{1/2}/c$ (the ratio of the flow speed to the speed of sound), the eigenvalues are real if the flow is supersonic ($M > 1$) when the equation is hyperbolic, complex if the flow speed is subsonic ($M < 1$) when the equation is elliptic, and real but coincident if the flow is transonic ($M = 1$) when the equation is degenerately parabolic.

Solutions 8.3

1. The eigenvalues and eigenvectors are

$$\lambda^{(1)} = 1, \quad \mathbf{1}^{(1)} = [1, 1], \quad \lambda^{(2)} = -1, \quad \mathbf{1}^{(2)} = [1, -1].$$

The Riemann invariants become

$$u_1 + u_2 = r(\beta) \text{ on } C^{(1)} \text{ characteristics: } \frac{dx}{dt} = 1$$

$$u_1 - u_2 = s(\alpha) \text{ on } C^{(2)} \text{ characteristics: } \frac{dx}{dt} = -1.$$

Thus the $C^{(1)}$ characteristics are given by $x = x_0 + t$ and the $C^{(2)}$ characteristics are given by $x = x_1 - t$. From the initial data $u(x, 0) = e^x$ and $u_2(x, 0) = e^{-x}$. Thus at $(x_0, 0)$ we have $u_1(x_0, 0) + u_2(x_0, 0) = \exp(x_0) + \exp(-x_0) = 2 \cosh x_0$, and in similar fashion we find that at $(x_1, 0)$ we have $u_1(x_1, 0) - u_2(x_1, 0) = 2 \sinh x_1$. Thus the Riemann invariants become $u_1(x, 0) + u_2(x, 0) = 2 \cosh x_0$ on the $C^{(1)}$ characteristic through $(x_0, 0)$ and $u_1(x, 0) - u_2(x, 0) = 2 \sinh x_1$ on the $C^{(1)}$ characteristic through $(x_1, 0)$. As $x_0 = x - t$ and $x_1 = x + t$ we find that $u_1(x, t) = \cosh(x - t) + \sinh(x + t)$ and $u_2(x, t) = \cosh(x - t) - \sinh(x + t)$.

3. Proceed as in Problem 1, but this time with the eigenvalues $\lambda^{(\pm)} = \pm 1$ and the corresponding eigenvectors $\mathbf{1}^{(\pm)} = [1, \pm 1]$. This leads to the results $u_1(x_0, 0) + u_2(x_0, 0) = 1 + \sin x_0$ on the $C^{(+)}$ characteristic through $(x_0, 0)$, and $u_1(x_1, 0) - u_2(x_1, 0) = 1 - \sin x_1$ on the $C^{(-)}$ characteristic through $(x_1, 0)$. As $x_0 = x - t$ and $x_1 = x + t$ the solution becomes $u_1(x, t) = 1 + \frac{1}{2}(\sin(x - t) - \sin(x + t))$ and $u_2(x, t) = 1 + \frac{1}{2}(\sin(x - t) + \sin(x + t))$.

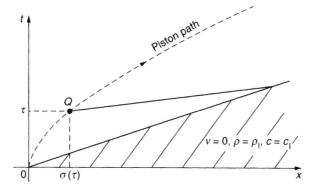

Figure 8.4.1

Solutions 8.4

1. The equation for v is $v_t + [c_I + \frac{1}{2}(\gamma - 1)v]v_x = 0$. When expressed in characteristic form this becomes $\frac{dv}{dt} = 0$ on $\frac{dx}{dt} = c_I + \frac{1}{2}(\gamma - 1)v$. Let the piston path be $x = \sigma(t)$, with $\sigma(0) = 0$ and $\sigma'(0) = 0$, so the piston is accelerated smoothly from rest. At point Q in Fig. 8.4.1 with coordinates $(\sigma(\tau), \tau)$ the piston speed will be $\sigma'(\tau)$. As $v = $ constant along characteristics and v is the piston speed at Q, the equation of the straight line characteristic through Q is obtained by integrating $\frac{dx}{dt} = c_I + \frac{1}{2}(\gamma - 1)v$. This yields $x = \sigma(\tau) + [c_I + \frac{1}{2}(\gamma - 1)\sigma'(\tau)](t - \tau)$.

The envelope of characteristics follows by eliminating τ between this equation and that obtained by differentiation with respect to τ, which gives $0 = \sigma'(\tau) + (t - \tau)\frac{1}{2}(\gamma - 1)\sigma''(\tau) - [c_I + \frac{1}{2}(\gamma - 1)\sigma'(\tau)]$.

Only the formation of the cusp is required, so we can take Q to be at the origin, when $x = 0$ and $\tau = 0$. It then follows from these two equations and the initial conditions $\sigma(0) = 0$ and $\sigma' = 0$ that the time t_P and position x_P of cusp formation is $t_P = \frac{2c_I}{(\gamma-1)\sigma''(0)} = 0$ and $x_P = c_I t_P$.

3. $U = \begin{bmatrix} i \\ v \end{bmatrix}$, $A = \begin{bmatrix} 0 & 1/L \\ 1/C & 0 \end{bmatrix}$, $B = \begin{bmatrix} R/C & 0 \\ 0 & G/C \end{bmatrix}$. The eigenvalues are $\lambda^{(+)} = c$, $\lambda^{(-)} = -c$, $c = 1/\sqrt{LC}$, while the corresponding eigenvectors are $\mathbf{l}^{(+)} = [LC, 1]$, $\mathbf{l} = [LC, -1]$.

When $R = G = 0$, $B = 0$ so the Riemann invariants become $(LC)i_\alpha + v_\alpha = -(Rci + \frac{Gv}{c})$ on the $C^{(+)}$ characteristics, and $(LC)i_\beta - v_\beta = -(Rci - \frac{Gv}{c})$ on the $C^{(-)}$ characteristics.

If $\frac{R}{L} = \frac{G}{C}$ these simplify to

$$\frac{\partial}{\partial \alpha}(Lci + v) = -\frac{R}{L}(Lci + v) \text{ on } C^{(+)} \text{ characteristics, and}$$

$$\frac{\partial}{\partial \alpha}(Lci + v) = -\frac{R}{L}(Lci + v) \text{ on } C^{(-)} \text{ characteristics.}$$

These can be integrated to give

$$L\dot{c}i + v = K(\beta)\exp\frac{-R\beta}{L} \text{ on } C^{(+)} \text{ characteristics, and}$$

$$L\dot{c}i - v = K(\alpha)\exp\frac{-R\alpha}{L} \text{ on } C^{(-)} \text{ characteristics.}$$

Thus

$$i = \frac{1}{2cL}\left[K(\beta)\exp\left(-\frac{R\alpha}{L}\right) + M(\alpha)\exp\left(-\frac{R\beta}{L}\right)\right]$$

and

$$v = \frac{1}{2}\left[K(\beta)\exp\left(-\frac{R\alpha}{L}\right) - M(\alpha)\exp\left(-\frac{R\beta}{L}\right)\right].$$

However, $\alpha = x + ct$ and $\beta = x - ct$, so as the functions K and M are arbitrary (because in a specific application they depend on the initial conditions, which may be imposed arbitrarily) the last results can be written as

$$i = \frac{1}{2cL}\exp\left(-\frac{2Rx}{L}\right)[f(x - ct) + g(x + ct)], \quad \text{and}$$

$$v = \frac{1}{2}\exp\left(-\frac{2Rx}{L}\right)[f(x - ct) - g(x + ct)],$$

where f and g are arbitrary twice differentiable functions of their arguments.

Solutions 8.5

1. The first equation is in conservation form when written $h_t + (uh)_x = 0$, so the jump condition is $s[[h]] = [[uh]]$. The second equation can be put in conservation form if it is multiplied by h and the result is added to u times the first equation to obtain $(uh)_t + (u^2h + \frac{1}{2}gh^2)_x = 0$. Thus the jump condition for this equation is $s[[uh]] = [[u^2h + \frac{1}{2}gh]] = 0$. In the problem $u_0 = 0$, so from the first jump condition $s(h_0 - h_1) = -u_1h_1$ so $u_1 = s(1 - \frac{h_0}{h_1})$. The second jump condition gives $s(-u_1h_1) = (-u_1^2h_1 + \frac{1}{2}g(h_0^2 - h_1^2)$. The required result for s follows by eliminating u_1 between these two equations and then taking the negative sign with the square root, because the bore moves to the *left*. Thus $s = -\sqrt{\frac{1}{2}g(\frac{h_1}{h_0})(h_0 + h_1)}$ and the speed of the water behind the bore is $u_1 = s(1 - \frac{h_0}{h_1})$.

3. The characteristics from $x < 0$ on the initial line have $\frac{dx}{dt} = 1$ while those from $0 \leq x \leq 1$ on the initial line have $\frac{dx}{dt} = 2$. Thus the characteristics diverge and there is a simple wave region in the wedge shown in Fig. 8.5.1

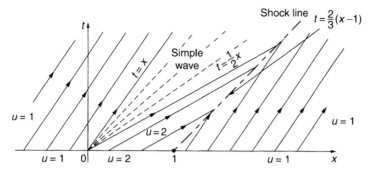

Figure 8.5.1

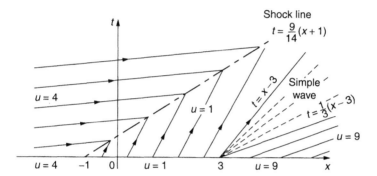

Figure 8.5.2

where the solution is $u = x/t$ for $1 \le \frac{x}{t} \le 2$. The characteristic to the left has the equation $t = x$ while that to the right has the equation $t = \frac{1}{2}x$. The characteristics from $x > 1$ on the initial line have $\frac{dx}{dt} = 1$, showing that adjacent characteristics converge, causing a shock to start from the point $(1, 0)$, as shown in Fig. 8.5.1. The conservation form of the equation is $u_t + \frac{1}{2}(u^2)_x = 0$, so the jump condition is $s[[u]] = \frac{1}{2}[[u^2]]$. Thus in this case the shock speed is given by $s(2 - 1) = \frac{1}{2}(4 - 1)$, so $s = \frac{3}{2}$. As in Fig. 8.5.2 the x axis is horizontal and the t axis is vertical, the slope of the shock in the (x, t) plane is $\frac{dt}{dx} = \frac{2}{3}$, so the equation of the shock is $t = \frac{2}{3}(x - 1)$. The shock intersects characteristics from the left where $\frac{1}{2}x = \frac{2}{3}(x - 1)$, and this occurs when $x_c = 4$ and $t_c = 2$. Thus the solution will only be valid for $0 \le t \le 2$ because of the interaction of a shock and a simple wave.

4. The characteristics from the initial line with $x < -1$ have $\frac{dx}{dt} = 2$, the characteristics from the initial line with $-1 \le x \le 3$ have $\frac{dx}{dt} = 1$, while the characteristics from the initial line with $x > 3$ have $\frac{dx}{dt} = 3$. Thus a shock starts from the point $(-1, 0)$ due to the convergence of characteristics, while a centered simple wave is located at $(3, 0)$, as shown in Fig. 8.5.2.

When written in conservation form the equation becomes $u_t + (\frac{2}{3}u^{3/2})_x = 0$, so the jump condition is $s[[u]] = \frac{2}{3}[[u^{3/2}]]$. Hence for the shock starting from $(-1, 0)$ we have $s(4-1) = \frac{2}{3}(4^{3/2} - 1)$, giving $s = \frac{14}{9}$. The equation of the shock line starting from $(-1, 0)$ in Fig. 8.5.2 is thus $t = \frac{9}{14}(x + 1)$, because in the (x, t) plane the x axis is horizontal and the t axis is vertical so the slope of the shock line is $\frac{dt}{dx} = \frac{9}{14}$. The centered simple wave located at $(3, 0)$ has the solution $u = \frac{x-3}{t}$ with $1 \le \frac{x-3}{t} \le 3$. The solution will be limited because the shock line $t = \frac{9}{14}(x + 1)$ will intersect the line $t = x - 3$, forming the left boundary of the simple wave region. This happens when $\frac{9}{14}(x + 1) = x - 3$, corresponding to $x_c = \frac{51}{5}$ and $t_c = \frac{36}{5}$. The solution is limited to $0 \le t < \frac{36}{5}$ because of the interaction of a shock and a simple wave.

Bibliography

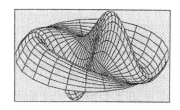

This short bibliography lists books on partial differential equations and boundary value problems written at about the same level as the present text, all of which provide additonal examples and useful background information. Books marked with an asterisk are written at a more advanced level, and while covering some of the topics contained in the present text, these serve both as reference works and as introductions to other more advanced aspects of the subject.

1. W. E. Boyce and R. C. DiPrima, *Elementary Differential Equations and Boundary Value Problems*, 4th ed., Wiley, New York, 1986.
2. J. W. Brown and R. V. Churchill, *Fourier Series and Boundary Value Problems*, 5th ed., McGraw-Hill, New York, 1993.
*3. H. S. Carslaw and J. C. Jaeger, *Conduction of Heat in Solids*, 2nd ed., Clarendon Press, Oxford, 1986.
4. P. DuChateau and D. Zachmann, *Applied Partial Differential Equations*, Harper and Row, New York, 1989.
*5. P. R. Garabedian, *Partial Differential Equations*, Wiley, New York, 1964.
*6. J. Kevorkian, *Analytical Solution Techniques*, 2nd ed., Springer-Verlag, New York, 2000.
7. R. Knobel, *An Introduction to the Mathematical Theory of Waves*, Student Mathematical Library Vol. 3, Amer. Mathe. Soc., Providence, RI, 2000.
8. G. L. Lamb, Jr., *Introductory Applications of Partial Differential Equations*, Wiley, New York, 1995.
*9. H. Levine, *Partial Differential Equations*, Studies in Advanced Mathematics Vol. 6, Amer. Math. Soc., Providence, RI, 1997.
10. P. V. O'Neil, *Beginning Partial Differential Equations*, Wiley, New York, 1999.

11. M. A. Pinsky, *Partial Differential Equations with Applications*, 3rd ed., McGraw-Hill, New York, 1998.

12. W. Strauss, *Partial Differential Equations: An Introduction*, Wiley, New York, 1992.

13. J. L. Troutman, *Boundary Value Problems of Applied Mathematics*, PWS, Boston, 1994.

14. E. Zauderer, *Partial Differential Equations of Applied Mathematics*, 2nd ed., Wiley, 1985.

Index

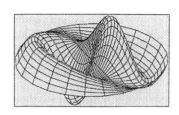

Printed and bound by CPI Group (UK) Ltd, Croydon, CR0 4YY

08/05/2025

01864827-0003